Alternative Splicing and Cancer

Editors

Muzafar A. Macha
Watson-Crick Centre for Molecular Medicine
Islamic University of Science and Technology
Kashmir, Jammu and Kashmir, India

Ajaz A. Bhat
Department of Human Genetics-Precision Medicine in Diabetes
Obesity and Cancer Research Program
Sidra Medicine, Doha, Qatar

Surinder Kumar Batra
Department of Biochemistry and Molecular Biology
University of Nebraska Medical Center
Omaha, NE, USA

CRC Press
Taylor & Francis Group
Boca Raton London New York

CRC Press is an imprint of the
Taylor & Francis Group, an **informa** business

A SCIENCE PUBLISHERS BOOK

Cover credit: Cover illustration reproduced by kind courtesy of the editors.

First edition published 2024
by CRC Press
2385 NW Executive Center Drive, Suite 320, Boca Raton FL 33431

and by CRC Press
4 Park Square, Milton Park, Abingdon, Oxon, OX14 4RN

CRC Press is an imprint of Taylor & Francis Group, LLC

Library of Congress Cataloging-in-Publication Data (applied for)

ISBN: 978-1-032-19659-6 (hbk)
ISBN: 978-1-032-19698-5 (pbk)
ISBN: 978-1-003-26039-4 (ebk)

DOI: 10.1201/9781003260394

Typeset in Times New Roman
by Radiant Productions

Preface

In the intricate tapestry of human biology, alternative splicing is a remarkable testament to the complexity and underlying elegance of genetic regulation. This multifaceted process allows for the diversification of proteins, serving as the cornerstone of cellular functionality. With a spotlight on its role in the genesis and progression of cancer, this edited volume, "Alternative Splicing and Cancer," provides an exhaustive and profound exploration of the subject.

The chapters in this volume span the spectrum of alternative splicing's implications in cancer, illuminating the path from fundamental understanding to innovative therapeutic applications. Together, they constitute an invaluable resource for scientists, researchers, and clinicians, offering insights, challenges, and possibilities that define this rapidly evolving field.

Chapter one sets the stage, introducing the multifaceted roles of alternative splicing in human diseases, serving as a springboard into the subsequent chapters. The second chapter then delves into the intricate mechanisms of RNA splicing, offering readers a detailed view of this complex and fascinating process.

Chapters three through eight explore the intriguing relationship between alternative splicing and cancer, painting a detailed picture of how deregulation in splicing can lead to the development and progression of cancer. We examine the splice variants and spliceosomes, oncogenic signaling pathways, the role of the splice variant of Pyruvate Kinase M2 in cancer metabolism, and how these cellular changes can lead to metastasis and evade immune surveillance.

In the last three chapters, we venture into the promising field of therapeutic intervention, exploring how we might exploit our understanding of alternative splicing to combat cancer. The chapters shed light on the diagnostic, prognostic, and therapeutic potentialities of alternative splicing and unravel its role in cancer drug resistance.

As we navigate the convoluted paths of alternative splicing, we encourage you, dear readers, to step into the complexity, explore the unknown, question the established, and contribute to the evolving narrative. Welcome to our journey through the intersection of alternative splicing and cancer. It is a journey of discovery, insight, and hope.

Contents

1

Alternative Splicing and Human Diseases

Mukesh Tanwar

1. Introduction

Alternative splicing (AS) allows a messenger RNA (mRNA) to control the production of various protein variations (isoforms) with distinct biological activities. It involves altering the splicing pattern of intron and exon components to change the mRNA coding sequence (Figure 1). It was originally described by Chow and colleagues as viral messenger RNA (mRNA) in 1977. Only roughly 10% of the human genome is made up of gene-coding exons, with the remainder being introns and untranslated regions (UTRs). As of October 28, 2021, the HGMD database contains 18,884 entries under the splicing head in the public entry domain and 27,959 items under the splicing head in the HGMD professional domain. These account for 8.98% of all disease-associated changes in the public entry domain and 8.63% in the HGMD professional domain. As most efforts are focused on identifying missense and nonsense variations, because splicing is so difficult to predict, this is likely an underestimate. The advancement of sequencing technology, like the latest next-generation sequencing (NGS) techniques [1], is likely to add many variants associated with different diseases to the HGMD. This, in turn, will increase our understanding of genotype-phenotype relationships. More than 90% of human gene transcripts are alternatively spliced, resulting in a variety of protein forms with distinct characteristics regarding molecular interactions, intracellular localization, and modulatory or suppressive activity [2].

Exon skipping, mutually exclusive exons, different donor (5'splice) sites, and intron retention are examples of AS [3]. Alternative polyadenylation sites and

Department of Genetics, Maharshi Dayanand University Rohtak (HR), INDIA-124001.
Email: mukeshtanwar@mdurohtak.ac.in; mukeshtanwar@gmail.com

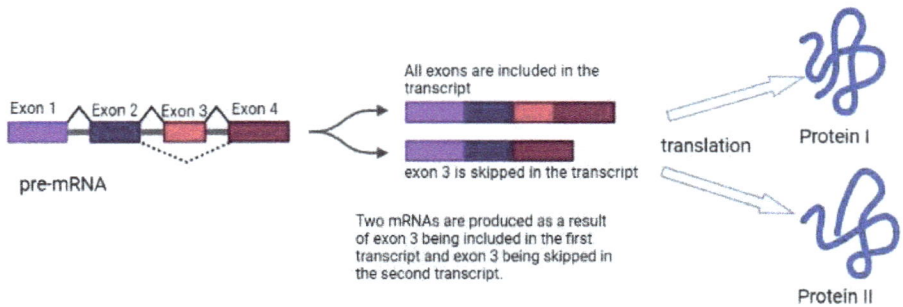

Figure 1: Showing the mechanism of splicing.

promoters, in addition to these AS mechanisms, can promote mRNA diversification [3–5]. Higher mammals, like humans and chimps, have more AS genes than mice, frogs, zebrafish, and fruit flies [6]. The literature indicates that AS variety decreases with increasing evolutionary distances from human beings to drosophila/fruit flies [6]. Exon skipping seems to be the most prevalent AS type in mammals (46–52%) and vertebrates (37–38%), while it is less common in invertebrates (20%). Despite its small genome size, the human proteome is extraordinarily diverse and rich, owing to species-specific variations in AS frequencies and mechanisms. The precise control of AS requires hundreds of components that facilitate appropriate site selection and subsequent splicing events [3]. Tazi and colleagues (2009) investigated the molecular pathways of pre-mRNA splicing modulation, namely the detection of splicing sites and their proper splicing [3]. Five short nuclear RNAs (U1, U2, U4, U5, and U6), essential complexes, and numerous governing components comprise the spliceosome. Numerous heterogeneous nuclear ribonucleoproteins and SR proteins govern splicing restriction and stimulation by binding intronic or exonic splicing suppressors, boosters, or suppressors in pre-mRNA AS [3, 7, 8]. Numerous causes, such as mutations in all these genes, contribute to abrupt AS and the development of numerous diseases.

Disruptions in the splicing process result in the death of live organisms, while minor changes in AS result in dysfunctional proteins that cause organ failure and a variety of disease conditions [9] (Table 1). After 10 years of discovery, the role of AS in human fibronectin genes was identified for the first time [10]. Numerous studies have been conducted to understand the significance of AS in the pathophysiology of the disease and to determine the diagnostic and therapeutic value of these events in a variety of conditions [3, 6, 11–15]. The role of AS in the disease manifestation of some major diseases is given below.

2. Implications of AS in Spinal Muscular Atrophy

Spinal muscular atrophy (SMA) is an overall terminology for a spectrum of disorders characterized by the degeneration of alpha motoneurons in the brainstem and spinal column. It is the second-most common recessive genetic condition. It is defined as progressive palsy caused by the degeneration of alpha-motor neurons. The viable

Table 1: Gene splicing in various diseases.

S. No.	Disease	Gene Name	Skipped Exon
1.	Spinal muscular atrophy	SMN2	Exon 7
2.	Cancer	COL12A1	Exon 2
3.	Cardiovascular disease	VEGF	Exon 6, 7, 8
4.	AD	MAPT	Exon 10
5.	PD	SNCA	Exon 5
		LRRK2	Exon 5
		PARK2/PRKN	Exon 4
		MAPT	Exon 10
6.	Frontotemporal dementia with parkinsonism-17	MAPT	Exon 10
7.	HD	MAP2	Exon 7–9
8.	Schizophrenia	DRD2	Exon 6
		DRD3	Exon 7

birth rate ranges from 1 in 6,000 to 10,000, and the frequency of the carrier is 1/40 [16]. SMA is caused by a mutation in the SMN1 gene, which codes for the SMN protein, which is required for small nuclear ribonucleoproteins (snRNP) formation. It is uncertain how SMN protein deficiency causes the disease and results in the death of specific motoneurons. SMN protein shortage resulted in cell-specific alterations to snRNAs and an overall decrease in snRNP assembly ability in the mouse model [17]. According to Zhang and colleagues (2008), several pre-mRNA splicing mechanisms are dysregulated across all investigated tissues. Several detected modifications correspond to a change in previously identified AS that was not standard. However, the majority of improper splicing events result in the generation of abnormal messenger RNAs. All of these imply that the combined impact of abnormal mRNA splicing may promote selective mortality of motoneurons, whereas alterations in neighboring cells may cause their death. Opera and colleagues (2008) reported six families in which eight female relatives were found to show no symptoms of SMA despite inheriting the identical SMN1 and SMN2 alleles as their afflicted brothers and sisters [18].

Actin-binding plasmin 3 (Plastin 3) was found to be a modulator. In SMN knockout mice, plastin 3 upregulation partially repaired the short neuronal axon length caused by SMA protein deficiency. According to these studies [18], splicing changes may not be the only cause of motoneurons' demise. A therapeutic approach would be to reestablish SMA protein synthesis, even if the cause of the disease's loss of SMA protein is not clearly grasped. In the human genome, the SMN2 gene is identical to SMN1. This new SMN2 results from a recent duplication in the human genome. Despite the fact that the genes' sequences are remarkably similar, a synonymous C > T change in exon 7 causes their splicing patterns to diverge, with exon 7 being largely omitted in the protein [19]. This exon-skipping event results in a protein that is shortened, unstable, and most likely non-functional. As a result, SMN2 cannot substitute for the absence of SMN1. In adults with SMA, at least one copy of SMN2 is preserved, as the absence of both SMN genes is lethal. Mice have a single SMN gene with a constitutively spliced exon 7. A homozygous deletion of this gene results in death. To explore how the human gene's splicing is regulated in mice, transgenic mice harboring the human gene were generated [20]. For unexplained causes, alpha motoneurons are by far the most severely damaged and perish, resulting in muscle wasting in SMA patients. SMA is classified into four categories (types I–IV), each with its distinct start and severity [21]. Because more SMN2 copies yield more SMN protein, symptoms are strongly correlated with SMN2 copy counts [21]. Because triggering SMN2 exon 7 increases SMN protein levels and may heal the condition, researchers are studying how exon 7 is managed [21]. The C > T transition at position 6 causes exon skipping, and there are now two hypothesized explanations. (i) The base change eliminates the exonic enhancer that connects SF2/ASF normally [22, 23], and (ii) the mutation results in the formation of a hnRNPA1 binding pocket that functions as a silencer [24, 25]. Both models can account for the majority of exon 7 skipping. Exon 7 inclusion is contingent upon the presence of a central tra2-beta1 enhancer region. The Tra2-beta1 protein is associated with the SR. Its activity is governed by the dephosphorylation of protein phosphatase 1 and unsurprisingly, exon 7 use is dependent on cellular PP1 activity [26].

SMA highlights several prevalent traits of misplacing illnesses. Splicing alterations can occur because of genome evolution, such as recent gene duplications that allow for recombination. A single mutation can have a big impact on alternative exon regulation, which is influenced by several factors and sequence properties. Finally, cellular signaling networks regulate splicing factor phosphorylation.

3. Implications of AS in Cancer

AS supports normal cell physiology via the regulation of activities, such as cell differentiation, proliferation, migration, as well as cell communication. Thus, faulty AS can disrupt normal cellular functioning and create cell development problems, like abnormal cell growth and proliferation. A recent review by Sterne-Weiler and Sanford (2014) established a correlation between typical AS episodes and cancer hallmarks, implying an unbreakable link between AS and cancer biology [9]. The "Cancer Genome Atlas (TCGA) project" just has unveiled a vast public repository of omics data derived from 33 unique forms of cancer. This enables the examination of AS patterns in a variety of malignancies. As a result, distinct AS isoforms associated with particular malignancies can be used as biomarkers and therapeutic targets [27].

The computational study can be utilized to identify potential AS genetic markers and cancer-causing differentially spiced genes utilizing the TCGA data [28]. Kim and colleagues (2018) constructed a network of cancer/AS patterns using TGCA data [6]. Employing data from the TGA, Kim and colleagues (2018) built a network of cancer/AS trends [6]. This network exhibited 33 distinct types of cancer, and the top 50 percent splice-in (PSI) rated genes for each of those cancers. The network includes 217 different genes that are all connected to a certain type of cancer. A total of 87% of 1,650 discovered AS genes were linked to more than one type of cancer, indicating that common AS alterations are linked to a wide range of cancers. In 31 distinct malignancies with a PSI value of more than 90, COL12A1 was shown to be the most frequently excluded exclusion gene [6]. Certain genes have PSI values that vary according to the kind of malignancy. In cholangiocarcinoma, Kim et al. discovered that the deletion of exon 2 was related to a PSI value of 85.9 but less than 70 in plenty of other cancers [6].

4. Implications of AS in Cardiovascular Disease

Heart development, contractility modulation, and numerous cardiac disorders are all linked to the AS of pre-mRNAs coding cytoskeletal and cell signaling proteins, as well as ion channels. A variety of hereditary cardiomyopathies can be caused by aberrant AS in different genes. By regulating AS, the correct amounts and functional activities of muscleblind-like proteins (MBNL), RNA-binding fox family proteins (RBFOX), and members of the CUG-binding protein (CUGBP) and ETR-like factors (CELF) families maintain healthy cardiogenesis. Adult cardiac splicing patterns can be remodeled to match those of an embryonic heart by manipulating the balance of these factors [29]. Adult cardiac failure is associated with such AS reversions to neonatal splicing patterns. Splicing alterations in sarcomeric genes, like cardiac troponin T, tropomyocin, etc., have been found to be linked to heart disease. Ischemic cardiomyopathy, aortic narrowing, and dilation of the cardiomyopathy are all associated with abnormalities in β-myosin heavy chain, cardiac troponin I, and filamin Cγ gene transcript splicing. Dilated cardiomyopathy is associated with abnormal AS isoforms, like tropomyocin [30], myosin binding protein C [31], EH-myomesin [32], LDB3 [33], or titin [34]. Mutated AS isoforms of ANK2 and ANK3 are linked to cardiac arrhythmias [35]. Multiple heart physiological processes are regulated by insulin-like growth factor (IGF-1), including energy metabolism, contractile activity, and apoptosis/autophagy. Cardiovascular function is also influenced by IGF-1-related AS. Activation of endothelial progenitor cells [36] and a decrease in oxidative stress [37] improve heart recovery after myocardial infarction and dilated cardiomyopathy. The Unfolded Protein Response (UPR) transcription factor, X-box binding protein 1 (XBP1), is one of the most versatile proteins in the heart. Vascular endothelial growth factor (VEGF) also regulates the AS of XBP1. As AS-mediated regulation of VEGF is itself AS regulated. When exons are skipped in the VEGFA gene, several VEGF isoforms can be generated. There are 13 distinct VEGF isoforms based on exon counts and sizes. Exons 1–5 are found in nearly all known VEGF isoforms, with exons 6, 7, and 8 frequently undergoing AS. Heparin-binding domains in exons 6 and 7 of the VEGF gene govern the isoforms' mobility. Exons 6 and 7 in the VEGF121 isoform are missing, making them more easily dispersed than other VEGF isoforms. To maintain heart health and prevent disease progression, the heparin-binding domains of VEGF are crucial. It has been shown that the expression of VEGF121 causes severe angiogenic abnormalities and cardiomyopathy in mice [38]. Exon 8 AS regulates the switch from pro-angiogenic VEGF xxx's to anti-angiogenic activities [39]. There are numerous activities for VEGF, and these findings show that AS isoform balance is critical for both angiogenesis and vascular pathology.

Antisense oligonucleotides (AON) are the main treatment strategy for inhibiting erroneous AS in some heart diseases. Administering AON treatments to control hyperlipidemia, diabetes, or hyperglycemia can help prevent heart disease. For example, mipomersen lowers LDL cholesterol levels by blocking the formation of the apolipoprotein B-100 AS isoform in the liver. Mipomersen binds directly to APOB mRNA and inhibits APOB production. Sarepta Therapeutics' Eteplirsen is an AON medication that induces exon 51 skipping in mutant dystrophin mRNA, which might enhance cardiac function. Although AON is in its infancy, it holds promise for additional heart conditions, such as hypertension, hypertrophy, DCM, and cardiac arrest [40, 41].

5. Implications of AS in Alzheimer's Disease

Alzheimer's disease (AD) is a chronic neurological condition that results in brain shrinkage and cell apoptosis. In most cases of dementia, AD is the underlying cause. Numerous protein alterations and AS events have been implicated in the progression of AD along with other molecular pathways. In AD, the proteins Fyn and tau and the peptide amyloid form a toxic trio. Previous investigations have established a relationship between alternative splicing-dependent regulation of Fyn protein and AD. The amyloid precursor protein (APP) is crucial for the progression and development of AD. AD is connected to changes in the APP isoform ratios. AS of exon 15 in the APP transcript has been shown to affect the synthesis and ration of betaA4 and p3. It is crucial for forming amyloid plaques in AD [42]. Additionally, Rockenstein and associates discovered that particular intron sequences contributed to the AS of APP pre-mRNA, which in turn contributed to the development of brain changes associated with AD [43].

Recent investigations have established a relationship between the regulation of the Fyn protein via AS and AD (Lee et al. 2016). According to Lee and colleagues (2017), upregulation of the FynT isoform in astrocytes in response to multiple inflammatory stimulants, such as TNF, may result in prolonged inflammatory responses, hence enhancing chronic neuroinflammation associated with AD [44].

Dysregulation in AS in tau exon 10 splicing mediated by protein kinase A (PKA) deficiency culminates in Alzheimer's neurofibrillary degeneration. By skipping exon 10 of tau's MAPT pre-mRNA, 3R-tau and 4R-tau are produced, and their equilibrium is maintained. AS regulates the proteins that regulate tau exon 10 splicing. So Glatz and colleagues discovered that the splicing patterns of tau, as well as pre-mRNAs for transformer-2 protein homolog beta (htra-beta/SFRS10) and the dual specificity protein kinase (CLK2), were considerably disturbed in AD patient's brains [45]. Apolipoprotein E receptor 2 (ApoER2/LRP8) AS regulation has been linked to AD. Postmortem brain samples from AD patients exhibit ApoER2/LRP8 exon 19 splicing imbalance [46].

Love et al. (2015) investigated the alternative splicing-dependent modulation of APP, tau, and apolipoprotein E genes in AD pathogenesis (APOE) [47]. Certain

APOE alleles and splicing events have already been related to an elevated risk of AD [46], but the significance of specific APOE isoforms in AD has remained unknown. Mills and colleagues discovered no differences in the APOE-005 expression in AD patients and healthy controls [48]. Rogaev and colleagues (1997) reported that the presenilin-1 gene (PSEN1) exhibits AS and has also been associated with an increased risk of developing early-onset AD [49]. PSEN1 and PSEN2 exon 9 AS are certainly implicated in the mechanisms underlying AD [50].

Additionally, AS of ion channel genes is also implicated in AD. Heinzen and colleagues (2007) recognized disease-associated AS in a variety of ion channels, including calcium, chloride, sodium, and potassium channels [51].

Additionally, global techniques have been used to detect AS variants related to AD. For example, Lai et al. (2014) reported 22 different candidate genes with differential AS events related to AD by using genome-wide profiling [52]. These genes are BEGAIN, BRWD1, CAMK2B, CYB5R3, EDF1, ELMO1, EPS15, FEZ1, FYN, GNAL, HK1, HPCAL1, KCNS3, KLC1, MAP4, NTRK2, PACSIN1, SCN4B, SIPA1L1, TKT, TRIM9, and ZNF365. In female individuals with Alzheimer's dementia, gender-specific alterations have already been reported. Sixty-two different estrogen receptor (ER) alpha messenger RNA splice isoforms have been found in distinct human brain regions associated with AD. AS isoforms of ER mRNA have been found to be dramatically diminished in the brains of AD patients, particularly in females [53].

6. Implications of AS in Parkinson's Disease

A neurodegenerative ailment known as Parkinson's disease (PD) is among the most prevalent among the aged. It accounts for 3% to 5% of cases in the elderly population [54]. Progressing loss of dopaminergic (DA) neurons in the *substantia nigra pars compacta* (SNc) and α-synuclein inclusions in the cytoplasm, or 'Lewy bodies', in remaining neuronal cells are the key pathogenic features of PD [55].

The AMPA '-amino-3-hydroxy-5-methyl-4-isoxazolepropionic acid' receptor is a vital component of the central nervous system's fast synaptic impulse transmission. It has been suggested that PD is partially caused by abnormal splicing of the AMPA receptor subunits [56]. Leukocytes of the body's immune system also have a role in the pathophysiology of PD. Six exons were found to be upregulated, and 12 exons were found to be down regulated in PD patients' leukocytes when AS patterns were examined. The most drastically downregulated exons were detected in pre-mRNAs for an ATP-binding protein and the LRRC 8C-associated protein (LRRC) 8C, and these AS alterations affect the NF-kB cascade and the immune response in PD [57].

6.1 Aberrant RNA Splicing of Alpha-Synuclein

The SNCA gene encodes alpha-synuclein (α-syn), the first gene linked to PD pathogenesis [58]. Lewy bodies in persisting substantia nigra cells are a neuropathological hallmark of PD, which is primarily composed of α-synuclein [59].

Five amphipathic helices are formed by six repetitive sequences on the N-terminus of α-syn and the acidic, glutamate-rich C-terminus acts as a chaperone to retain the protein solubility. Accumulating data indicates that the dosage and expression levels of α-syn involved in the etiology of PD. The SNCA gene has seven exons, five of which code for proteins. The alternatively spliced isoforms of SNCA have been studied to validate their disease-associated expression patterns [60]. The complete SNCA 140 and the shortened versions SNCA126 (with exon 3 deleted), SNCA112 (with exon 5 deleted), and SNCA98 (with exon 3 and 5 deleted) are the four major isoforms of SNCA [61, 62]. Exon 5 deletion (SNCA112) in the SNCA 3' regions predicts functional impairments linked to Lewy Body disease. The deletion dramatically shortens the acidic C-terminus and promotes α-syn aggregation, which may lead to the formation of Lewy bodies [63, 64].

6.2 Aberrant RNA Splicing of Leucine-Rich Repeat Kinase 2

Dardarin (leucine-rich repeat kinase 2) is a member of the Ras/GTPase superfamily's ROCO group. It is represented by the LRRK2 gene. This family of multi-domain proteins includes the human ROCO proteins (ROC-COR supra-domain) [65]. The human ROCO protein family includes LRRK1, LRRK2, DAPK1, and MASL1 [66]. LRRK2's function is unknown. It could be a cytoplasmic kinase implicated in Parkinson's pathogenesis [67, 68]. The LRRK2 gene contains 51 exons and encodes a predicted 2,527 amino acid protein. LRRK2 mutations are the prevalent hereditary cause of PD condition. Di Fonzo and colleagues (2006) conducted a comprehensive analysis of LRRK2 in a large sample of families with PD [69]. They sequenced the complete LRRK2 in 60 autosomal dominant PD probands. Three uncertain intronic variants were identified. Parkinson's patients had a considerably greater allelic frequency of IVS30 +12delT than controls, and two other intronic changes (IVS4-38A>G, IVS5+33T>C) were observed in patients only. Johnson and colleagues (2007) sequenced all the exons of LRRK2 in familial cases of PD and revealed a 4-bp intronic deletion in exon 19 (IVS20+4delGTAA) caused abnormal splicing *in vitro* [70].

Increased LRRK2 expression changes the splicing of tau and α-syn genes related to Parkinson's phenotype. Kalivendi et al. (2012) have suggested that alternative splicing-mediated modulation of α-syn is involved in PD [71]. The oxidant-induced 112-syn isoform, created by exon 5 deletion, was reported to contribute to the development of PD by accelerating dopamine neuron cell death [72]. PD patients have AS dysregulation in the splicing factor serine/arginine repetitive matrix protein 2 (SRRM2) [73]. All of these studies highlight the importance of aberrant AS in PD's etiology.

6.3 Aberrant RNA Splicing in PARK2/PRKN

PARK2 has 12 exons that collectively encode 465 amino acids of cytosolic ubiquitin-E3-ligase, the parkin protein. Parkin's N-terminus has a ubiquitin-like motif (represents a unique substrate protein), and the C-terminus contains two C3HC4 RING finger domains (E2 interaction sites) (in-between RING finger domain)

[74, 75]. Parkin aids in the transfer of ubiquitin to substrate proteins, marking them for proteasomal degradation [75]. Numerous proteins have been identified as parkin targets, for example, synphilin-1 and α-synuclein [76]. PARK2 generates seven protein isoforms, comprising transcript variations (TVs) "1, 2, 3, 6, 7, 11, and 12" [77]. Their differential expression levels are associated with the genesis of Lewy bodies. The differential expression levels of TV3 (in-frame deletions of exons 3–5) and TV12 (in-frame deletions of exons 2–7) were increased in PD but not in controls [60, 76].

D'Agata and Cavallaro (2004) studied parkin variants in adult rats and human fetal brain cDNA species [77]. These isoforms have one or two C3HC4 ring fingers, an in-between RING finger motif, and a thiol protease functional group. The structural variety of parkin alternatively spliced isoforms may have significant consequences for the pathogenesis of autosomal recessive juvenile parkinsonism. Tan and colleagues (2005) discovered an alternatively spliced variant of parkin (E4SV) in sporadic PD patients [78].

Exon 4 (122 bp) loss resulted in a frameshift at the juncture of exons 3, and also the addition of a stop codon upstream of exon 3. The shortened protein was found to be devoid of the two-RING finger motif entirely consequently, enzymatic function. Sporadic PD patients have higher E4SV expression than healthy controls. Additionally, E4SV expression increased with the age in patients but not in the controls [78].

6.4 Aberrant RNA Splicing in Synphilin-1

The SNCAIP gene encodes a presynaptic protein called synphilin-1. It is found around synaptic terminals [60]. Synphilin-1 may be necessary for the formation of Lewy bodies, as co-expression of parkin, synphilin-1, and a-synuclein results in an increase in the formation of cytoplasmic aggregates mimicking LBs [79, 80]. Additionally, synphilin-1 is a parkin substrate and appears to mediate synaptic activity via the ubiquitin-proteasome mechanism [79].

Synphilin-1 has two major functional domains: one that interacts with parkin and another with α-synuclein. The transcript variations demonstrate a variable decrease in both domains. At least eight alternatively spliced protein isoforms are derived from SNCAIP, and four of these isoforms have reduced C-terminal. Synphilin-1 isoforms with shorter lengths may demonstrate more interactions than longer proteins. Their differential expression has been shown to vary in LB illnesses [76]. Synphilin-1A is present in PD [76, 81]. Synphilin-1A overexpression saturates the proteasome, increases aggregation, and directly increases protein inclusion, synthesis, and neurotoxicity, hinting that this plays a role in neuronal degradation [76, 80]. Additionally, synphilin-1B was found to be highly increased in PD patients [60].

6.5 Aberrant RNA Splicing in Microtubule-Associated (MT) Protein

Microtubule-associated protein (MAPT), commonly known as tau protein, is encoded by the MAPT gene. Its primary role is to assist in the assembly and stability of the microtubule. MAPT mutations affect protein functionality or transcriptional regulation in frontotemporal dementia with parkinsonism-chromosome 17 types (FTDP-17) and result in insoluble filament cytosolic inclusions. Tau's capacity to bind to microtubules and/or promote microtubule assembling is sometimes reduced because of mutations in MAPT's exons. Six tau isoforms are created from the AS of exons 2, 3, and 10 in the central nervous system of the adult human. Four microtubule-binding domains are encoded by exons 9 to 12. There are three isoforms of microtubule-binding protein, each with four repeats of exon-10, which contains the 31 amino acid repetition (4R tau). Three tau isoforms with three repetitions each are found in the other three tau isoforms (3R tau). Increased tau mutations in FTDP-17 generate increased 4R tau production, as shown in postmortem tau samples from FTDP-17 patients [82, 83]. According to Hutton (2001), tau pathology can be avoided if 3R/4R isoforms ratio tau is kept constant at 1:1 [84]. As the RNA stem-loop becomes unstable because of mutations at the 3'-end of exon 10, resulting from the last two exon-10 bps and the neighbouring 16-base intron, the presence of E10 is becoming increasingly common. This stem-loop provides an ideal binding site for the U1 small nuclear RNA, hence increasing the effectiveness of exon-10. This loop prevents the splicing machinery from accessing exon-10 in healthy persons, hence inhibiting mRNA inclusion. It has also been found that all of these reported alterations have improved splice site sequence compatibility with U1 small nuclear RNA's 5' end [85]. There are cis-acting components in E1exon-10 that have an impact on splice site selection [86].

According to D'Souza and colleagues (2000), there are three cis-acting regulatory factors that can enhance/reduce the alternative splicing of exon-10 in tau [86]. As a result, the inclusion or exclusion of exon-10 from the tau transcript can be caused by either the strengthening or destruction of an ESE (N279K, delK280). The FTDP-17 variant L284L resulted in an overabundance of exon-10 isoforms in the brain. Intronic regions bordering the 5'-splice location of exon-10 are related to a third element that inhibits AS, and FTDP-17 mutants promote the incorporation of exon-10 into the transcript [85, 86].

7. Implications of AS in Various Tauopathies

Tauopathies are a set of CNS conditions characterized by intracellular buildup of the microtubule-affiliated protein tau. Tau interacts with microtubules via repeating microtubule regions. Exon 10 produces one of these microtubule-binding sites. Adults have species-specific splicing. Exon 10 is alternately spliced in adults in humans, whereas it is constitutively utilized in mice. Exon use is regulated during development in both species [3].

Dominant Tau gene mutations associated with frontotemporal dementia with parkinsonism have been related to chromosome 17. Splicing regulation of exon 10 is affected by the bulk of the mutations. The 3R and 4R repetition tau isoforms are connected to exon 10 mutations in FTDP-17 [87, 88].

According to these findings, splicing mutations promote neuropathology by altering the 3R/4R isoforms ratio. N279K is the most well-understood mutation [89].

That is what causes this mutation: a change in the DNA sequence. A tra2-beta1 binding site is formed by the GAAGAAG sequence. SMN2's exon 7 contains two partially overlapping GAAG binding sites, just like this mutant variant. Cotransfections tests demonstrated that tra2-beta1 facilitated the incorporation of exon 10 into reporter gene constructs by increasing the affinity of the mutation *in vitro* [90, 91].

According to *in vitro* experiments [82], the asparagine to lysine mutation has no effect on tau-tubulin-binding, tau aggregate formation, or microtubule assembly. These findings indicate that a shift in 3R/4R ratio is a significant underlying cause of FTDP-17. Other tauopathies, such as AD, also include abnormal intracellular tau aggregations [82]. The 4R to 3R tau pre-mRNA ratio was examined in several studies to see if it changed in AD. Since only postmortem human tissue can be analyzed, the investigation's findings were unsatisfactory. A few studies did not find statistical variations in the 4R/3R ratio between different regions [92, 93]. In one study [94], there was no apparent link between AD and regional disparities. AD may be caused or contributed to by various alterations in the selection of AS sites [45]. This shows that splice variants can also affect the proportion of protein isoforms in the body, which in turn can lead to disease. The regulation of splicing, which is occasionally complicated by species-specific differences in splicing, can be better understood by examining genetic mutants. Nevertheless, animal models that mimic human pathology can be created despite these difficulties.

8. Implications of AS in Huntington's Disease

Huntington's disease (HD) is a degenerative neurological condition caused by an increase in the Huntingtin (HTT) gene's glutamine-coding CAG region. Healthy humans have fewer than 36 CAG repeats in exon 1 of HTT (on average 15–25), whereas afflicted patients possess repeat lengths of more than 39 [95]. The expanded polyglutamine stretch at the amino–terminal causes HTT proteins to aggregate and toxicity because of gain-of-function [96]. HD predominantly affects striatal and cortical neurons, and patients may exhibit depression distress, cognitive impairments, motor dysfunction, sleep difficulty, and weight loss [97, 98].

Huntingtin functions as a transcription modulator in part [99], and while the function of wild-type HTT remains uncertain, enlarged polyglutamine HTT demonstrates a transformed DNA interaction [100], resulting in transcriptional dysbiosis of a number of genes, including neurotransmitter receptors and neuropeptides [95, 101, 102]. Expanded polyQ-containing proteins are toxic, interfering with multiple important cellular functions, including transcription [103], mitochondrial function [104], and proteostasis mechanisms [105]. Here we are focusing on the pathogenic pathways of mutant Huntington protein that affect the neuronal cytoskeleton. The microtubule-associated proteins (or MAPs) MAP1A, MAP1B, MAP2, and tau [106] are indeed the proteins that contribute to stabilizing microtubules. The enlarged PolyQ-Htt interferes with motor proteins that move cargo along microtubules, such as dynein–dynactin, which is a hallmark pathogenic mechanism of enlarged PolyQ-Htt [107, 108].

AS has been identified as a novel pathogenic mechanism in HD [6, 109, 110] that influences microtubule-associated protein molecules such as tau and MAP2, impairing their function and resulting in a toxic gain of functionality, leading to the classification of HD as a *tauopathy*.

8.1 Aberrant RNA Splicing in Serine and Arginine-Rich Splicing Factor 6

Serine and Arginine Rich Splicing Factor 6 (SRSF6) regulates AS in HD [111]. CUG repeats interact with certain splicing factors and intriguingly CAG repeats have been demonstrated to mirror CUG repeats in terms of AS dysregulation [112]. Sathasivam and colleagues (2013) bioinformatically predicted that the splicing factor SRSF6 (also referred to as SRp55) interacts with CAG repeats using the six-mer YRCRKM (T/C, A/G, C, A/G, C, A/G, G/T) [113]. This unanimity has been developed and expanded to a 9-mer pattern, with the final eight nucleotides matching the CAGCAGCA pattern perfectly [114].

SRSF6 expression is changed in HD patients and mouse model brains [110]. The volume of SRSF6 protein in Huntington's patients' striatum is increased, associated with the increased molecular weight band, which may indicate an increase in its phosphorylation level. Indeed, phospho-SR antibodies detected increased amounts of phospho-SRSF6 at the predicted molecular weight of 55 kDa in the striatum [110] and cortex [115] of Huntington patients, as well as in mouse model (R6/1) for HD [110].

In R6/1 mice, SRSF6 accumulates in nuclear inclusion bodies, where it colocalizes with Htt. Additionally, SRSF6 aggregates in human HD striatal Htt-inclusion complexes and different cytoplasm and nucleus stores with a range of morphologies have been found, comprising inclusion-, foci-, thread-, and rodlike aggregation. As a result, two SRSF6-mediated splicing alterations linked with HD have been identified, resulting in a very harmful N-terminal variant of mutant HTT [113].

8.2 Aberrant RNA Splicing in Microtubule-Associated Protein 2

Dendritic anomalies in striatal medium spiny neurons are the primary indicators of HD. Microtubule-associated protein 2 (MAP2) is the primary microtubule-associated protein in dendrites, and it is also the most abundant. The microtubule-binding protein 2 is the nearest homolog to MAP Tau [116] as Tau is the primary protein associated with microtubules that are expressed especially in axons. The microtubule-binding protein 2 is a modulator of the neuronal cytoskeleton that interacts with microtubules and F-actin, helping in their nucleation and stability. It has been postulated that MAP2 plays a crucial part in stabilizing microtubules in developed dendrites and helps to move organelles and scaffolding inside this neuronal segment [116, 117].

Multiple isoforms of MAP2 are classified into two groups: those with a high molecular weight (HMW MAP2) and those with a low molecular weight (LMW

MAP2). As a result, the N-and C-terminal sections of HMW and LMW MAP2 isoforms are identical, while exons 7–9 are omitted in LMW. As a result, it lacks a central domain. MAP2 AS is modulated during neurodevelopment and is required for the protein's subcellular localization. On the other hand, LMW isoforms are widely dispersed across neuronal regions throughout development and, with the exception of the cerebellum, are expressed at a low level in adult brains [118]. These isoforms are downregulated following the initial phases of neuronal development when high molecular weight variants (HMW-MAP2) containing the exons E7–E9 are favored. In light of two pathogenic mis-splicing instances that resulted in an extremely toxic N-terminal form of mutated Huntington and a detrimental imbalance in MAP tau isoforms with three or four tubulin-binding domains/motifs; it has recently been postulated that splicing modification contributes to HD. The SR splicing factor SRSF6 is hypothesized to be the target of both splicing events in HD.

Recently, it was shown that HD patients have substantial change in the SR splicing factor SRSF6 [111]. It has been reported that HD patients' striatum may exhibit altered MAP2 AS. Cabrera et al. (2016) detected HMW MAP2 isoforms (> 225 kDa) in controls but not in HD [115]. HD patients had a slight drop in the volume of LMW MAP2 isoforms (between 70–75 kDa). Cabrera et al. (2016) noticed a significantly elevated LMW/HMW ratio in patients with HD. All of this indicates that altered AS in the MAP2 contributes to dendritic atrophy in patients with HD [115].

9. Implications of AS in Schizophrenia

Schizophrenia is characterized by cognitive, perceptual, emotional, language, self-perception, and behavioral abnormalities. Schizophrenia is a long-lasting neurological condition that affects around 20 million individuals worldwide [119]. Humans have five G-protein-coupled dopaminergic receptors (DRD1–DRD5), which are divided into two classes: the D1 class, which includes DRD1 and DRD5, and the D2 class, which includes DRD2, DRD3, and DRD4.

9.1 D2 Class Dopaminergic Receptor RNA Splicing Anomaly

9.1.1 D2 Receptors

Research on D2 class receptors in the striatum and other parts of the brain (such as the thalamus and anterior cingulate cortex) in schizophrenia has produced mixed results. When it comes to schizophrenia's unpleasant symptoms, dopamine communication in the prefrontal cortex is mostly coordinated by DRD1. Schizophrenia medication therapies have traditionally focused on D2 class DA receptors. Schizophrenia has been linked to two aberrant splicing processes in D2 class receptors. Either exon 6 of the DRD2 gene is omitted or included, resulting in the two alternatively spliced DA receptors known as D2S and D2L. D2S and D2L can be found in the pre- and postsynaptic membranes, respectively [120]. In the classic antipsychotic drug haloperidol, D2S serves as a presynaptic auto-receptor and inhibits D1 receptor-mediated actions; D2L is the target of this medication. Zhang et al. (2007) showed that schizophrenia patients' striatum and prefrontal cortex (PFC) had dramatically lower D2S expression, meaning the D2L/D2S ratio was elevated. According to Zhang and colleagues, intron 5 and intron 6 single nucleotide polymorphisms "rs2283265 and rs1076560" were found to be significantly associated with variations in the D2L/D2S expression ratio. According to bioinformatics studies, splicing factors, including the signal-recognition particle (SRP) proteins, attach to these intronic regions. SRP55, SC35, and SRP40's binding would be affected if these SNPs were present, and vice versa. In the striatum, a reduction in D2S isoform levels is predicted to promote medium spiny neuron excitability by decreasing glutamate inhibition [121, 122].

9.1.2 D3 Receptors

Schizophrenia may be caused in part by mutations in the DRD3 gene, which codes for a D2 class DA receptor. Patients with chronic schizophrenia have a higher prevalence of an alternate DRD3 isoform that omits 98 nucleotides from exon 7 because of intra-exonic splicing ($DRD3_{nf}$). Despite the presence of GU and AG motifs immediately preceding and following the splice site, this intra-exonic variation deviates from the conventional GU-AG splicing pattern. Due to this splicing, protein synthesis is interrupted after the addition of 55 amino acids, which are not seen in the normal DRD3 protein. Increased $DRD3_{nf}$ splicing in SZ would result in a decrease in full-length DRD3 expression at a specific $DRD3/DRD3_{nf}$ ratio. The $DRD3-DRD3_{nf}$ interaction has been found to produce DRD3 mislocalization, as well as the $DRD3-DRD3_{nf}$ interaction [123].

9.2 Aberrant RNA Splicing of GABAA Receptor Genes

Most GABA signalling is mediated by type A, B, and C receptors. The $GABA_B$ receptor is metabotropic, while $GABA_A$ and $GABA_C$ are ionotropic (chloride channels). Schizophrenia has been linked to GABAergic neurotransmission dysfunction (124–126). Schizophrenia patients have also been found to have abnormal splicing in the $GABA_A$ receptor genes. The human $GABA_A$ receptor is a pentameric construct made up of at least nineteen subunits, each of which is encoded by a distinct gene (α, β, γ, θ, ε, δ, and π).

9.2.1 Aberrant RNA Splicing in GABAA Receptor β2 Subunits

The GABAA receptor β subunits consist of three parts (GABAA β1–β3). AS-variants of the $GABA_A$ β2 gene have been reported in schizophrenia patients. The inclusion or absence of exon 10 of the GABRB2 gene results in this AS. The extended and abbreviated variants were designated as $β_{2L}$ and $β_{2s}$, respectively [127, 128]. Zhao and associates (2006) found a significant decrease in the relative expression of the $β_{2L}$ type in the dorsolateral PFC of SZ patients [129]. Exon 10 of GABRB2 gene contains a calmodulin-dependent protein kinase II phosphorylation site. This enzyme is also suspected of being involved in schizophrenia. Another GABRB2 isoform is generated in which exon 10 is skipped, and exon 11 is spliced intra-exonally, leading to a 64-bp loss and frameshift (β_{2S2}) (Zhao et al. 2009). In comparison to controls, β_{2S2} expression was significantly lower in the DLPFC of schizophrenia patients. In schizophrenia men, two SNPs, rs2546620 and rs1872669, were shown to be strongly linked with b2S2 expression levels.

9.2.2 Aberrant RNA Splicing in GABAA Receptor γ2 Subunits

The GABA$_A$ receptor γ_2 subunit GABRG2 gene generates two different isoforms, γ_{2L} and γ_{2S}. Huntsman and colleagues (1998) found that SZ patients' DLPFC had 51% fewer γ_{2S} transcripts than healthy controls [130]. The γ_{2L}/γ_{2S} ratio increased because of a non-significant drop (16.9%) in γ_{2L} levels in schizophrenia patients. The eight amino acids in γ_{2L} but not γ_{2S} include a phosphorylation site for protein kinase C (Ser343). This phosphorylation inhibits GABA-activated currents at GABAA receptors [131]. The elevated γ_{2L}/γ_{2S} ratio in schizophrenia likely reduces GABA$_A$ activation.

9.3 Aberrant RNA Splicing of Glutamate Receptor Genes

There are two types of glutamate receptors: Ionotropic and metabotropic [132–134]. G-protein coupled metabotropic glutamate receptors (mGluRs; GRMs) are categorized into three subtypes [134]. The mGluRs 1 and 5, the mGluRs 2 and 3, and the mGluRs 4, 6, 7, and 8 are found in groups I, II, and III. These mGluRs can be expressed in a number of differently spliced configurations [134]. Ionotropic glutamate receptors are two types of ligand-gated ion channels: NMDA and non-NMDA receptors. NMDA receptors are composed of the NR1, NR2A-D, and NR3 subunits; AMPA receptors are composed of the GluR1–4 subunits; and kainate receptors are composed of the GluR5–7, KA1, and KA2 subunits. Additionally, non-NMDA receptors are classified as amino-3-hydroxy-5-methyl-4-isoxazole-propionate (AMPA) and kainate receptors [132, 133]. In patients with schizophrenia, abnormal AS of mGluR3 (GRM3) and NR1 (GRIN1) has been reported. Schizophrenia has been linked to specific SNPs or haplotypes of GRM3 in humans [135]. The schizophrenia-associated SNP rs2228595 in exon 3 of GRM3 has been shown to boost GRM3Δ4 expression in the DLPFC of SZ patients [136]. GRM3 exon 4 produces a membrane-spanning domain, implying that the GRM3Δ4 isoform may act as a distinct glutamate receptor.

Ca2+ channels are allosteric and ligand-gated tetrameric NMDA receptors. The NMDA receptor has several GRIN1 subunits and one NR2 (GRIN2) subunit [133]. Exons 5 and 21 are alternate 3' terminal exons of the cassette type, whereas exons 22 and 22' are alternative 3' terminating exons. By combining these alternate exons, eight isoforms can be created. Two instances of GRIN1 AS have been observed in SZ patients [137]. Numerous studies have established that the SNP rs175174 in intron 4 of the zinc finger, DHHC domain containing 8, is related to SZ (ZDHHC8). Mukai and colleagues found that the SNP rs175174 regulates the intron 4-retaining isoform's synthesis [138]. Mukai and coworkers (2004) found that the SZ SNP rs175174 regulates the intron 4-retaining isoform [138]. This variant encodes a small protein with a stop codon in the retained intron 4. The palmitoylation of proteins in the postsynaptic density may create abnormalities in the glutamate-signalling system in a mouse model [139]. This protein is palmitoylated and binds AMPA and NMDA receptors in postsynaptic membranes [140].

References

[1] Goodwin, S., McPherson, J.D. and McCombie, W.R. (2016, May). Coming of age: Ten years of next-generation sequencing technologies. Nat. Rev. Genet. 17(6): 333–51.

[2] Graveley, B.R. (2001, Feb). Alternative splicing: Increasing diversity in the proteomic world. Trends Genet TIG 17(2): 100–7.

[3] Tazi, J., Bakkour, N. and Stamm, S. (2009, Jan). Alternative splicing and disease. Biochim. Biophys. Acta 1792(1): 14–26.

[4] Keren, H., Lev-Maor, G. and Ast, G. (2010, May). Alternative splicing and evolution: Diversification, exon definition and function. Nat. Rev. Genet. 11(5): 345–55.

[5] Tian, B. and Manley, J.L. (2017, Jan). Alternative polyadenylation of mRNA precursors. Nat. Rev. Mol. Cell Biol. 18(1): 18–30.

[6] Kim, H.K., Pham, M.H.C., Ko, K.S., Rhee, B.D. and Han, J. (2018, Jul). Alternative splicing isoforms in health and disease. Pflugers Arch. 470(7): 995–1016.

[7] Smith, C.W. and Valcárcel, J. (2000, Aug). Alternative pre-mRNA splicing: The logic of combinatorial control. Trends Biochem. Sci. 25(8): 381–8.

[8] Wang, Z. and Burge, C.B. (2008, May). Splicing regulation: From a parts list of regulatory elements to an integrated splicing code. RNA N Y N 14(5): 802–13.

[9] Sterne-Weiler, T. and Sanford, J.R. (2014, Jan). Exon identity crisis: Disease-causing mutations that disrupt the splicing code. Genome Biol. 15(1): 201.

[10] Mardon, H.J., Sebastio, G. and Baralle, F.E. (1987, Oct). A role for exon sequences in alternative splicing of the human fibronectin gene. Nucleic Acids Res. 15(19): 7725–33.

[11] Bergsma, A.J., van der Wal, E., Broeders, M., van der Ploeg, A.T. and Pim Pijnappel, W.W.M. (2018). Alternative splicing in genetic diseases: Improved diagnosis and novel treatment options. Int. Rev. Cell Mol. Biol. 335: 85–141.

[12] Chen, J., Liu, Y., Min, J., Wang, H., Li, F., Xu, C. et al. (2021). Alternative splicing of lncRNAs in human diseases. Am. J. Cancer Res. 11(3): 624–39.

[13] Dery, K.J., Gusti, V., Gaur, S., Shively, J.E., Yen, Y. and Gaur, R.K. (2009). Alternative splicing as a therapeutic target for human diseases. Methods Mol. Biol. Clifton NJ 555: 127–44.

[14] Montes, M., Sanford, B.L., Comiskey, D.F. and Chandler, D.S. (2019, Jan). RNA splicing and disease: Animal models to therapies. Trends Genet TIG 35(1): 68–87.

[15] Ohe, K. and Hagiwara, M. (2015, Apr). Modulation of alternative splicing with chemical compounds in new therapeutics for human diseases. ACS Chem. Biol. 10(4): 914–24.

[16] Pearn, J. (1980, Apr). Classification of spinal muscular atrophies. Lancet Lond. Engl. 1(8174): 919–22.

[17] Zhang, Z., Lotti, F., Dittmar, K., Younis, I., Wan, L., Kasim, M. et al. (2008, May). SMN deficiency causes tissue-specific perturbations in the repertoire of snRNAs and widespread defects in splicing. Cell 133(4): 585–600.

[18] Oprea, G.E., Kröber, S., McWhorter, M.L., Rossoll, W., Müller, S., Krawczak, M. et al. (2008, Apr). Plastin 3 is a protective modifier of autosomal recessive spinal muscular atrophy. Science 320(5875): 524–7.

[19] Wirth, B., Brichta, L. and Hahnen, E. (2006). Spinal muscular atrophy and therapeutic prospects. Prog. Mol. Subcell Biol. 44: 109–32.

[20] Monani, U.R., Sendtner, M., Coovert, D.D., Parsons, D.W., Andreassi, C., Le, T.T. et al. (2000, Feb). The human centromeric survival motor neuron gene (SMN2) rescues embryonic lethality in Smn(-/-) mice and results in a mouse with spinal muscular atrophy. Hum. Mol. Genet. 9(3): 333–9.

[21] Feldkötter, M., Schwarzer, V., Wirth, R., Wienker, T.F. and Wirth, B. (2002, Feb). Quantitative analyses of SMN1 and SMN2 based on real-time light Cycler PCR: Fast and highly reliable carrier testing and prediction of severity of spinal muscular atrophy. Am. J. Hum. Genet. 70(2): 358–68.

[22] Cartegni, L., Hastings, M.L., Calarco, J.A., de Stanchina, E. and Krainer, A.R. (2006, Jan). Determinants of exon 7 splicing in the spinal muscular atrophy genes, SMN1 and SMN2. Am. J. Hum. Genet. 78(1): 63–77.

[23] Cartegni, L. and Krainer, A.R. (2002, Apr). Disruption of an SF2/ASF-dependent exonic splicing enhancer in SMN2 causes spinal muscular atrophy in the absence of SMN1. Nat. Genet. 30(4): 377–84.

[24] Kashima, T., Rao, N., David, C.J. and Manley, J.L. (2007, Dec). hnRNP A1 functions with specificity in repression of SMN2 exon 7 splicing. Hum. Mol. Genet. 16(24): 3149–59.

[25] Kashima, T. and Manley, J.L. (2003, Aug). A negative element in SMN2 exon 7 inhibits splicing in spinal muscular atrophy. Nat. Genet. 34(4): 460–3.

[26] Novoyatleva, T., Heinrich, B., Tang, Y., Benderska, N., Butchbach, M.E.R., Lorson, C.L. et al. (2008, Jan). Protein phosphatase 1 binds to the RNA recognition motif of several splicing factors and regulates alternative pre-mRNA processing. Hum. Mol. Genet. 17(1): 52–70.

[27] Ryan, M., Wong, W.C., Brown, R., Akbani, R., Su, X., Broom, B. et al. (2016, Jan). TCGA Splice Seq a compendium of alternative mRNA splicing in cancer. Nucleic Acids Res. 44(D1): D1018–1022.

[28] Sveen, A., Kilpinen, S., Ruusulehto, A., Lothe, R.A. and Skotheim, R.I. (2016, May). Aberrant RNA splicing in cancer; expression changes and driver mutations of splicing factor genes. Oncogene 35(19): 2413–27.

[29] Kalsotra, A., Xiao, X., Ward, A.J., Castle, J.C., Johnson, J.M., Burge, C.B. et al. (2008, Dec). A postnatal switch of CELF and MBNL proteins reprograms alternative splicing in the developing heart. Proc. Natl. Acad. Sci. USA 105(51): 20333–8.

[30] Rajan, S., Jagatheesan, G., Karam, C.N., Alves, M.L., Bodi, I., Schwartz, A. et al. (2010, Jan). Molecular and functional characterization of a novel cardiac-specific human tropomyosin isoform. Circulation 121(3): 410–8.

[31] Watkins, H., Conner, D., Thierfelder, L., Jarcho, J.A., MacRae, C., McKenna, W.J. et al. (1995, Dec). Mutations in the cardiac myosin binding protein-C gene on chromosome 11 cause familial hypertrophic cardiomyopathy. Nat. Genet. 11(4): 434–7.

[32] Schoenauer, R., Emmert, M.Y., Felley, A., Ehler, E., Brokopp, C., Weber, B. et al. (2011, Mar). EH-myomesin splice isoform is a novel marker for dilated cardiomyopathy. Basic. Res. Cardiol. 106(2): 233–47.

[33] Arimura, T., Inagaki, N., Hayashi, T., Shichi, D., Sato, A., Hinohara, K. et al. (2009, Jul). Impaired binding of ZASP/Cypher with phosphoglucomutase 1 is associated with dilated cardiomyopathy. Cardiovasc. Res. 83(1): 80–8.

[34] Makarenko, I., Opitz, C.A., Leake, M.C., Neagoe, C., Kulke, M., Gwathmey, J.K. et al. (2004, Oct). Passive stiffness changes caused by upregulation of compliant titin isoforms in human dilated cardiomyopathy hearts. Circ. Res. 95(7): 708–16.

[35] Cunha, S.R., Le Scouarnec, S., Schott, J.J. and Mohler, P.J. (2008, Dec). Exon organization and novel alternative splicing of the human ANK2 gene: Implications for cardiac function and human cardiac disease. J. Mol. Cell. Cardiol. 45(6): 724–34.

[36] Santini, M.P., Lexow, J., Borsellino, G., Slonimski, E., Zarrinpashneh, E., Poggioli, T. et al. (2011, Jul). IGF-1Ea induces vessel formation after injury and mediates bone marrow and heart cross-talk through the expression of specific cytokines. Biochem. Biophys. Res. Commun. 410(2): 201–7.

[37] Vinciguerra, M., Santini, M.P., Martinez, C., Pazienza, V., Claycomb, W.C., Giuliani, A. et al. (2012, Feb). mIGF-1/JNK1/SirT1 signaling confers protection against oxidative stress in the heart. Aging Cell 11(1): 139–49.

[38] Cáceres, J.F. and Kornblihtt, A.R. (2002, Apr). Alternative splicing: Multiple control mechanisms and involvement in human disease. Trends Genet TIG 18(4): 186–93.

[39] Dehghanian, F., Hojati, Z. and Kay, M. (2014, Oct). New insights into VEGF-A alternative splicing: Key regulatory switching in the pathological process. Avicenna J. Med. Biotechnol. 6(4): 192–9.

[40]   Athyros, V.G., Kakafika, A.I., Tziomalos, K., Karagiannis, A. and Mikhailidis, D.P. (2008, Jul). Antisense technology for the prevention or the treatment of cardiovascular disease: The next blockbuster? Expert. Opin. Investig. Drugs 17(7): 969–72.

[41]   Phillips, M.I., Costales, J., Lee, R.J., Oliveira, E. and Burns, A.B. (2015). Antisense therapy for cardiovascular diseases. Curr. Pharm. Des. 21(30): 4417–26.

[42]   Hartmann, T., Bergsdorf, C., Sandbrink, R., Tienari, P.J., Multhaup, G., Ida, N. et al. (1996, May). Alzheimer's disease betaA4 protein release and amyloid precursor protein sorting are regulated by alternative splicing. J. Biol. Chem. 271(22): 13208–14.

[43]   Rockenstein, E.M., McConlogue, L., Tan, H., Power, M., Masliah, E. and Mucke, L. (1995, Nov). Levels and alternative splicing of amyloid beta protein precursor (APP) transcripts in brains of APP transgenic mice and humans with Alzheimer's disease. J. Biol. Chem. 270(47): 28257–67.

[44]   Lee, C., Low, C.Y.B., Wong, S.Y., Lai, M.K.P. and Tan, M.G.K. (2017, Mar). Selective induction of alternatively spliced FynT isoform by TNF facilitates persistent inflammatory responses in astrocytes. Sci. Rep. 7: 43651.

[45]   Glatz, D.C., Rujescu, D., Tang, Y., Berendt, F.J., Hartmann, A.M., Faltraco, F. et al. (2006, Feb). The alternative splicing of tau exon 10 and its regulatory proteins CLK2 and TRA2-BETA1 changes in sporadic Alzheimer's disease. J. Neurochem. 96(3): 635–44.

[46]   Hinrich, A.J., Jodelka, F.M., Chang, J.L., Brutman, D., Bruno, A.M., Briggs, C.A. et al. (2016, Apr). Therapeutic correction of ApoER2 splicing in Alzheimer's disease mice using antisense oligonucleotides. EMBO Mol. Med. 8(4): 328–45.

[47]   Love, J.E., Hayden, E.J. and Rohn, T.T. (2015, Aug). Alternative splicing in Alzheimer's Disease. J. Park. Dis. Alzheimers Dis. 2(2): 6.

[48]   Mills, J.D., Sheahan, P.J., Lai, D., Kril, J.J., Janitz, M. and Sutherland, G.T. (2014, Oct). The alternative splicing of the apolipoprotein E gene is unperturbed in the brains of Alzheimer's disease patients. Mol. Biol. Rep. 41(10): 6365–76.

[49]   Rogaev, E.I., Sherrington, R., Wu, C., Levesque, G., Liang, Y., Rogaeva, E.A. et al. (1997, Mar). Analysis of the 5' sequence, genomic structure, and alternative splicing of the presenilin-1 gene (PSEN1) associated with early onset Alzheimer disease. Genomics 40(3): 415–24.

[50]   Rogaev, E.I., Sherrington, R., Wu, C., Levesque, G., Liang, Y., Rogaeva, E.A. et al. (1997, Mar). Analysis of the 5' sequence, genomic structure, and alternative splicing of the presenilin-1 gene (PSEN1) associated with early onset Alzheimer disease. Genomics 40(3): 415–24.

[51]   Heinzen, E.L., Yoon, W., Weale, M.E., Sen, A., Wood, N.W., Burke, J.R. et al. (2007). Alternative ion channel splicing in mesial temporal lobe epilepsy and Alzheimer's disease. Genome Biol. 8(3): R32.

[52]   Lai, M.K.P., Esiri, M.M. and Tan, M.G.K. (2014, Dec). Genome-wide profiling of alternative splicing in Alzheimer's disease. Genomics Data 2: 290–2.

[53]   Ishunina, T.A. and Swaab, D.F. (2012, Feb). Decreased alternative splicing of estrogen receptor-α mRNA in the Alzheimer's disease brain. Neurobiol Aging 33(2): 286–296.e3.

[54]   Dexter, D.T. and Jenner, P. (2013, Sep). Parkinson disease: From pathology to molecular disease mechanisms. Free Radic. Biol. Med. 62: 132–44.

[55]   Heller, J., Dogan, I., Schulz, J.B. and Reetz, K. (2014, Feb). Evidence for gender differences in cognition, emotion and quality of life in Parkinson's disease? Aging Dis. 5(1): 63–75.

[56]   Kobylecki, C., Crossman, A.R. and Ravenscroft, P. (2013, Sep). Alternative splicing of AMPA receptor subunits in the 6-OHDA-lesioned rat model of Parkinson's disease and L-DOPA-induced dyskinesia. Exp. Neurol. 247: 476–84.

[57]   Soreq, L., Bergman, H., Israel, Z. and Soreq, H. (2012). Exon arrays reveal alternative splicing aberrations in Parkinson's disease leukocytes. Neurodegener Dis. 10(1-4): 203–6.

[58]   Polymeropoulos, M.H., Lavedan, C., Leroy, E., Ide, S.E., Dehejia, A., Dutra, A. et al. (1997, Jun). Mutation in the alpha-synuclein gene identified in families with Parkinson's disease. Science 276(5321): 2045–7.

[59]   Spillantini, M.G., Schmidt, M.L., Lee, V.M., Trojanowski, J.Q., Jakes, R. and Goedert, M. (1997, Aug). Alpha-synuclein in Lewy bodies. Nature 388(6645): 839–40.

[60]   Beyer, K., Domingo-Sàbat, M., Humbert, J., Carrato, C., Ferrer, I. and Ariza, A. (2008, Jul). Differential expression of alpha-synuclein, parkin, and synphilin-1 isoforms in Lewy body disease. Neurogenetics 9(3): 163–72.

[61] Campion, D., Martin, C., Heilig, R., Charbonnier, F., Moreau, V., Flaman, J.M. et al. (1995, Mar). The NACP/synuclein gene: Chromosomal assignment and screening for alterations in Alzheimer disease. Genomics 26(2): 254–7.

[62] Uéda, K., Saitoh, T. and Mori, H. (1994, Dec). Tissue-dependent alternative splicing of mRNA for NACP, the precursor of non-A beta component of Alzheimer's disease amyloid. Biochem. Biophys. Res. Commun. 205(2): 1366–72.

[63] Baba, M., Nakajo, S., Tu, P.H., Tomita, T., Nakaya, K., Lee, V.M. et al. (1998, Apr). Aggregation of alpha-synuclein in Lewy bodies of sporadic Parkinson's disease and dementia with Lewy bodies. Am. J. Pathol. 152(4): 879–84.

[64] Lee, H.J., Choi, C. and Lee, S.J. (2002, Jan). Membrane-bound alpha-synuclein has a high aggregation propensity and the ability to seed the aggregation of the cytosolic form. J. Biol. Chem. 277(1): 671–8.

[65] Bosgraaf, L. and Van Haastert, P.J.M. (2003, Dec). Roc, a Ras/GTPase domain in complex proteins. Biochim. Biophys. Acta 1643(1-3): 5–10.

[66] Lewis, P.A. (2009, Mar). The function of ROCO proteins in health and disease. Biol Cell. 101(3): 183–91.

[67] Paisán-Ruíz, C., Jain, S., Evans, E.W., Gilks, W.P., Simón, J., van der Brug, M. et al. (2004, Nov). Cloning of the gene containing mutations that cause PARK8-linked Parkinson's disease. Neuron 44(4): 595–600.

[68] Zimprich, A., Biskup, S., Leitner, P., Lichtner, P., Farrer, M., Lincoln, S. et al. (2004, Nov). Mutations in LRRK2 cause autosomal-dominant parkinsonism with pleomorphic pathology. Neuron 44(4): 601–7.

[69] Di Fonzo, A., Tassorelli, C., De Mari, M., Chien, H.F., Ferreira, J., Rohé, C.F. et al. (2006, Mar). Comprehensive analysis of the LRRK2 gene in sixty families with Parkinson's disease. Eur. J. Hum. Genet. EJHG 14(3): 322–31.

[70] Johnson, J., Paisán-Ruíz, C., Lopez, G., Crews, C., Britton, A., Malkani, R. et al. (2007). Comprehensive screening of a north american parkinson's disease cohort for LRRK2 mutation. Neurodegener. Dis. 4(5): 386–91.

[71] Elliott, D.A., Kim, W.S., Gorissen, S., Halliday, G.M. and Kwok, J.B.J. (2012, Jul). Leucine-rich repeat kinase 2 and alternative splicing in Parkinson's disease. Mov. Disord Off J. Mov. Disord. Soc. 27(8): 1004–11.

[72] Kalivendi, S.V., Yedlapudi, D., Hillard, C.J. and Kalyanaraman, B. (2010, Feb). Oxidants induce alternative splicing of alpha-synuclein: Implications for Parkinson's disease. Free Radic. Biol. Med. 48(3): 377–83.

[73] Shehadeh, L.A., Yu, K., Wang, L., Guevara, A., Singer, C., Vance, J. et al. (2010, Feb). SRRM2, a potential blood biomarker revealing high alternative splicing in Parkinson's disease. PloS One 5(2): e9104.

[74] Imai, Y., Soda, M. and Takahashi, R. (2000, Nov). Parkin suppresses unfolded protein stress-induced cell death through its E3 ubiquitin-protein ligase activity. J. Biol. Chem. 275(46): 35661–4.

[75] Shimura, H., Hattori, N., Kubo, Si, Mizuno, Y., Asakawa, S., Minoshima, S. et al. Familial Parkinson disease gene product, parkin, is a ubiquitin-protein ligase. Nat. Genet. 25(3): 302–5.

[76] Humbert, J., Beyer, K., Carrato, C., Mate, J.L., Ferrer, I. and Ariza, A. (2007, Jun). Parkin and synphilin-1 isoform expression changes in Lewy body diseases. Neurobiol. Dis. 26(3): 681–7.

[77] Dagata, V. and Cavallaro, S. (2004, Sep). Parkin transcript variants in rat and human brain. Neurochem. Res. 29(9): 1715–24.

[78] Tan, E.K., Shen, H., Tan, J.M.M., Lim, K.L., Fook-Chong, S., Hu, W.P. et al. (2005, Dec). Differential expression of splice variant and wild-type parkin in sporadic Parkinson's disease. Neurogenetics 6(4): 179–84.

[79] Chung, K.K., Zhang, Y., Lim, K.L., Tanaka, Y., Huang, H., Gao, J. et al. (2001, Oct). Parkin ubiquitinates the alpha-synuclein-interacting protein, synphilin-1: Implications for Lewy-body formation in Parkinson disease. Nat. Med. 7(10): 1144–50.

[80]   Engelender, S., Kaminsky, Z., Guo, X., Sharp, A.H., Amaravi, R.K., Kleiderlein, J.J. et al. (1999, May). Synphilin-1 associates with alpha-synuclein and promotes the formation of cytosolic inclusions. Nat. Genet. 22(1): 110–4.

[81]   Eyal, A., Szargel, R., Avraham, E., Liani, E., Haskin, J., Rott, R. et al. (2006, Apr). Synphilin-1A: An aggregation-prone isoform of synphilin-1 that causes neuronal death and is present in aggregates from alpha-synucleinopathy patients. Proc. Natl. Acad. Sci. USA 103(15): 5917–22.

[82]   Hong, M., Zhukareva, V., Vogelsberg-Ragaglia, V., Wszolek, Z., Reed, L., Miller, B.I. et al. (1998, Dec). Mutation-specific functional impairments in distinct tau isoforms of hereditary FTDP-17. Science 282(5395): 1914–7.

[83]   Hutton, M., Lendon, C.L., Rizzu, P., Baker, M., Froelich, S., Houlden, H. et al. (1998, Jun). Association of missense and 5'-splice-site mutations in tau with the inherited dementia FTDP-17. Nature 393(6686): 702–5.

[84]   Hutton, M. (2001, Jun). Missense and splice site mutations in tau associated with FTDP-17: Multiple pathogenic mechanisms. Neurology 56(11 Suppl 4): S21–25.

[85]   D'Souza, I., Poorkaj, P., Hong, M., Nochlin, D., Lee, V.M., Bird, T.D. et al. (1999, May). Missense and silent tau gene mutations cause frontotemporal dementia with parkinsonism-chromosome 17 type, by affecting multiple alternative RNA splicing regulatory elements. Proc. Natl. Acad. Sci. USA 96(10): 5598–603.

[86]   D'Souza, I. and Schellenberg, G.D. (2000, Jun). Determinants of 4-repeat tau expression. Coordination between enhancing and inhibitory splicing sequences for exon 10 inclusion. J. Biol. Chem. 275(23): 17700–9.

[87]   Gallo, J.M., Noble, W. and Martin, T.R. (2007, Jul). RNA and protein-dependent mechanisms in tauopathies: Consequences for therapeutic strategies. Cell Mol. Life Sci. CMLS 64(13): 1701–14.

[88]   Andreadis, A. (2006). Misregulation of tau alternative splicing in neurodegeneration and dementia. Prog. Mol. Subcell. Biol. 44: 89–107.

[89]   Hasegawa, M., Smith, M.J., Iijima, M., Tabira, T. and Goedert, M. (1999, Jan). FTDP-17 mutations N279K and S305N in tau produce increased splicing of exon 10. FEBS Lett. 443(2): 93–6.

[90]   Jiang, Z., Tang, H., Havlioglu, N., Zhang, X., Stamm, S., Yan, R. et al. (2003, May). Mutations in tau gene exon 10 associated with FTDP-17 alter the activity of an exonic splicing enhancer to interact with Tra2 beta. J. Biol. Chem. 278(21): 18997–9007.

[91]   Wang, J., Gao, Q.S., Wang, Y., Lafyatis, R., Stamm, S. and Andreadis, A. (2004, Mar). Tau exon 10, whose missplicing causes frontotemporal dementia, is regulated by an intricate interplay of cis elements and trans factors. J. Neurochem. 88(5): 1078–90.

[92]   Boutajangout, A., Boom, A., Leroy, K. and Brion, J.P. (2004, Oct). Expression of tau mRNA and soluble tau isoforms in affected and non-affected brain areas in Alzheimer's disease. FEBS Lett. 576(1-2): 183–9.

[93]   Connell, J.W., Rodriguez-Martin, T., Gibb, G.M., Kahn, N.M., Grierson, A.J., Hanger, D.P. et al. Quantitative analysis of tau isoform transcripts in sporadic tauopathies. Brain Res. Mol. Brain Res. 137(1-2): 104–9.

[94]   Ingelsson, M., Ramasamy, K., Cantuti-Castelvetri, I., Skoglund, L., Matsui, T., Orne, J. et al. (2006, Oct). No alteration in tau exon 10 alternative splicing in tangle-bearing neurons of the Alzheimer's disease brain. Acta Neuropathol (Berl). 112(4): 439–49.

[95]   Ross, C.A. and Shoulson, I. (2009, Dec). Huntington disease: Pathogenesis, biomarkers, and approaches to experimental therapeutics. Parkinsonism Relat. Disord. 15 Suppl 3: S135–138.

[96]   Williams, A.J. and Paulson, H.L. (2008, Oct). Polyglutamine neurodegeneration: Protein misfolding revisited. Trends Neurosci. 31(10): 521–8.

[97]   Hodges, A., Strand, A.D., Aragaki, A.K., Kuhn, A., Sengstag, T., Hughes, G. et al. (2006, Mar). Regional and cellular gene expression changes in human Huntington's disease brain. Hum. Mol. Genet. 15(6): 965–77.

[98]   Thu, D.C.V., Oorschot, D.E., Tippett, L.J., Nana, A.L., Hogg, V.M., Synek, B.J. et al. (2010, Apr). Cell loss in the motor and cingulate cortex correlates with symptomatology in Huntington's disease. Brain J. Neurol. 133(Pt 4): 1094–110.

[99] Benn, C.L., Sun, T., Sadri-Vakili, G., McFarland, K.N., DiRocco, D.P., Yohrling, G.J. et al. (2008, Oct). Huntingtin modulates transcription, occupies gene promoters *in vivo*, and binds directly to DNA in a polyglutamine-dependent manner. J. Neurosci. Off. J. Soc. Neurosci. 28(42): 10720–33.

[100] Kegel, K.B., Meloni, A.R., Yi, Y., Kim, Y.J., Doyle, E., Cuiffo, B.G. et al. (202, Mar). Huntingtin is present in the nucleus, interacts with the transcriptional corepressor C-terminal binding protein, and represses transcription. J. Biol. Chem. 277(9): 7466–76.

[101] Augood, S.J., Faull, R.L. and Emson, P.C. (1997, Aug). Dopamine D1 and D2 receptor gene expression in the striatum in Huntington's disease. Ann. Neurol. 42(2): 215–21.

[102] Cha, J.H., Kosinski, C.M., Kerner, J.A., Alsdorf, S.A., Mangiarini, L., Davies, S.W. et al. (1998, May). Altered brain neurotransmitter receptors in transgenic mice expressing a portion of an abnormal human huntington disease gene. Proc. Natl. Acad. Sci. USA 95(11): 6480–5.

[103] McCampbell, A. and Fischbeck, K.H. (2001, May). Polyglutamine and CBP: Fatal attraction? Nat. Med. 7(5): 528–30.

[104] Yano, H., Baranov, S.V., Baranova, O.V., Kim, J., Pan, Y., Yablonska, S. et al. (2014, Jun). Inhibition of mitochondrial protein import by mutant huntingtin. Nat. Neurosci. 17(6): 822–31.

[105] Ortega, Z. and Lucas, J.J. (2014). Ubiquitin-proteasome system involvement in Huntington's disease. Front. Mol. Neurosci. 7: 77.

[106] Matus, A. (1988). Microtubule-associated proteins: Their potential role in determining neuronal morphology. Annu. Rev. Neurosci. 11: 29–44.

[107] Gauthier, L.R., Charrin, B.C., Borrell-Pagès, M., Dompierre, J.P., Rangone, H., Cordelières, F.P. et al. (2004, Jul). Huntingtin controls neurotrophic support and survival of neurons by enhancing BDNF vesicular transport along microtubules. Cell 118(1): 127–38.

[108] Li, S.H., Gutekunst, C.A., Hersch, S.M. and Li, X.J. (1998, Feb). Interaction of huntingtin-associated protein with dynactin P150Glued. J. Neurosci. Off. J. Soc. Neurosci. 18(4): 1261–9.

[109] Faulty splicing in Huntington's disease | Nature Reviews Neuroscience [Internet]. [cited 2022 Aug 27]. Available from: https://www.nature.com/articles/nrn3807.

[110] Fernández-Nogales, M., Santos-Galindo, M., Hernández, I.H., Cabrera, J.R. and Lucas, J.J. (2016, Nov). Faulty splicing and cytoskeleton abnormalities in Huntington's disease. Brain Pathol. Zurich. Switz. 26(6): 772–8.

[111] Vuono, R., Winder-Rhodes, S., de Silva, R., Cisbani, G. and Drouin-Ouellet. J. (2015, Jul). Registry Investigators of the European Huntington's Disease Network, et al. The role of tau in the pathological process and clinical expression of Huntington's disease. Brain J. Neurol. 138(Pt 7): 1907–18.

[112] Mykowska, A., Sobczak, K., Wojciechowska, M., Kozlowski, P. and Krzyzosiak, W.J. (2011, Nov). CAG repeats mimic CUG repeats in the misregulation of alternative splicing. Nucleic Acids Res. 39(20): 8938–51.

[113] Sathasivam, K., Neueder, A., Gipson, T.A., Landles, C., Benjamin, A.C., Bondulich, M.K. et al. (2013, Feb). Aberrant splicing of HTT generates the pathogenic exon 1 protein in Huntington disease. Proc. Natl. Acad. Sci. USA 110(6): 2366–70.

[114] Jensen, M.A., Wilkinson, J.E. and Krainer, A.R. (2014, Feb). Splicing factor SRSF6 promotes hyperplasia of sensitized skin. Nat. Struct. Mol. Biol. 21(2): 189–97.

[115] Cabrera, J.R. and Lucas, J.J. (2017, Mar). MAP2 splicing is altered in Huntington's disease. Brain Pathol. Zurich. Switz. 27(2): 181–9.

[116] Dehmelt, L. and Halpain, S. (2005). The MAP2/Tau family of microtubule-associated proteins. Genome Biol. 6(1): 204.

[117] Arnold, S.E., Lee, V.M., Gur, R.E. and Trojanowski, J.Q. (1991, Dec). Abnormal expression of two microtubule-associated proteins (MAP2 and MAP5) in specific subfields of the hippocampal formation in schizophrenia. Proc. Natl. Acad. Sci. USA 88(23): 10850–4.

[118] Ferhat, L., Represa, A., Ferhat, W., Ben-Ari, Y. and Khrestchatisky, M. (1998, Jan). MAP2d mRNA is expressed in identified neuronal populations in the developing and adult rat brain and its subcellular distribution differs from that of MAP2b in hippocampal neurones. Eur. J. Neurosci. 10(1): 161–71.

[119] GBD. (2017). Disease and Injury Incidence and Prevalence Collaborators. Global, regional, and national incidence, prevalence, and years lived with disability for 354 diseases and injuries for 195 countries and territories, 1990–2017: A systematic analysis for the Global Burden of Disease Study 2017. Lancet Lond. Engl. 392(10159): 1789–858.

[120] Usiello, A., Baik, J.H., Rougé-Pont, F., Picetti, R., Dierich, A., LeMeur, M. et al. (2000, Nov). Distinct functions of the two isoforms of dopamine D2 receptors. Nature 408(6809): 199–203.

[121] Centonze, D., Grande, C., Usiello, A., Gubellini, P., Erbs, E., Martin, A.B. et al. (2003, Jul). Receptor subtypes involved in the presynaptic and postsynaptic actions of dopamine on striatal interneurons. J. Neurosci. Off. J. Soc. Neurosci. 23(15): 6245–54.

[122] Centonze, D., Gubellini, P., Usiello, A., Rossi, S., Tscherter, A., Bracci, E. et al. (2004). Differential contribution of dopamine D2S and D2L receptors in the modulation of glutamate and GABA transmission in the striatum. Neuroscience 129(1): 157–66.

[123] Karpa, K.D., Lin, R., Kabbani, N. and Levenson, R. (2000, Oct). The dopamine D3 receptor interacts with itself and the truncated D3 splice variant d3nf: D3-D3nf interaction causes mislocalization of D3 receptors. Mol. Pharmacol. 58(4): 677–83.

[124] Guidotti, A., Auta, J., Davis, J.M., Dong, E., Grayson, D.R., Veldic, M. et al. (2005, Jul). GABAergic dysfunction in schizophrenia: New treatment strategies on the horizon. Psychopharmacology (Berl). 180(2): 191–205.

[125] Kantrowitz, J., Citrome, L. and Javitt, D. (2009, Aug). GABA(B) receptors, schizophrenia and sleep dysfunction: A review of the relationship and its potential clinical and therapeutic implications. CNS Drugs 23(8): 681–91.

[126] Ong, J. and Kerr, D.I.B. (2005). Clinical potential of GABAB receptor modulators. CNS Drug Rev. 11(3): 317–34.

[127] McKinley, D.D., Lennon, D.J. and Carter, D.B. (1995, Jan). Cloning, sequence analysis and expression of two forms of mRNA coding for the human beta 2 subunit of the GABAA receptor. Brain Res. Mol. Brain Res. 28(1): 175–9.

[128] Mueller, T.M., Remedies, C.E., Haroutunian, V. and Meador-Woodruff, J.H. (2015, Aug). Abnormal subcellular localization of GABAA receptor subunits in schizophrenia brain. Transl. Psychiatry 5: e612.

[129] Zhao, C., Xu, Z., Chen, J., Yu, Z., Tong, K.L., Lo, W.S. et al. (2006, Dec). Two isoforms of GABA(A) receptor beta2 subunit with different electrophysiological properties: Differential expression and genotypical correlations in schizophrenia. Mol. Psychiatry 11(12): 1092–105.

[130] Huntsman, M.M., Tran, B.V., Potkin, S.G., Bunney, W.E. and Jones, E.G. (1998, Dec). Altered ratios of alternatively spliced long and short gamma2 subunit mRNAs of the gamma-amino butyrate type A receptor in prefrontal cortex of schizophrenics. Proc. Natl. Acad. Sci. USA 95(25): 15066–71.

[131] Krishek, B.J., Xie, X., Blackstone, C., Huganir, R.L., Moss, S.J. and Smart, T.G. (1994, May). Regulation of GABAA receptor function by protein kinase C phosphorylation. Neuron 12(5): 1081–95.

[132] Marek, G.J., Behl, B., Bespalov, A.Y., Gross, G., Lee, Y. and Schoemaker, H. (2010, Mar). Glutamatergic (N-Methyl-d-aspartate Receptor) hypofrontality in schizophrenia: Too little juice or a miswired brain? Mol. Pharmacol. 77(3): 317–26.

[133] Dingledine, R., Borges, K., Bowie, D. and Traynelis, S.F. (1999, Mar). The glutamate receptor ion channels. Pharmacol. Rev. 51(1): 7–61.

[134] Niswender, C.M. and Conn, P.J. (2010). Metabotropic glutamate receptors: Physiology, pharmacology, and disease. Annu. Rev. Pharmacol. Toxicol. 50: 295–322.

[135] Egan, M.F., Straub, R.E., Goldberg, T.E., Yakub, I., Callicott, J.H., Hariri, A.R. et al. (2004, Aug). Variation in GRM3 affects cognition, prefrontal glutamate, and risk for schizophrenia. Proc. Natl. Acad. Sci. USA 101(34): 12604–9.

[136] Sartorius, L.J., Weinberger, D.R., Hyde, T.M., Harrison, P.J., Kleinman, J.E. and Lipska, B.K. (2008, Oct). Expression of a GRM3 splice variant is increased in the dorsolateral prefrontal cortex of individuals carrying a schizophrenia risk SNP. Neuropsychopharmacol. Off. Publ. Am. Coll. Neuropsychopharmacol. 33(11): 2626–34.

[137] Le Corre, S., Harper, C.G., Lopez, P., Ward, P. and Catts, S. (2000, Apr). Increased levels of expression of an NMDARI splice variant in the superior temporal gyrus in schizophrenia. Neuroreport 11(5): 983–6.

[138] Mukai, J., Liu, H., Burt, R.A., Swor, D.E., Lai, W.S., Karayiorgou, M. et al. (2004, Jul). Evidence that the gene encoding ZDHHC8 contributes to the risk of schizophrenia. Nat Genet. 36(7): 725–31.

[139] Mukai, J., Dhilla, A., Drew, L.J., Stark, K.L., Cao, L., MacDermott, A.B. et al. (2008, Nov). Palmitoylation-dependent neurodevelopmental deficits in a mouse model of the 22q11 microdeletion. Nat. Neurosci. 11(11): 1302–10.

[140] el-Husseini A. el-Din and Bredt, D.S. (2002, Oct). Protein palmitoylation: A regulator of neuronal development and function. Nat. Rev. Neurosci. 3(10): 791–802.

2

Mechanism of RNA Splicing

Tabasum Ashraf, Humaira Shah, Rouf Maqbool,
Auqib Manzoor and *Ashraf Dar**

1. Introduction

In living organisms, the genetic information encrypted in the DNA is transcribed into the related molecules called messenger RNAs (mRNAs), which are then translated into proteins. However, in higher organisms, such as plants and animals, the coding regions in the genes (called exons) are quite often intervened by non-coding regions (called introns). During transcription, the entire gene is copied into the primary transcript or pre-mRNA, which contains both introns and exons. The introns are then excised out from the pre-mRNA, and exons are joined together into a contiguous sequence or mature mRNA by a well-organized mechanistic process called RNA splicing [1]. The mature mRNA forms the template for ribosome-mediated translation or protein synthesis.

The 5' and 3' ends of introns are marked by short conserved sequences known as 5' and 3' splice sites (Figure 1). Most commonly, a splice site in a removable intron begins with dinucleotide, GU at the 5' end, and AG at the 3' end [2]. These consensus sequences are critical for splicing. Very rarely, splice sites begin with AU and AC at 5' and 3' ends. Any change or mutation in a single residue inhibits splicing. In addition, another very important sequence located anywhere between 18 to 40 nucleotides upstream of the 3' end of an intron is called a branch site (BS) [3]. A typical BS sequence is represented by YNYYRAY, which is loosely conserved except for the presence of adenine (A) all the time. Here, Y, N, and R denote a pyrimidine, any nucleotide, and a purine, respectively.

Splicing is a multistep process carried out by the catalytic activity of a multiprotein ribonucleoprotein complex known as spliceosome (Figure 1) [4]. The process initiates by the cleavage of pre-mRNA at the 5' end of the intron by the

Department of Biochemistry, University of Kashmir, Srinagar-190006, Jammu and Kashmir, India.
* Corresponding author: ashrafdar@kashmiruniversity.ac.in

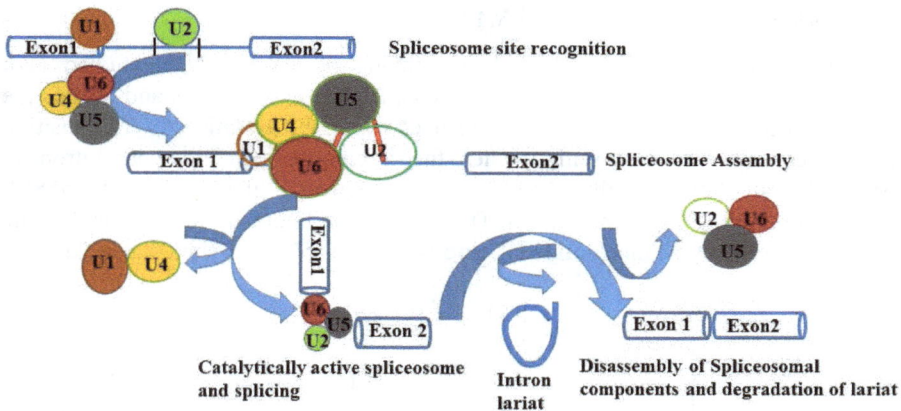

Figure 1: Overview of splicing.
RNA splicing comprising of two transesterification reactions is catalysed by a huge molecular machine called the spliceosome, which assembles anew on the substrate pre-mRNAs from small nuclear RNAs (snRNAs) and different proteins in a stepwise manner that undergoes different compositional and structural rearrangements to generate an active site for catalytic reaction of splicing.

spliceosome [5]. Next, the guanine nucleotide from the 5' cut-end base pairs with the adenine nucleotide of the conserved branch point downstream [6]. This results in the formation of a looped structure called a lariat. Finally, the 3' end of the intron is brought into proximity of the lariat, cut, and ligated to the 5' end by transesterification reaction. The adjoining exons are joined together by covalent linkages, and the resulting lariat is released.

The eukaryotic genes with long introns, in addition to splice sites, contain exonic splicing enhancers (ESEs). The ESEs are present in exons which help position the splicing apparatus to the correct site [7]. Most commonly, the splicing occurs between exons in a single pre-mRNA transcript but occasionally, trans-splicing occurs where exons from two different pre-mRNAs are ligated together to form a mature mRNA [8].

In human cells, the number of proteins is far greater than the number of protein-coding genes (~ 25,000) [9]. The alternative splicing in 95% of the total protein-coding gene transcripts and the presence of 8 introns, on average, in a gene greatly expands the proteome produced from a limited number of genes [10]. This chapter provides detailed information about the splicing machinery or spliceosome, the molecular mechanism of splicing reaction, and the current understanding of the regulation of splicing in eukaryotic cells.

2. The Chemistry of RNA Splicing

The splicing reactions involve recognizing the consensus sequences at the ends and within the intron. These sequences serve as a landing platform for spliceosome assembly. The spliceosome machinery removes the introns and ligates the exons in two transesterification reactions.

2.1 *Splice Sites (SS) in Pre-mRNA*

The biochemical mechanism by which splicing occurs has been studied in a number of systems and is now fairly well characterized. The exons and introns are distinguished by specific nucleotide sequences within the pre-mRNAs. The position of these conserved sites determines where the splicing reaction will occur. Introns in pre-mRNA are defined by three short consensus sequences; the 5' splice site (5' SS), the 3' splice site (3' SS), and the BS. The exon-intron boundary at the 5' end of the intron is marked by a dinucleotide GU and is called as 5' splice site (Figure 2a). The

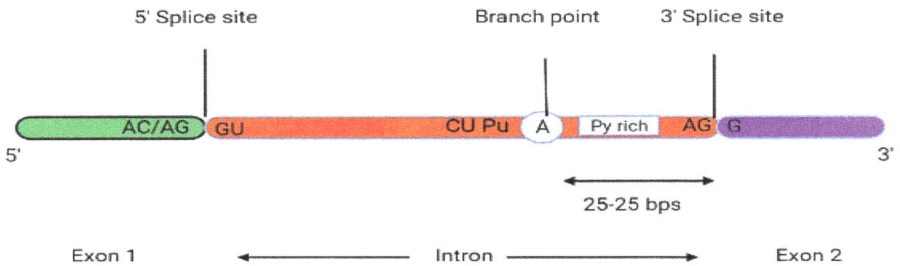

Figure 2a: Consensus sequences at intron-exon border.
The 5' SS, BS (or BP), and 3' SS sequences are three short conserved sequences that define introns. (This figure has been generated in Biorender).

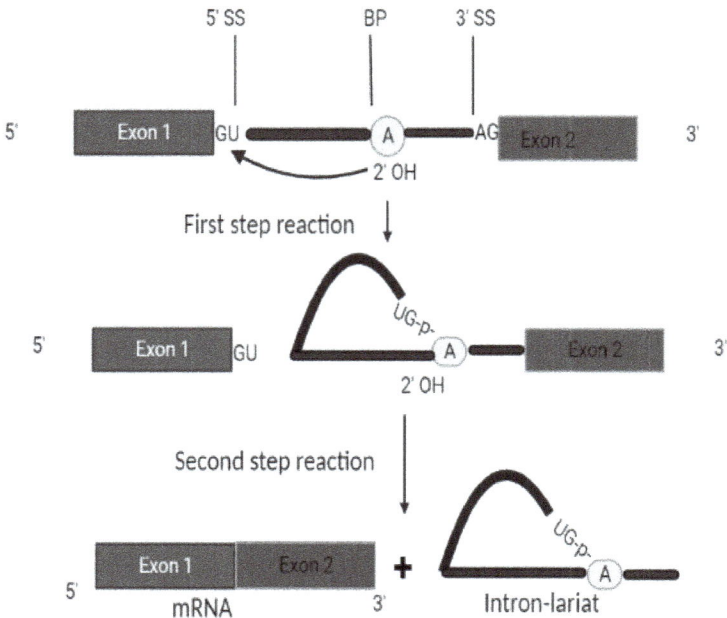

Figure 2b: Two transesterification steps of splicing.
Branching and exon ligation, two transesterification events catalysed at a single active site, are used to eliminate introns. (This figure has been generated in Biorender).

intron-exon boundary at the 3' end of the intron is marked by the 3' SS comprising dinucleotide AG [11]. A third sequence important for splicing reaction is the BS which is located within the intron about 18–40 nucleotides upstream from its 3' end, often followed by a pyrimidine-rich tract (PPT). These consensus sequences are critical for splicing because changing or mutating one of the conserved nucleotides results in the inhibition of splicing. The branch site always contains an adenine but is otherwise loosely conserved [12]. A typical sequence is YNYYRAY, where Y indicates a pyrimidine, N denotes any nucleotide, R denotes any purine, and A denotes adenine. Rarely, alternate splice site sequences are found that begin with the dinucleotide AU and end with AC; these are spliced through a similar mechanism. These conserved signalling sequences are all positioned within the intron to be excised out. These special sequences are recognized by specific components of the spliceosome.

2.2 Exons Have a Role in Splicing

How are the introns cleaved from between two successive exons, which are then ligated together precisely to make a correct reading frame for translation? It is intriguing that correct splice sites at the intron ends are recognized during splicing, while the cryptic sites (mRNA sequences that have the potential for interacting with the spliceosome) are avoided.

Besides introns, exons (50–250 nt) are also recognized and play a role in splicing events. Through a process called exon definition, the splice sites on either side of an exon can stimulate each other's recognition [13]. Several studies have reported various non-splice-site sequences that help in the recognition of exons. For example, ESEs are present in exons and are involved in the stimulation of splicing at adjacent exons. Exons lacking these elements are usually excluded from the mature mRNA [14]. Other sequences include exonic splicing silencer elements (ESS), as well as intronic splicing enhancers and silencers (ISE and ISS), that help in the recognition of correct splice sites. These RNA elements (ESEs, ESS, ISE, and ISS) bind proteins that regulate the spliceosomal assembly either positively or negatively, among which most studied are SR proteins [15]. SR proteins, which bind to an internal exon, block the splicing of the flanking exons and therefore repress the splicing.

2.3 Introns are Removed from Pre-mRNA by Two Transesterification Reactions

Introns are removed from primary transcripts through two successive transesterification reactions in which phosphodiester bonds within the pre-mRNA are broken, and new ones are created (Figure 2b). In the first reaction, termed branching, the 2' hydroxyl group of the conserved adenosine of the BS initiates the splicing reaction [16]. It acts as a nucleophile and attacks the phosphate group in the sugar-phosphate in the conserved G at the 5' splice junction [17]. This reaction is of SN2 type and proceeds through a pentavalent phosphorus intermediate. As a result of this first reaction, the phosphodiester bond between the sugar and the phosphate at the 5' exon is cleaved, creating a break in the backbone of the RNA molecule. The cut 5' end of the intron attaches to the conserved branch point region downstream through the pairing of G and A nucleotides from the 5' end and the BS, respectively, to form a looped structure known as a lariat [18]. Thus, the products in the first cleavage are the free 5' exon and the RNA lariat. Note that the ribose of the branch point adenosine has, in addition to 5' and 3' backbone linkages, a third newly created phosphodiester bond that extends from 2'-OH.

In the second transesterification reaction, termed exon ligation, the newly released 3' OH of the 5' exon reverses its role and becomes a nucleophile that attacks the phosphoryl group of the downstream exon at the 3' SS. As a result of this second reaction, two adjoining 5' and 3' exons are covalently ligated together with the liberation of the intron as a leaving group in the shape of a lariat. ATP and magnesium (Mg^{2+}) are found to be indispensable for the *in vitro* splicing of adenovirus pre-mRNA [19]. A two-metal-ion mechanism was proposed by Steitz and Steitz to explain the splicing catalysis in which two metal ions stabilize the pentavalent transition states in the spliceosomal active site [20]. One of the two metal ions (M1) stabilizes the leaving group, the 3' OH of the last 5' exon nucleotide, and the other metal ion (M2) activates the attacking nucleophile, the 2' OH group of the BS adenosine (A), in the first phosphoryl transfer process. To allow the binding of the 3' SS to the active site, one of the initial reaction's reactants, the BS adenosine, must leave the active site. M1 activates the 3' OH of the 5' exon in the second phosphoryl transfer event (exon ligation), whereas M2 stabilizes the leaving group in the intron lariat. Since M1 and M2 are coordinated by RNA, the spliceosome is categorized as a ribozyme.

In conclusion, RNA splicing occurs through two consecutive transesterification reactions. There is no net gain or loss in the number of chemical bonds, two phosphodiester bonds are made, and two are broken. Since splicing involves the reshuffling of bonds only, there is no net energy input needed for the chemistry of this process. Thus, in principle, splicing could be fairly reversible and go in either a forward or reverse direction, which does not happen practically in cells. To ensure the forward direction of this process, the products are immediately removed from the nucleus. The mature mRNA is exported to the cytoplasm for translation into proteins, and the intron lariat is degraded or converted to snoRNAs.

3. Spliceosomal Machinery and its Assembly

In cells, a spliceosome does not exist as a pre-assembled complex. The spliceosome assembly occurs fresh and is coupled to the splicing of pre-mRNA. The catalytic complex is built up from RNA and protein components in a stepwise manner via several intermediary non-catalytic complexes [21].

3.1 Composition of Spliceosome

The spliceosome is a multi-megadalton ribonucleoprotein catalytic complex present in the cells. It is made up of five different small ribonuclear protein subunits (snRNPs) [22, 23]. The snRNPs, in turn, are composed of five different small nuclear RNAs (snRNAs), U1, U2, U4, U5, and U6, complexed with approximately 100 proteins.

The U1, U2, U4, and U5 are transcribed by RNA polymerase II, whereas U6 is transcribed by RNA polymerase III [24]. At the 3' ends of U1, U2, U4, and U5 snRNAs lie Uridine-rich sequences called Sm sites [25]. A ring is formed around these Sm sites by the assembly of seven homologous proteins called Sm proteins (B/B', D3, D2, D1, E, F, and G) [26]. On the hand, uridine rich sequence (at 3' end) of U6 passes through a pre-assembled ring formed by seven paralogous LSm2-8 proteins [27]. Each of these snRNAs binds to a specific set of additional proteins to form snRNP particle or 'snrup'. The snRNP particles contain a single snRNA molecule except U4/U6, which is built on two snRNA molecules [28]. Several non-snRNP-associated proteins and protein complexes, including splicing factors and 8 ATP-depend helicases, also play a role in splicing reaction. The snRNAs within spliceosome help in substrate recognition and play an essential role in catalytic reactions [29].

The snRNPs and associated proteins assemble on pre-mRNA and form a multi-megadalton ribonucleoprotein complex or spliceosome. The assembly of the complex occurs afresh on the pre-mRNA template for every round of splicing and involves controlled association and dissociation of snRNPs. The spliceosome is not present in the cell as a pre-assembled enzyme complex but is formed anew on substrate pre-mRNAs from small nuclear RNAs (snRNAs) and different proteins in a stepwise manner, which undergoes different compositional and structural rearrangements to generate an active site for catalytic reaction of splicing [30]. The rearrangement of snRNPs and association of various trans-acting factors with pre-mRNA is necessary for correctly identifying and pairing the splice sites among a myriad of identical sequences. The rearrangement of spliceosome machinery further ensures the correct alignment of the splice sites for transesterification reactions to take place within atomic distances, despite these sites being located tens of thousands of bases far away.

The configurational and architectural dynamics of the spliceosome offer the splicing complex its accuracy and flexibility during substrate-dependent complex assembly, catalytic activation, and active site remodelling [31]. Proteomic analysis, genetic approaches, and biochemical procedures of human and yeast spliceosomes have shed light on their complexity, and ten distinct intermediates of the spliceosome during the assembly and catalytic process have been isolated and purified. These intermediates are E, A, pre-B, B, B^{act}, B*, C, C*, P, and ILS (Figure 3). Their assembly and role will be discussed later on [32].

3.1.1 Major (U-2) and Minor (U-12) Spliceosomes

In eukaryotic organisms, two types of spliceosomes, U-2 and U-12, are present. U2-dependent spliceosome or major spliceosome is one of the most abundant structures in the cell and excises about 99.5% of introns in metazoans [33]. U2-dependent spliceosomes assemble on and splice out U2-type introns. These introns are characterized by poorly conserved sequences at 5' and 3' ends. The variability and flexibility in splice site choices by U2-dependent spliceosomes promote alternative splicing in a significant number of genes in the cell [34].

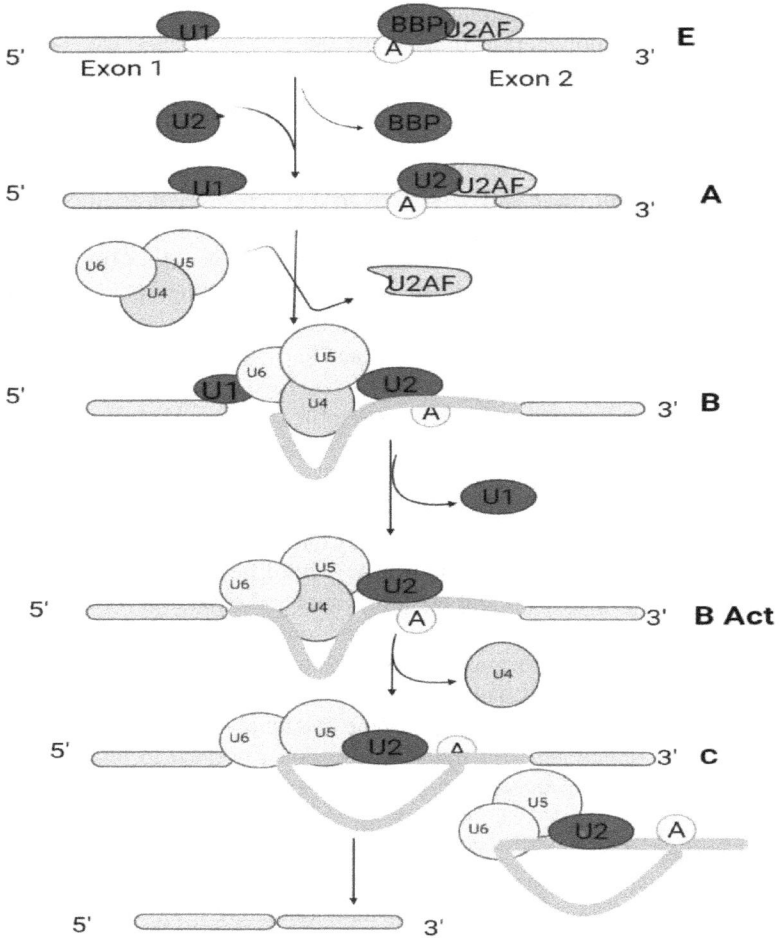

Figure 3: Steps of the spliceosomal assembly and catalysis of splicing.
Components of Spliceosome machinery assemble at sites of transcription. At each step, spliceosome components enter or exit the complex, resulting in structural changes that are required for the splicing reaction to proceed. (This figure has been generated in Biorender).

The U12-dependent spliceosome, or minor spliceosome, has U11/U12 and U4atac/U6atac snRNPs as main components, which are compositionally different but functionally similar to the components of the major spliceosome [35]. This spliceosome excises nearly 0.5% of the total number of introns by recognizing and utilizing significantly conserved sequences at the 5' and 3' ends of the U12-type intron.

U-12 spliceosome is assembled from compositionally different but functionally related U11/U12 and U4atac/U6atac snRNPs, with the U5 snRNP shared by both types of spliceosomal complexes. The overall splicing mechanism and spliceosome assembly are similar between the two complexes regardless of their compositional

Table 1:   List of snRNPs and associated proteins.

snRNP	Sm Proteins	Associated Proteins	Other Core Proteins Associated with snRNA
U1	B, D3, G, E, F, D1, and D2	Urn1, Snu56Prp39, Prp42, Snu71, Prp40 and Nam8	Mud1(U1A), Snp1(U1-70K), and Yhc(U1C)
U2	B, D3, G, E, F, D1, and D2	U2AF65, U2AF35, BBP	U2A, Prp11(SF3A66), Prp21(SF3A120), Rds3 (SF3B14B), Ysf3(SF3B10)U2B, Prp9(SF3B60), Snu17(SF3B14), AHsh49(SF3B49), Hsh155 (SF3B155), Cus1(SF3B145), and Rse1(SF3B130)
U4–U6	U4: B, D3, G, E, F, D1, and D2 U6: Lsm2–8		Prp4, Prp3, Snu13, and Prp31
U5	B, D3, G, E, F, D2, and D1	Spp2, Yju2, Cbc2, Prp2, Prp38, Snu23	Snu114, U5-40K, Dib1, Brr2, Prp6, Prp8, and Prp28
U4–U6.U5	Two sets of B, D3, G, E, F, D1 and D2U6: Lsm2–8	Yju2, Snu66, Sad1, Snu23, Prp2, Prp38 and Spp2	Prp31, Prp4, Prp3, Prp8, Snu13Prp6, Brr2, Prp28 Snu114, U5–40K, Dib1, and snRNP27
U11/U12	B, D3, G, E, F, D2, and D1	SNRNP35, PDCD7, Urp, ZCRB1, ZMAT5	SF3b complex, ZMAT5, SNRNP25, ZCRB1
U4/U6	Same as that of U4/U6 of the major spliceosome	Same as that of U4/U6 major spliceosome	

differences. The details of various protein factors associated with snRNPs in U-2 and U-12 type spliceosomes are given in Table 1.

In U-2 type spliceosome, the 5' SS and BS in the intron are identified initially by separate U1 and U2 snRNPs [36]. In contrast, U11/U12 di-snRNP complex is involved in initial recognition in U-12 spliceosome. Also, the polypyrimidine tract (PPT) and 3'SS in U2-type spliceosome are recognized by the U2AF1/U2AF2 protein complex, whereas in U12-spliceosome 3'SS is recognized by the ZRSR2 protein. This is followed by the entry of U4/U6 U5 tri-snRNP in the U-2 spliceosome or U4atac/U6atac.U5 tri-snRNP in the U-12 spliceosome [37]. Finally, in both cases, there is the generation of catalytic complexes, which excise the respective introns from substrate pre-mRNAs.

3.2 Assembly of Spliceosome

The assembly of the spliceosome in a stepwise process. It involves the ordered interaction of snRNAs and snRNPs and many other splicing factors. There are four stages in spliceosome assembly, including (i) recognition of substrate and assembly, (ii) activation, (iii) catalysis, and (iv) disassembly. All these stages involve rearrangement in RNA-RNA and RNA-Protein interactions modulated by RNA helicases that facilitate kinetic proofreading during splicing events [38].

3.2.1 Recognition of Substrate and Formation of E-Complex

In the first stage of spliceosome assembly, the intron on pre-mRNA is recognized and marked by the U1 snRNP [39]. The process begins with the recruitment of U1 snRNP, where the 5' end of the U1 snRNA base pairs with the 5' splice site (SS) of the intron to form the U1 snRNA:5' SS duplex [40]. This interaction is stabilized by protein components in U1 snRNP. The non-snRNP protein factors, such as SF1/BBP and U2 auxiliary factor (U2AF), bind to the BS and the polypyrimidine tract (PPT) next to BS, respectively, to form an early assembly intermediate called as E-complex or cross-intron spliceosome complex [41].

3.2.2 Recognition of BS, 3'SS and Formation of A-Complex

Formation of the E complex is followed by ATP-dependent stable association of U2 snRNP with BS to form A-complex or prespliceosome [42]. U2AF is a heteromeric dimer consisting of 65 kDa and 35 kDa subunits. The 65 kDa subunit binds to PPT, whereas the 35 kDa subunit recognizes and contacts 3'SS. In the following step, the snRNA component of the U2 snRNP base pairs with BS. SF1/BBP are displaced, and the branch point adenosine is contacted by the U2-associated SF3b complex [43]. The U2:BS duplex is stabilized by SF3a and SF3b and the arginine-serine-rich domain of the 65 kDa subunit of U2AF. The prespliceosome attains a bipartite structure with limited contacts between U1 and U2 snRNPs.

3.2.3 Assembly of Pre-Catalytic B-Complex

Following the demarcation of 5'SS and BS by A-complex, pre-assembled U4/U6.U5 tri-snRNP (formed from U4/U6 and U5 snRNPs) is recruited to generate a fully assembled pre-catalytic complex called B-complex. Both A-complex and U4/U6 [44]. U5 tri snRNP complex is held together by base pairing between the 5' end of the U2 snRNA and 3' end of the U6 snRNA [45]. The B-complex contains all the snRNPs but remains inactive catalytically because the active site is still not generated, and the branching in pre-mRNAs is prevented by U1 and U2 snRNPs [46]. Within the B-complex, the interactions between components of A-complex and tri-snRNP are weak, and the splice sites remain far apart from the active-site cavity. These interactions are later stabilized during various steps of catalytic activation [47].

3.2.4 Active Site Generation and Activation of Spliceosome

Various compositional and conformational rearrangements in RNA-RNA and RNA-protein interactions lead to the destabilization of the U1 and U4 snRNPs by RNA helicase, BRR2 to form catalytically active spliceosome with fully formed catalytic center and is called the B^{act}-complex [48]. The B^{act}-complex is then transformed by the DEAH-box ATPase, Prp2, to form the catalytically active B*-complex, which catalyzes the branch helix docking and branching reaction [49]. This produces the C-complex, which is compositionally similar to the B-complex but differs only in the linkage of the phosphate group of the first intron nucleotide. The C-complex holds the cleaved 5'exon and the intron lariat-3'exon [50]. The C-complex is remodeled by the helicase, PRP16, to exon-ligation conformation—i.e., C*-complex to carry out the next step, which is exon ligation. This step results in the formation of post-catalytic spliceosome or P-complex [51].

3.2.5 Release of the Ligated Exons

The ligated exons are released by DEAH-box ATPase, prp22 [52]. Prp22 docks within the exon on nucleotides upstream of the 3'SS before exon ligation [53] and downstream of the 3'SS after exon ligation. After ATP hydrolysis, Prp22 translocates along the mRNA from 3' to 5', which eventually results in the release of mRNA and loss of exon-ligation factors from the spliceosome to form intron-lariat-spliceosome (ILS).

3.2.6 Disassembly of Spliceosome

At the end of the second catalytic step, ILS is dissociated to release and decay intron lariat by the action of helicase, Prp43. The mRNA is released as a mature product and, after additional remodelling, U2, U5, U6 snRNPs, and associated factors are also released for new cycle of splicing [54].

4. Self-Splicing

The predominant class of splicing is nuclear pre-mRNAs splicing, which involves a well-dedicated spliceosomal machinery. There are two other classes of splicing found in rare genes, mediated by group I and group II self-splicing introns. These introns are called self-splicing introns because they can catalyze the chemistry of their excision from their parent RNA molecule. The self-splicing introns are not enzymes as they catalyze only one round of RNA processing and are called ribozymes. Group II introns work similarly as that of nuclear pre-mRNAs, with the only difference being that spliceosomal machinery is not involved; instead, the RNA enzyme encoded by intron acts as a catalyst. However, the group I introns excise by a different mechanism. The nucleophile that attacks the 5' end of the intron is a free G nucleoside instead of BS A residue (Figure 4). The first transesterification reaction takes place in the same way as that of nuclear pre-mRNA splicing, which results in lariat formation by fusing the G to the 5' end of the intron. In the second transesterification reaction, the released 3' end of the exon attacks the 3' splice site and thus results in the ligation of two successive exons and liberation of a linear intron instead of a lariat intron.

5. Fidelity Mechanisms in Pre-mRNA Splicing

Fidelity in splicing is critical for maintaining functions and survival of cells, whereas variations in it result in deletions, insertions, or frameshift mutations in mRNA. Loss of specificity during splicing has been associated with the etiology and progression

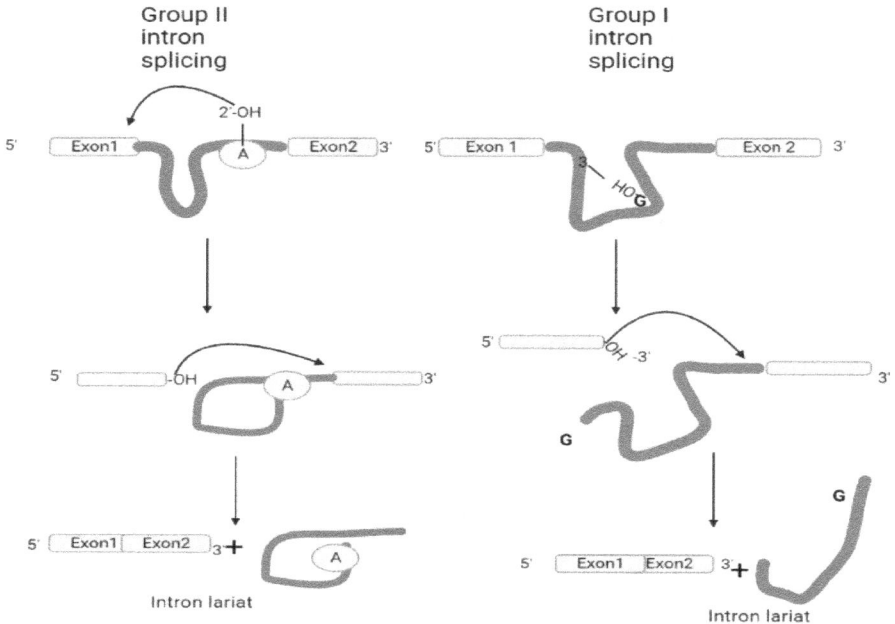

Figure 4: Comparison between splicing mechanism of group II and group I introns. To excise the intron and ligate the exons, a sequence of transesterification processes are used in both cases. A guanosine cofactor initiates the reaction in group I introns, while an internal adenosine initiates the process in group II introns. Nuclear pre-mRNA splicing follows the same pattern as group II intron splicing; however, it takes place on a massive ribonucleoprotein complex. (This figure has been generated in Biorender).

of various diseases. Spliceosome ensures the specificity of pre-mRNA splicing via thermodynamic and kinetic mechanisms by discriminating against suboptimal substrates during the early assembly and later catalytic stages [55]. According to the thermodynamic mechanism, which is best characterized during catalytic steps of splicing, the spliceosome selectively sequesters suboptimal substrates in stable and non-productive conformations that are in equilibrium with the spliceosome's catalytic conformations [56]. The kinetic mechanism, also called kinetic proofreading, mediated by DEAD/H-box ATPases, works by promoting an optimal substrate and antagonizing the suboptimal substrates through branches in the splicing pathway and competition with the productive step [57, 58] (Figure 5). Eight DEAD- and DEAH/RHA-box ATPases play a role in these ATP-dependent rejection steps by mediating RNP rearrangements and splicing of only optimal substrates [59]. Five DEAD- and DEAH ATPases, Prp28, Prp5, Prp22, Prp43, and Prp16 enhance splicing specificity by rejecting sub-optimal substrates [60, 61]. Prp16, which is conserved in all eukaryotes, mediates the rearrangement of splicing intermediates for the ligation of exons and thus facilitates the splicing of only optimal substrates [62, 63]. Prp16 also rejects suboptimal substrates that contain BS that are not similar to consensus sequences [64]. Further, *Prp16* mutants have increased levels of abnormal BS intermediates

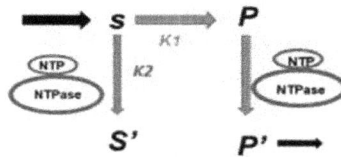

Figure 5: General outlook for energy dependent mechanism of kinetic proofreading of fidelity by DEAD/h BOX ATPases.
The specificity of step S-P is increased by the action of NTPases by competing with S-S'. K1 shows some processes like cleavage of 5' splice site, and K2 shows rate rejection of suboptimal substrate. Specificity is increased when k1/k2 ratio is high for an optimal substrate than a suboptimal substrate.

and aberrant mRNAs. Prp5, A DEAD-box ATPase, also ensures fidelity during early intron recognition by associating U2 snRNA with optimal BS [65]. Fidelity is also increased by a DExD/H-box ATPase, Prp22p, which not only mediates the release of spliced mRNA from spliceosome but also inhibits the ligation of intermediates with abnormal 3' SS, 5' SS, and BS [66, 67]. Mutations that lead to compromised ATPase activities in PRP5, PRP16, or PRP22 permit enhanced use of suboptimal substrates *in vivo,* indicating that their ATPase activity is important for their fidelity function [68, 69]. The spliceosome disassembly pathway mediated by Prp43 also plays a role in maintaining fidelity by releasing stalled suboptimal substrates in the spliceosome complex. Prp43 mediates the release of not only pre-mRNA discarded by Prp16 but also intermediates rejected before 3' splice site cleavage.

6. Alternative Splicing

The one gene, one mRNA, and one protein hypothesis do not answer the discrepancy between the number of protein-coding genes (~ 25,000 in humans) and the proteins they produce (> 90,000) [70]. The genes of higher eukaryotes are spliced in alternative ways to produce two or more protein products [71]. More than 95% of human genes undergo splicing in a developmental, tissue-specific, or signal transduction-dependent manner [72]. Alternative splicing is a well-controlled process that modifies pre-mRNA molecules preceding translation and gives rise to a number of mRNAs from a single gene by arranging coding sequences from recently spliced RNA sequences into different combinations. This complex process is guided by the functional coupling between transcription and splicing where diverse interacting elements work together including cis-acting elements and trans-acting factors [73]. Besides contributing to proteomic diversity, alternative splicing plays a role in the regulation of gene expression. As a result, different proteins are generated in different cell types or in response to different signals.

A simple example of alternative splicing is from the mammalian muscle protein troponin T, which contains five exons, as shown in Figure 6a. Its pre-mRNA is spliced to form two alternative mRNAs with four exons. They are spliced differentially with the elimination of different exons from each of the mRNAs, thus producing two products that differ in one exon.

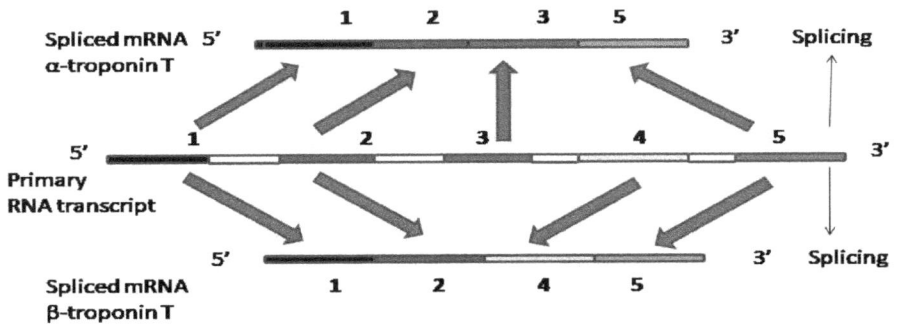

Figure 6a: Splicing variants of troponin T gene.
The troponin T gene encodes five exons that generate two splice variants by alternative splicing, as shown.

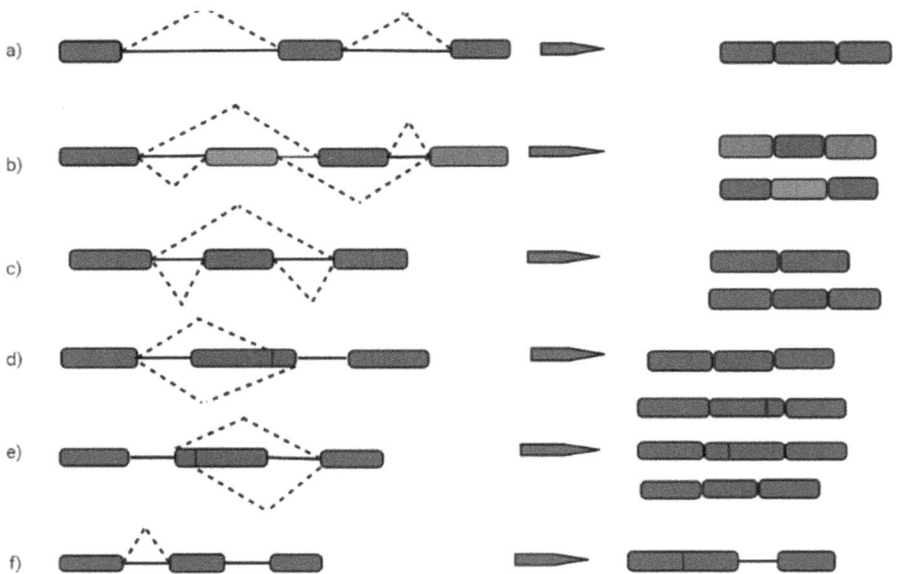

Figure 6b: Five main types of alternative splicing.
(a) Constitutive splicing; (b) mutually exclusive exons; (c) cassette alternative exon; (d) alternative 3' splice site; (e) alternative 5' splice site; and (f) intron retention (This figure has been generated in Biorender).

6.1 Types of Alternative Splicing and Mechanism

Based on the mechanism of splicing reactions, seven main types of alternative splicing have been described [74]. Of this, cassette-type alternative exon accounts for half of the total alternative splicing events in cells (Figure 6b).

6.1.1 Cassette-Type Alternative Exon

In the human genome, cassette exon splicing, also known as exon skipping, accounts for 50 to 60% of all alternative splicing and is the most prevalent form of alternative splicing [75]. Here, an intervening exon between the two exons from the mature mRNA sequence may be either included or skipped, resulting in two distinct protein isoforms (Figure 6b).

6.1.2 Intron Retention

This is mostly found in lower metazoans. This is the most uncommon way of alternative splicing in mammals and is primarily positioned in human transcripts in UTRs [76]. Intron retention is mostly associated with weaker splice sites, short intron length, and the regulation of cis-regulatory sites [77]. In this type of splicing, intron may be spliced out or retained and is differentiated from exon skipping as the retained sequence is not flanked by introns.

6.1.3 Alternative 5' and 3' Splice Site

An alternative 5' splice junction known as a donor site is utilized to create 3' extension of a predicate exon; similarly, 3' splice junction (acceptor site) creates 5' boundary of a proximate exon [78]. An alternative selection of these splice sites within exons leads to minuscule substitutes in the coding regions, thus adding a layer of complexity with absolute alternate exons [79].

6.1.4 Mutually Exclusive Exons

ME splicing is a form of alternative splicing in which only one exon is carried by the mature mRNA from a set of two or more variants. Originating from exon duplication events, ME exons encode interchangeable peptide segments that can modulate protein function [80]. Mutually exclusive election of alternative exons is ensured by following mechanisms so that when one is selected, the other is not.

6.1.4.1 Steric Hindrance

If the splicing site within the intron separating two exons are too close together, splicing factors cannot bind to both sides simultaneously. Therefore binding of U1 snRNP to the 5' splice site will inhibit the binding of U2 snRNP on the BS within the same intron. Alternative splicing in the troponin T gene is made mutually exclusive by this mechanism.

6.1.4.2 Combinations of major and minor splice sites

The minor spliceosome does not recognize the SS that is recognized by the major spliceosome. None of these spliceosomes can remove an intron with a combination of SSs (splice sites), therefore, ensuring a fair selection of SS. The Human JNK1 gene is spliced by this mechanism.

6.1.4.3 Nonsense—Mediated Decay

NMD is another extensive and complicated mechanism, ranging from yeast to humans, to achieve a higher level of gene expression control post-transcriptionally.

This mechanism ensures that transcripts that have only one or the other exon (never both) survive. If a message includes both exons, the mRNA produced will contain a premature terminating codon and, as a result, will be destroyed by NMD.

7. Regulation of Splicing

Splicing is an essential post-transcriptional mechanism for reprogramming gene expression profiles and broadening transcriptomic and proteomic diversity in eukaryotic cells. However, its dysregulation results in aberrant biological actions causing scores of pathological conditions in humans [81]. Regulation of splicing involves RNA sequence elements and protein regulators. The RNA sequence elements comprise cis-regulatory elements, which are divided into splicing enhancers, viz., ESEs and ISEs and splicing silencers, viz., exonic splicing silencers (ESSs) and intronic splicing silencers (ISSs) depending on their position and function [82, 83]. Splicing enhancers activate while splicing silencers repress the splicing at the nearby splice sites. A majority of ESEs are bound by members of the SR (Ser-Arg) protein family [84], whereas most of ESSs and ISSs are bound by heterogeneous nuclear RNPs (hnRNPs) [85]. ISEs have been poorly characterized compared to the other three types of elements, although several proteins, including hnRNP F, hnRNP H, neurooncological ventral antigen 1 (NOVA1), NOVA2, and FOX1 and FOX2 (also known as RBM9), have been recently shown to bind ISEs and stimulate splicing [84]. Recent studies have demonstrated that splicing is regulated at the stages of splice site recognition [86], different stages of spliceosome assembly [87], and even during conformational changes between the two steps of transesterification [88]. Additionally, there are evidence of transcription-coupled regulation of splicing [89].

Collaborative interactions between these elements either promote or inhibit spliceosomal assembly at splice sites. Constitutive splicing is dominated by enhancing elements, whereas alternative splicing is significantly controlled by silencers [90].

7.1 Activators of Splicing

Members of the SR protein family play crucial roles in directing splicing machinery to different splice sites under different conditions. The family includes the proteins, SRp20, SRp30c, 9G8, SRp40, SRp55, SRp70, ASF/SF2, and SC35 [91]. SR proteins bind ESEs through their RNA recognition motifs (RRMs) and other proteins of splicing machinery through their arginine-serine-rich (RS) domain (Figure 7a). They facilitate splice site recognition by recruiting U1 snRNP to the 5' splice site and the U2AF complex and U2 snRNP to the 3' splice site [92]. Splicing-enhancing complexes may also be formed by SR proteins interacting with other positively regulatory factors, such as transformer 2 (TRA2) and the SR-related nuclear matrix proteins SRm160 (also known as SRRM1) and SRm300. Table 2 contains various cellular proteins that have been described to activate splicing [93].

The *Drosophilla* half-pint protein is an activator that promotes a particular alternative splicing event in a set of pre-mRNAs in the fly ovary by binding

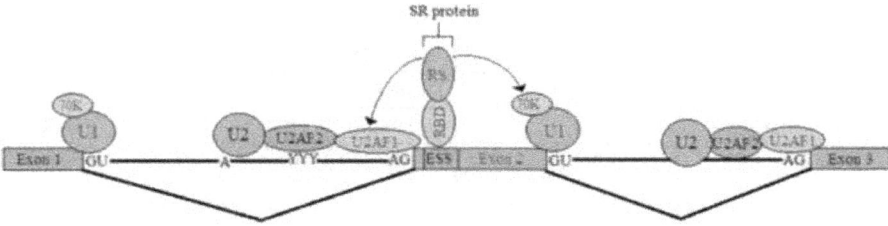

Figure 7a: Positive regulators of splicing.
ESEs bind to Ser–Arg proteins to stimulate U2AF binding to the upstream 3' SS or U1 small nuclear ribonucleoprotein (snRNP) binding to the downstream 5' SS.

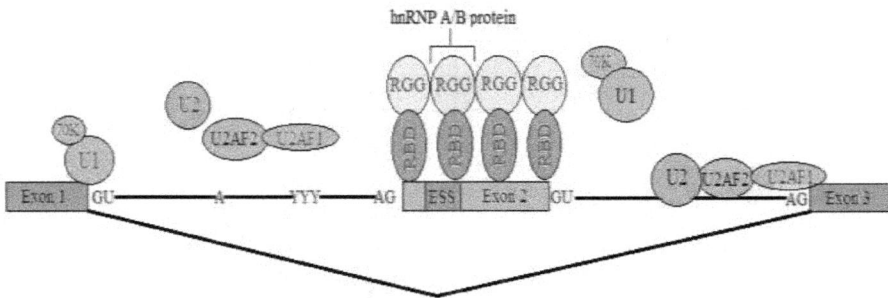

Figure 7b: Negative regulators of splicing.
Various heterogeneous nuclear ribonucleoproteins (hnRNPs) typically recognise ESSs to inhibit splicing.

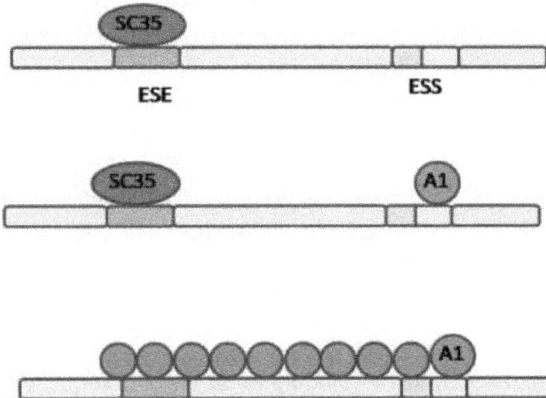

Figure 7c: Mechanism of inhibition splicing of HIV Tat Exon 3 by hnRNPA1.
SC35 is an activator of splicing and promotes exonic exclusion by binding to ESE. A1 acts as an inhibitor and after sitting on ESS, it spreads to ESE through cooperative binding, thus inhibiting the binding of SC35 to ESE.

to sites near the 3' splice site of specific exons and recruiting the U2AF splicing factor. Microarray-based expression profiles from mouse, chimpanzee, and human tissues revealed the role of core spliceosomal proteins (CSPs) in the regulation of splicing.

Table 2: Positive regulators of splicing.

Positive Regulator	Domains	Target Gene	Target Sequence
Tra2	RRM, RS1 and RS2	Dsx and Fru	GAAARGARR
SRp20 (SFRS3)	RRM and RS	CALCA and INSR	GCUCCUCUUC
ASF/SF2 (SFRS1)	RRM, RRMH, and RS	CAMK2D, HIV RNAs and GRIA1–GRIA4	RGAAGAAC
SC35 (SFRS2)	RRM and RS	AChE and GRIA1–GRIA4	UGCUGUU
SRp40 (SFRS5)	RRM, RRMH, and RS	hIPK3, PRKCB and FN1	AGGAGAAGGGA
SRp55 (SFRS6)	RRM, RRMH, and RS	TNNT2 and CD44	GGCAGCACCUG
9G8 (SFRS7)	RRM, zinc finger, and RS	TAu, GNRh, and SFRS7	(GAC)n
SRp75 (SFRS4)	RRM, RRMH, and RS	FN1, E1A and CD45	GAAGGA
SRp38 (FUSIP1)	RRM and RS	GRIA2 and TRD	AAAGACAAA
RNPS1	RRM and Ser-rich	TRA2B	Not determined
CAPER (RBM39)	RRM and RS	VEGF	Not determined

7.2 Inhibitors of Splicing

The best characterized exonic and intronic silencers or repressors are bound by hnRNPs (Figure 7b). The hnRNP proteins are a large group of molecules whose association with unspliced mRNA precursors (hnRNA) makes them distinctive [94]. They are highly conserved from nematodes to mammals and are known to play critical roles in the negative regulation of various stages of splicing, ranging from binding to ESS and exclusion of SR proteins. The well-studied among these proteins, HnRNP A1, counters the action of SR proteins by binding to ESS within an exon of HIV tat pre-mRNA and prevents the inclusion of that exon in the final mRNA [95]. After binding to the ESS site, HnRNP A1 spreads through cooperative binding until it occludes the nearby enhancer site (ESE) and blocks the binding of the activator SC35 (SR protein). The blocking is not direct in this case; the two binding sites do not overlap, rather hnRNP1 promotes the binding of additional hnRNPA1 molecules to adjacent sequences, spreading over the enhancer site (Figure 7c).

Inhibition of RNA splicing can also occur by sterically blocking by polypyrimidine-tract binding protein (PTB; also known as PTB1 and hnRNP I), which binds the polypyrimidine tract and blocks the binding of U2AF to regulated exons [96]. Additionally, tissue-specific splicing factors FOX1 and FOX2 inhibit the formation of the E-complex by binding to an intronic sequence to prevent SF1 from binding to the BS of CAlCA (calcitonin-related polypeptide-α) pre-mRNA [97]. Furthermore, splicing inhibitors also sterically block the binding of activators to enhancers. Hu/ELAV family proteins inhibit the binding of U1 snRNP by competing with TIA1 to an AU-rich sequence downstream of the exon 23a 5' SS of neurofibromatosis type 1 pre-mRNA. In Table 3, there is a list of proteins that promote repression or negative regulation of splicing.

Table 3: Negative regulators of splicing.

Negative Regulators	Binding Domains	Target Genes	Binding Sequence
PTB	RRM	*TNTT2, nPTB, SRC CALCA,* and *GRIN3B*	UCUU and CUCUCU
SRRF	RRM	*CD45*	C and A rich
hnRNP A1	RRM, RGG and G	*SMN2* and *RAS*	UAGGGA/U
hnRNP H	RRM, RGG	HIV *tat* and *BCL2L1*	GYR and GY GGGA and G rich
AUF1	RRM	*APP*	U rich
hnRNP A2	RRM, RGG and G	HIV *tat* and *IKBKAP*	(UUAGGG)n
hnRNP F	RRM, RGG and GY	*PLP, SRC,* and *BCL2L2*	GGGA and G rich
hnRNP L	RRM	*NOS* and *CD45*	C and A rich
hnRNP G	RRM and SRGY	*SMN2* and *TMP1*	CC(A/C) and AAGU

7.3 Multifactorial Regulation of Splicing

The excision of individual introns is regulated by the combinatorial effects of both activators and silencers [98]. Whether an exon is to be included or excluded from an mRNA is decided by the concentration or activity of each type of regulator, often by SR proteins and hnRNPs [99]. For instance, in the case of the splicing of α-tropomyosin exon 2SR, protein 9G8 (also known as SFRS7), hnRNP F and hnRNP H compete for binding to the same element, and their antagonizing effects determine the activation or repression of splicing of this exon [100]. Additionally, the activation or repression of cis-acting elements and their cognate binding proteins depends on their position relative to regulated exons. For instance, NOVA1 interacts with an ISE in the pre-mRNA of GABRG2 (GABA A receptor 2) and promotes exon 9 inclusion, but upon binding to ESS in the exon 4 of its own pre-mRNA, it prevents the inclusion of exon 4 [101].

Alternative splicing is crucial in determining tissue specificity. Recent high-throughput investigations have found that 50% or more of alternative splicing isoforms are expressed differently across human tissues, showing that most alternative splicing is tissue-specific [102]. The brain is the most functionally varied of all human tissues with the highest number of tissue-specific alternative splicing isoforms. Although no single determinant of tissue specificity of splicing has been identified in any system, there are splicing regulatory proteins whose expression is confined to distinct cell types (Table 4). Examples of such factors include the Neuronal PTB protein and ETR3 variations such as NAPOR. The neuronal proteins Nova-1 and Nova-2 are the best-investigated splicing regulators with precise tissue-specific expression. These are differently expressed in the brain tissues of post-natal mice, with Nova-2 predominantly being expressed in the neocortex and hippocampus, while NOVA-1 is primarily expressed in the hindbrain and spinal cord [103] (Table 4).

Table 4: Tissue specific splicing factors.

Splicing Factor	Characteristic Domains	Target Gene	Target Tissue
NoVA1	KH, YCAY	*NOVA1, GLyRA2,* and *GABRG2*	Hindbrain and spinal cord
NoVA2	KH YCAY	*KCNJ, APLP2, GPhN, JNK2, NEO, GRIN1,* and *PLCB4*	Neocortex and hippocampus
nPTB	RRM	*SRC, MEF2, NASP, ATP2B1, SPAG9,* and *GLyRα2,*	Neural progenitor cells and N2A neuroblastoma cells
Fox1	RRM	*EWSR1, ACTN, FGFR2, SRC,* and *FN1*	Neurons and cardiac muscles
Fox2	RRM	*SRC, FGFR2, FN1,* and *EWS*	Neurons and cardiac muscles
RBM35a	RRM	*ENAh, CD44, CTNND1,* and *FGFR2*	Epithelial cells
TIA1	RRM	*CALCA CD95, FGFR2, MyPT1, COL2A1, IL8, VEGF, NF1,* and *TIAR*	Testes, brain, and spleen
MBNL	CCCH zinc finger domain YGCU(U/G)Y	*CLCN1, INSR, TNTT2* and *TNNT3*	Ovaries, muscles, and uterus

7.4 Regulation of Alternative Splicing Decides the Sex in Flies

The sex of Drosophila is determined by the regulation of alternative splicing of the sex-lethal (Sxl) gene, and the sex of the given fly depends on which of the two alternative isoforms of this mRNA is produced. The functional protein of this gene is exclusively present in females but not in males, which is ensured by the autoregulation of the splicing of its own message. Sxl protein negatively regulates the splicing of the Tra gene by binding to the pyrimidine tract at 3' SS, producing its functional protein only in females [104]. This Tra protein itself positively regulates the splicing of mRNA of double-sex (Dsx) gene by binding to the enhancer site of one of its exons [105]. mRNA of the Dsx gene is spliced differently in response to Tra protein, producing an alternative isoform of this protein with a stretch of 30 amino acids at the carboxy-terminal end that distinguishes it from the isoform produced in males in the absence of Tra protein [106]. This female isoform of the Dsx gene acts as an activator for genes required for development into a female fly and represses the genes needed for the development of a male fly.

References

[1] Chow, L.T., Gelinas, R.E., Broker, T.R. and Roberts, R.J. (1977). An amazing sequence arrangement at the 5' ends of adenovirus 2 messenger RNA. Cell 12: 1–8.
[2] Nilsen, T.W. (1998). RNA–RNA interactions in nuclear pre-mRNA splicing. In RNA Structure and Function, 279–307.
[3] Berglund, J.A., Chua, K., Abovich, N., Reed, R. and Rosbash, M. (1997). The splicing factor BBP interacts specifically with the pre-mRNA branchpoint sequence UACUAAC. Cell 89: 781–787.
[4] Wahl, M.C., Will, C.L. and Luhrmann, R. (2009). The spliceosome: Design principles of a dynamic RNP machine. Cell 136: 701–718.
[5] Zamore, P.D. and Green, M.R. (1989). Identification, purification, and biochemical characterization of U2 small nuclear ribonucleoprotein auxiliary factor. Proc. Natl Acad. Sci. 86: 9243–9247.

[6] Rino, J. and Carmo-Fonseca, M. (2009). The spliceosome: A self organized macromolecular machine in the nucleus? Trends Cell Biol. 19: 375.

[7] Plaschka, C., Lin, P.C., Charenton, C. and Nagai, K. (2018). Prespliceosome structure provides insights into spliceosome assembly and regulation. Nature 559(7714): 419–22.

[8] Human Genome Sequencing Consortium I and International Human Genome Sequencing C: Finishing the euchromatic sequence of the human genome. (2004). Nature 431: 931–945.

[9] Nilsen, T.W. and Graveley, B.R. (2010). Expansion of the eukaryotic proteome by alternative splicing. Nature 463: 457–463.

[10] Qin, D., Huang, L., Wlodaver, A., Andrade, J. and Staley, J.P. (2016). Sequencing of lariat termini in *S. cerevisiae* reveals 5_ splice sites, branch points, and novel splicing events. RNA 22(2): 237–53.

[11] Sheth, N., Roca, X., Hastings, M.L., Roeder, T., Krainer, A.R. and Sachidanandam, R. (2006). Comprehensive splice site analysis using comparative genomics. Nucleic Acids Res. 34(14): 3955–67.

[12] Berget, S.M. (1995). Exon recognition in vertebrate splicing. J. Biol. Chem. 270: 2411–2414.

[13] Wahl, M.C., Will, C.L. and Luhrmann, R. (2009). The spliceosome: Design principles of a dynamic RNP machine. Cell 136: 701–718.

[14] Long, J.C. and Caceres, J.F. (2009). The SR protein family of splicing factors: Master regulators of gene expression. Biochem. J. 417: 15–27.

[15] Rodriguez, J.R., Pikielny, C.W. and Rosbash, M. (1984). *In vivo* characterization of yeast mRNA processing intermediates. Cell 39(3, Part 2): 603–10.

[16] Kastner, B., Will, C.L., Stark, H. and Lührmann, R. (2019). Structural insights into nuclear pre-mRNA splicing in higher eukaryotes. Cold Spring Harbor Perspectives in Biology 11(11): a032417.

[17] Ruskin, B., Krainer, A.R., Maniatis, T. and Green, M.R. (1984). Excision of an intact intron as a novel lariat structure during pre-mRNA splicing *in vitro*. Cell 38(1): 317–31.

[18] Hardy, S.F., Grabowski, P.J., Padgett, R.A. and Sharp, P.A. (1984). Cofactor requirements of splicing of purified messenger RNA precursors. Nature 308: 375–377.

[19] Steitz, T.A. and Steitz, J.A. (1993). A general two-metal-ion mechanism for catalytic RNA. Proc. Natl Acad. Sci. USA 90: 6498–6502.

[20] Steitz, J.A., Dreyfuss, G., Krainer, A.R., Lamond, A.I., Matera, A.G. and Padgett, R.A. (2008). Where in the cell is the minor spliceosome? Proceedings of the National Academy of Sciences 105(25): 8485–8486.

[21] Brody, E. and Abelson, J. (1985). The 'spliceosome': Yeast pre-messenger RNA associates with a 40S complex in a splicing-dependent reaction. Science 228: 963–967.

[22] Lerner, M.R., Boyle, J.A., Mount, S.M., Wolin, S.L. and Steitz, J.A. (1980). Are snRNPs involved in splicing? Nature 283: 220–224.

[23] Reddy, R. and Busch, H. (1988). Small nuclear RNAs: RNA sequences, structure, and modifications. pp. 1–37. *In*: Birnstiel, M.L. (ed.). Structure and Function of Major and Minor Small Nuclear Ribonucleoprotein Particles. Berlin: Springer.

[24] Bringmann, P. and Lührmann, R. (1986). Purification of the individual snRNPs U1, U2, U5 and U4/U6 from HeLa cells and characterization of their protein constituents. EMBO J. 5(13): 3509–16.

[25] Séraphin, B. (1995). Sm and Sm-like proteins belong to a large family: Identification of proteins of the U6 as well as the U1, U2, U4 and U5 snRNPs. EMBO J. 14(9): 2089–98.

[26] Zhou, L., Hang, J., Zhou, Y., Wan, R. and Lu, G. (2013). Crystal structures of the Lsm complex bound to the 3 end sequence of U6 small nuclear RNA. Nature 506(7486): 116–20.

[27] Achsel, T., Brahms, H., Kastner, B., Bachi A., Wilm, M. and Lührmann, R. (1991). A doughnut-shaped heteromer of human Sm-like proteins binds to the 3-end of U6 snRNA, thereby facilitating U4/U6 duplex formation *in vitro*. EMBO J. 18(20): 5789–802.

[28] Galej, W.P., Oubridge, C., Newman, A.J. and Nagai, K. (2013). Crystal structure of Prp8 reveals active site cavity of the spliceosome. Nature 493(7434): 638–43.

[29] Kastner, B., Will, C.L., Stark, H. and Lührmann, R. (2019). Structural insights into nuclear pre-mRNA splicing in higher eukaryotes. Cold Spring Harbor Perspectives in Biology 11(11): 032417.

[30] Will, C.L. and Lührmann, R. (2011). Spliceosome structure and function. Cold Spring Harbor Perspectives in Biology 3(7): a003707.

[31]   Wahl, M.C., Will, C.L. and Luhrmann, R. (2009). The spliceosome: Design principles of a dynamic RNP machine. Cell 136: 701–718.

[32]   Akinyi, M.V. and Frilander, M.J. (2021). At the intersection of major and minor spliceosomes: Crosstalk mechanisms and their impact on gene Expression. Frontiers in Genetics 12: 700744.

[33]   Patel, A.A. and Steitz, J.A. (2003). Splicing double: Insights from the second spliceosome. Nat. Rev. Mol. Cell Biol. 4(12): 960–70.

[34]   Staley, J.P. and Guthrie, C. (1998). Mechanical devices of the spliceosome: Motors, clocks, springs, and things. Cell 92(3): 315–326.

[35]   Reddy, R. and Busch, H. (1988). Small nuclear RNAs: RNA sequences, structure, and modifications. pp. 1–37. *In*: Birnstiel, M.L. (ed.). Structure and Function of Major and Minor Small Nuclear Ribonucleoprotein Particles. Berlin: Springer.

[36]   Wilkinson, M.E., Charenton, C. and Nagai, K. (2020). RNA splicing by the spliceosome. Annual Review of Biochemistry 89: 359–388.

[37]   Plaschka, C., Lin, P.C., Charenton, C. and Nagai, K. (2018). Prespliceosome structure provides insights into spliceosome assembly and regulation. Nature 559(7714): 419–22.

[38]   Gilbert, W. (1978). Why genes in pieces? Nature 271: 501.

[39]   Wong, M.S., Kinney, J.B. and Krainer, A.R. (2018). Quantitative activity profile and context dependence of all human 5 splice sites. Mol. Cell 71(6): 1012–13.

[40]   Hui, J., Hung, L.H., Heiner, M., Schreiner, S., Neumüller, N., Reither, G., Haas, S.A. and Bindereif, A. (2005). Intronic CA-repeat and CA-rich elements: A new class of regulators of mammalian alternative splicing. EMBO J. 24: 1988–1998.

[41]   Frendewey, D. and Keller, W. (1985). Stepwise assembly of a pre-mRNA splicing complex requires U-snRNPs and specific intron sequences. Cell 42(1): 355–67.

[42]   Valcárcel, J., Gaur, R.K., Singh, R. and Green, M.R. (1996). Interaction of U2AF65 RS region with pre-mRNA branch point and promotion of base pairing with U2 snRNA [corrected]. Science (New York, N.Y.) 273(5282): 1706–1709.

[43]   Gozani, O., Feld, R. and Reed, R. (1996). Evidence that sequence-independent binding of highly conserved U2 snRNP proteins upstream of the branch site is required for assembly of spliceosomal complex A. Genes & Development 10(2): 233–243.

[44]   Kastner, B., Will, C.L., Stark, H. and Lührmann, R. (2002). Structural insights into nuclear pre-mRNA splicing in higher eukaryotes. Cold Spring Harb. Perspect. Biol. In Press.

[45]   Boesler, C., Rigo, N., Anokhina, M.M., Tauchert, M.J., Agafonov, D.E., Kastner, B., Urlaub, H., Ficner, R., Will, C.L. and Lührmann, R. (2016). A spliceosome intermediate with loosely associated tri-snRNP accumulates in the absence of Prp28 ATPase activity. Nature Communications 7: 11997.

[46]   Bai, R., Wan, R., Yan, C., Lei, J. and Shi, Y. (2018). Structures of the fully assembled Saccharomyces cerevisiae spliceosome before activation. Science (New York, N.Y.) 360(6396): 1423–1429.

[47]   Will, C.L. and Lührmann, R. (2011). Spliceosome structure and function. Cold Spring Harbor Perspectives in Biology 3(7): a003707.

[48]   Bessonov, S., Anokhina, M., Krasauskas, A., Golas, M.M., Sander, B., Will, C.L., Urlaub, H., Stark, H. and Lührmann, R. (2010). Characterization of purified human B act spliceosomal complexes reveals compositional and morphological changes during spliceosome activation and first step catalysis. RNA (New York, N.Y.) 16(12): 2384–2403.

[49]   Galej, W.P., Oubridge, C., Newman, A.J. and Nagai, K. (2013). Crystal structure of Prp8 reveals active site cavity of the spliceosome. Nature 493(7434): 638–43.

[50]   Fica, S.M., Tuttle, N., Novak, T., Li, N.S. and Lu, J. (2013). RNA catalyses nuclear pre-mRNA splicing. Nature 503(7475): 229–34.

[51]   Galej, W.P. (2018). Structural studies of the spliceosome: Past, present and future perspectives. Biochemical Society Transactions 46(6): 1407–1422.

[52]   Schwer, B. (2008). A conformational rearrangement in the spliceosome sets the stage for Prp22-dependent mRNA release. Molecular Cell 30(6): 743–754.

[53]   Fica, S.M., Oubridge, C., Galej, W.P., Wilkinson, M.E., Bai, X.C., Newman, A.J. and Nagai, K. (2017). Structure of a spliceosome remodelled for exon ligation. Nature 542(7641): 377–380.

[54]   Patel, A.A. and Steitz, J.A. (2003). Splicing double: Insights from the second spliceosome. Nat. Rev. Mol. Cell. Biol. 4(12): 960–70.

[55]   Koodathingal, P. and Staley, J.P. (2013). Splicing fidelity: DEAD/H-box ATPases as molecular clocks. RNA Biol. 10(7): 1073–1079.

[56] Smith, D.J., Query, C.C. and Konarska, M.M. (2008). Nought may endure but mutability: Spliceosome dynamics and the regulation of splicing. Mol Cell. 30(6): 657–66.

[57] Semlow, D.R. and Staley, J.P. (2012). Staying on message: Ensuring fidelity in pre-mRNA splicing. Trends in Biochemical Sciences 37(7): 263–273.

[58] Mayas, R.M., Maita, H. and Staley, J.P. (2006). Exon ligation is proofread by the DExD/H-box ATPase Prp22p. Nature Structural & Molecular Biology 13(6): 482–490.

[59] Koodathingal, P. and Staley, J.P. (2013). Splicing fidelity: DEAD/H-box ATPases as molecular clocks. RNA Biol. 10(7): 1073–1079.

[60] Mayas, R.M., Maita, H., Semlow, D.R. and Staley, J.P. (2010, Jun). Spliceosome discards intermediates via the DEAH box ATPase Prp43p. Proc. Natl. Acad. Sci. USA 107(22): 10020–5.

[61] Yang, F., Wang, X.Y., Zhang, Z.M., Pu, J., Fan, Y.J., Zhou, J., Query, C.C. and Xu, Y.Z. (2013). Splicing proofreading at 5' splice sites by ATPase Prp28p. Nucleic Acids Res. 41(8): 4660–70.

[62] Schwer, B. and Guthrie, C. (1991). PRP16 is an RNA-dependent ATPase that interacts transiently with the spliceosome. Nature 349(6309): 494–9.

[63] Schwer, B. and Guthrie C. (1992). A conformational rearrangement in the spliceosome is dependent on PRP16 and ATP hydrolysis. EMBO J. 11(13): 5033–9.

[64] Burgess, S.M. and Guthrie, C. (1993). A mechanism to enhance mRNA splicing fidelity: The RNA-dependent ATPase Prp16 governs usage of a discard pathway for aberrant lariat intermediates. Cell 73(7): 1377–91.

[65] Perriman, R. and Ares, M. Jr. (2010). Invariant U2 snRNA nucleotides form a stem loop to recognize the intron early in splicing. Mol. Cell 38(3): 416–27.

[66] Wagner, J.D., Jankowsky, E., Company, M., Pyle, A.M. and Abelson, J.N. (1998). The DEAH-box protein PRP22 is an ATPase that mediates ATP-dependent mRNA release from the spliceosome and unwinds RNA duplexes. EMBO J. 17(10): 2926–37.

[67] Schwer, B. and Gross, C.H. (1998). Prp22, a DExH-box RNA helicase, plays two distinct roles in yeast pre-mRNA splicing. EMBO J. 17(7): 2086–94.

[68] Mayas, R.M., Maita, H. and Staley, J.P. (2006). Exon ligation is proofread by the DExD/H-box ATPase Prp22p. Nature Structural & Molecular Biology 13(6): 482–490.

[69] Xu, Y.Z. and Query, C.C. (2007). Competition between the ATPase Prp5 and branch region-U2 snRNA pairing modulates the fidelity of spliceosome assembly. Molecular Cell 28(5): 838–849.

[70] Human Genome Sequencing Consortium I and International Human Genome Sequencing C: Finishing the euchromatic sequence of the human genome. (2004). Nature 431: 931–945.

[71] Koch, L. (2017). Alternative splicing: a thermometer controlling gene expression. Nat. Rev. Genet. 18(9): 515.

[72] Nilsen, T.W. and Graveley, B.R. (2010). Expansion of the eukaryotic proteome by alternative splicing. Nature 463: 457–463.

[73] Moldon, A. (2008). Promoter-driven splicing regulation in fission yeast. Nature 455: 997–1000.

[74] Blencowe, B.J. (2006). Alternative splicing: new insights from global analyses. Cell 126: 37–47.

[75] Kim, E., Magen, A. and Ast, G. (2007). Different levels of alternative splicing among eukaryotes. Nucleic Acids Res. 35: 125–131.

[76] Galante, P.A., Sakabe, N.J., Kirschbaum-Slager, N. and de Souza, S.J. (2004). Detection and evaluation of intron retention events in the human transcriptome. RNA 10: 757–765.

[77] Sakabe, N.J. and de Souza, S.J. (2007). Sequence features responsible for intron retention in human. BMC Genomics 8: 59.

[78] Dredge, B.K. and Darnell, R.B. (2003). Nova regulates GABAA receptor γ2 alternative splicing via a distal downstream UCAU-rich intronic splicing enhancer. Mol. Cell. Biol. 23: 4687–4700.

[79] Di Giammartino, D.C., Nishida, K. and Manley, J.L. (2011). Mechanisms and consequences of alternative polyadenylation. Mol. Cell 43: 853–866.

[80] Goldstrohm, A.C., Greenleaf, A.L. and Garcia-Blanco, M.A. (2001). Co-transcriptional splicing of pre-messenger RNAs: Considerations for the mechanism of alternative splicing. Gene 277: 31–47.

[81] Cooper, T.A., Wan, L. and Dreyfuss, G. (2009). RNA and disease. Cell 136(4): 777–93.

[82] Wang, Z., Xiao, X., Van Nostrand, E. and Burge, C.B. (2006). General and specific functions of exonic splicing silencers in splicing control. Mol. Cell 23: 61–70.

[83] Wang, Z. and Burge, C.B. (2008). Splicing regulation: From a parts list of regulatory elements to an integrated splicing code. RNA 14: 802–813.

[84]   Long, J.C. and Caceres, J.F. (2009). The SR protein family of splicing factors: Master regulators of gene expression. Biochem. J. 417: 15–27.

[85]   Ule, J. (2006). An RNA map predicting Nova-dependent splicing regulation. Nature 444: 580–586.

[86]   Lallena, M.J., Chalmers, K.J., Llamazares, S., Lamond, A.I. and Valcarcel, J. (2002). Splicing regulation at the second catalytic step by Sex-lethal involves 3' splice site recognition by SPF45. Cell 109: 285–296.

[87]   House, A.E. and Lynch, K.W. (2006). An exonic splicing silencer represses spliceosome assembly after ATP-dependent exon recognition. Nature Struct. Mol. Biol. 13: 937–944.

[88]   Sheth, N., Roca, X., Hastings, M.L., Roeder, T., Krainer, A.R. and Sachidanandam, R. (2006). Comprehensive splicesite analysis using comparative genomics. Nucleic Acids Res. 34(14): 3955–67.

[89]   Montes, M., Becerra, S., Sanchez-Alvarez, M. and Sune, C. (2012). Functional coupling of transcription and splicing. Gene 501: 104–117.

[90]   Lerner, M.R. and Steitz, J.A. (1979). Antibodies to small nuclear RNAs complexed with proteins are produced by patients with systemic lupus erythematosus. PNAS 76(11): 5495–99.

[91]   Long, J.C. and Caceres, J.F. (2009). The SR protein family of splicing factors: Master regulators of gene expression. Biochem. J. 417: 15–27.

[92]   Ghosh, G. and Adams, J.A. (2011). Phosphorylation mechanism and structure of serine-arginine protein kinases. FEBS J. 278: 587–597.

[93]   Longman, D. (2011). Multiple interactions between SRm160 and SR family proteins in enhancer dependent splicing and development of *C. elegans*. Curr. Biol. 11: 1923–1933.

[94]   Krecic, A.M. and Swanson, M.S. (1999). Functional roles of non-coding Y RNAs. Curr. Opin. Cell Biol. 11: 363–71.

[95]   Tange, T.O., Damgaard, C.K., Guth, S., Valcarcel, J. and Kjems, J. (2001). The hnRNP A1 protein regulates HIV-1 *tat* splicing via a novel intron silencer element. EMBO J. 20: 5748–5758.

[96]   Spellman, R. and Smith, C.W. (2006). Novel modes of splicing repression by PTB. Trends Biochem. Sci. 31: 73–76.

[97]   Zhou, H.L. and Lou, H. (2008). Repression of prespliceosome complex formation at two distinct steps by Fox-1/ Fox-2 proteins. Mol. Cell. Biol. 28: 5507–5516.

[98]   Black, D.L. (2003). Mechanisms of alternative pre-messenger RNA splicing. Annu. Rev. Biochem. 72: 291–336.

[99]   Mayeda, A., Helfman, D.M. and Krainer, A.R. (2001). Modulation of exon skipping and inclusion by heterogeneous nuclear ribonucleoprotein A1 and pre-mRNA splicing factor SF2/ASF. Mol. Cell. Biol. 13: 2993–3001.

[100] Expert-Bezancon, A. (2004). hnRNP A1 and the SR proteins ASF/SF2 and SC35 have antagonistic functions in splicing of β-tropomyosin exon 6B. J. Biol. Chem. 279: 38249–59.

[101] Dredge, B.K. and Darnell, R.B. (2003). Nova regulates GABAA receptor γ2 alternative splicing via a distal downstream UCAU-rich intronic splicing enhancer. Mol. Cell. Biol. 23: 4687–4700.

[102] Wang, Z. and Burge, C.B. (2008). Splicing regulation: from a parts list of regulatory elements to an integrated splicing code. RNA 14: 802–813.

[103] Dredge, B.K., Stefani, G., Engelhard, C.C. and Darnell, R.B. (2005). Nova autoregulation reveals dual functions in neuronal splicing. EMBO J. 24: 1608–1620.

[104] González, A.N., Lu, H. and Erickson, J.W. (2008). A shared enhancer controls a temporal switch between promoters during Drosophila primary sex determination. Proc. Natl. Acad. Sci. USA 105: 18436–18441.

[105] Wojtowicz, W.M., Flanagan, J.J., Millard, S.S., Zipursky, S.L. and Clemens, J.C. (2004). Alternative splicing of Drosophila Dscam generates axon guidance receptors that exhibit isoform-specific homophilic binding. Cell 118: 619–633.

[106] Chertemps, T., Duportets, L., Labeur, C., Ueda, R. and Takahashi, K. (2007). A female-biased expressed elongase involved in long-chain hydrocarbon biosynthesis and courtship behavior in Drosophila melanogaster. Proc. Natl. Acad. Sci. USA 104: 4273–4278.

3

Alternative Splicing

A New Cancer Hallmark

Mohd Younis Rather,[1] *Ajaz Ahmad Waza,*[1,*]
Yasmeena Hassan[2] and *Sabhiya Majid*[1,3]

1. Introduction

The blueprint for constructing an organism resides within the hereditary genetic material, typically referred to as DNA. This DNA comprises linear sequences of four distinct nucleotides, serving as the driving force behind protein synthesis, the essential components responsible for the cell's functionality. Nonetheless, an intermediary molecule is indispensable for transmitting information from DNA to proteins, and this intermediary is known as RNA. This fundamental principle is acknowledged as the central tenet of biology, wherein DNA sequences are converted into RNA sequences, ultimately culminating in the creation of proteins. These proteins play pivotal roles in providing structural integrity and Executing enzymatic functions within cells and organisms. While extensive research over the past six decades has uncovered intricate details adding complexity to this process, the fundamental "DNA-RNA-protein" flow remains a cornerstone in molecular biology [1].

The process of copying RNA from one of the two DNA strands is termed transcription, and consequently, the produced RNA molecules are also known as transcripts. Transcription of DNA is executed by RNA polymerases. Conversely, translation is the process through which the information encoded in "messenger" RNA molecules is transformed into proteins. Facilitating this vital cellular function are ribosomes, which are complex assemblies consisting of both nuclear proteins and RNA molecules [1, 2].

[1] Multidisciplinary Research Unit, Department of Health Research, Government Medical College, Srinagar, J&K, India.
[2] Mader-e-Meharban Institute of Nursing Sciences & Research (MMINSR), Sher-I- Kashmir Institute of Medical Sciences (SKIMS), Soura Srinagar, J&K, India.
[3] Department of Biochemistry, Government Medical College Srinagar, J&K, India.
* Corresponding author: ajazahmad09@gmail.com

The genome represents the entirety of an organism's genetic material, encompassing DNA that contains genes responsible for encoding various proteins, as well as intergenic regions. These intergenic regions were once believed to lack functional elements but are now acknowledged for their crucial roles in gene regulation and the generation of transcripts that do not undergo protein translation [3]. Throughout this information flow, numerous modifications occur at each step. In the realm of eukaryotic genes, there exists a combination of coding (exons) and non-coding (introns) sequences. Exons exist solely in mature mRNAs, which are subsequently translated into proteins, while introns are specific to premature RNAs (pre-mRNAs) and must be excised to enable mRNA translation into proteins. This process of intron removal and exon fusion is referred to as pre-mRNA splicing [3, 4]. Splicing constitutes a major component of pre-mRNA processing, unfolding within the nucleus of all eukaryotic cells and orchestrated by the spliceosome, a highly dynamic and intricate ribonucleoprotein complex [4, 5].

Decades of research have unveiled and expanded our understanding of the intriguing functions underpinning the necessity to eliminate segments of primary transcripts. Although the origins of introns remain a subject of debate [6], several compelling reasons support the existence of splicing. Foremost among these is the potential for intricate regulation; numerous non-coding RNAs with regulatory roles, such as snoRNAs (small nucleolar RNAs) and miRNAs (micro RNAs), emerge from excised introns [7]. Additionally, a fascinating class of RNAs known as circular RNAs can also originate from intronic sequences and be linked to biologically significant functions [7]. Furthermore, it has been postulated that one of the benefits of RNA splicing is to defend against parasitic nucleic acids lacking optimal splicing signals, resulting in the stalling and degradation of spliceosomes [8].

In higher eukaryotes, there exists the capacity for entire exons or portions thereof to be either included or excluded in the mature mRNA molecules. This flexibility in RNA processing is termed alternative splicing (as depicted in Figure 1). Consequently, a single gene can give rise to multiple distinct mature transcripts. This feature bestows a significant advantage upon eukaryotic genomes, as it diversifies the array of proteins that can be generated from the same genetic material, thereby greatly enhancing their informational richness [9]. Functional disparities among proteins derived from alternatively spliced transcripts, known as isoforms, encompass variations in the existence or lack of protein domains and regions of polypeptides that lack a defined or structured conformation play pivotal roles in protein-protein interactions [10].

Nevertheless, in certain instances, alternative splicing may not result in variations at the protein level; instead, it impacts the stability of the alternative transcripts. For instance, alternative splicing can be associated with nonsense-mediated decay (NMD). In this process, transcripts containing premature stop codons, as observed in alternatively spliced RNAs, are degraded after the initial round of translation. Alternatively, in some situations, the functional differences between isoforms are dictated by the unique binding patterns of RNA-binding proteins (RBPs). These RBPs oversee various aspects of RNA metabolism, including mRNA stability and translation efficiency. Collectively, these examples illustrate that alternative splicing plays a multifaceted role in gene regulation and protein diversity [6, 11].

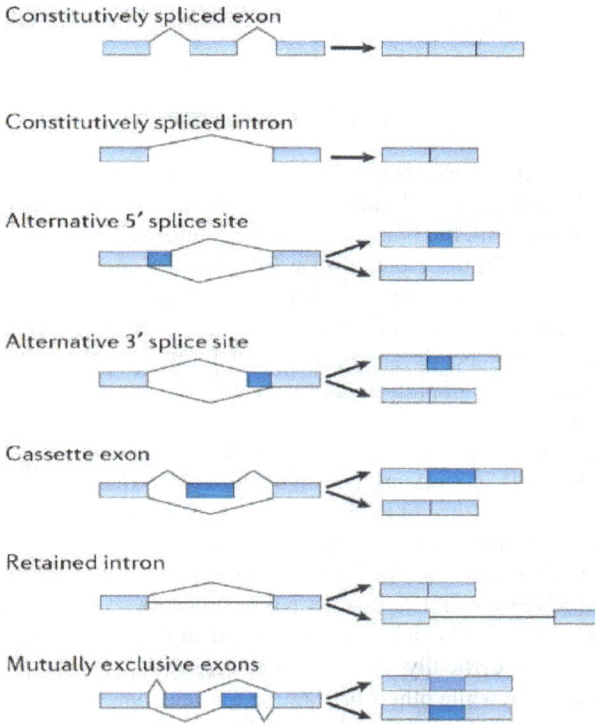

Figure 1: Types of Alternative Splicing.
Splicing within a region is labeled as constitutive when it adheres to a uniform pattern in every cell. In this context, introns that are consistently spliced and exons that are consistently included in the mature transcripts are termed constitutive elements. Conversely, alternative splicing arises from the potential to generate diverse splicing variants. This variability can manifest through the utilization of alternative splice sites (both 5' and 3'), the omission of exons (referred to as cassette exon or exon skipping events), the retention of introns within mature transcripts, or the occurrence of mutually exclusive exons. Complex events can also arise from combinations of these different splicing patterns. Splice sites that are only employed under specific, non-physiological conditions are often termed cryptic splice sites (adapted from [15]).

These factors encompass the unique requirements of cells found in various tissues or developmental stages. These requirements entail not only the activation or suppression of distinct genes via transcriptional regulation but also the attainment of a distinct equilibrium of RNA and protein isoforms, which translates to different transcripts or proteins originating from the same gene [12, 13]. Consequently, despite the fact that all cells within a complex organism possess identical genomes, their transcriptomes and proteomes can exhibit substantial disparities, with these variations stemming from both transcriptional and post-transcriptional processes.

Alternative splicing appears to have held a pivotal role in the course of evolution. For instance, despite human and *C. elegans* worm genomes containing roughly similar gene counts (around 20,000), the greater complexity observed in humans can be partly attributed to the heightened occurrence of alternative splicing [8]. In fact, over 95% of human genes are capable of undergoing alternative splicing [7, 14] (as depicted in Figure 1).

2. The Spliceosome

Splicing catalysis unfolds in a two-step process (as depicted in Figure 2). Initially, the nucleotide located at the 5' end of the intron forms a covalent bond, a 2'-5' phosphodiester bond, with the branch adenosine found within the branch point (BP) sequence. This results in the creation of a free 5' exon and a lariat-shaped intron. In the second step, the free 5' exon is connected (ligated) to the 3' exon, and the lariat structure is removed. Each of these steps involves trans-esterification reactions wherein phosphodiester bonds are broken and formed, with one electron donor oxygen facilitating a nucleophilic assault on a phosphate group within the RNA backbone [11, 16, 17].

The molecular machinery performing RNA splicing is known as the spliceosome [18]. This is a remarkably dynamic and intricate assembly composed of non-coding RNAs and proteins, distinguished by its unique ability to assemble sequentially on each intron prior to splicing. The splicing machinery consists of five snRNPs (small nuclear ribonucleoproteins): U1, U2, U4, U5, and U6 snRNPs. Each of these snRNPs contains a distinct RNA molecule (U1, U2, U4, U5, and U6 snRNAs, known for their U-rich sequences) and multiple polypeptides. Some of these polypeptides, specifically the Sm core, are shared across all spliceosomal snRNPs except for U6, while others are unique to each complex. The machinery can assemble on each intron by recognizing specific sequence signals. At the intronic 5' boundary (the 5' splice site or 5'ss), there is a GURAGU consensus, and three elements are located at the 3' boundary (the 3' splice site or 3'ss). These elements consist of a branch point sequence (BP), characterized by a degenerate YUNAY consensus in higher eukaryotes (where Y represents a pyrimidine and N represents A, G, C, or U), a stretch of pyrimidines known as the polypyrimidine tract (Py-tract), and an AG dinucleotide [5, 16] (as depicted in Figure 3).

In the early phases of spliceosome assembly, the 5' splice site sequences are recognized by U1 snRNP through base-pairing interactions between the 5' end of U1 snRNA and the 5'ss sequence. At the same time, proteins SF1, U2AF65 (or U2AF2), and U2AF35 (or U2AF1) identify the BP sequence, the polypyrimidine tract, and the AG dinucleotide, respectively. The coordinated binding of these proteins initiates the formation of the early spliceosomal complex E (as shown in Figure 3) and facilitates the recruitment of U2 snRNP at the BP sequence [9]. The control of U1 and U2 snRNPs recruitment plays a crucial role in the context of alternative splicing [6]. The formation of the U2 small nuclear ribonucleoprotein (snRNP) complex entails base-pairing interactions between the U2 snRNA component of the snRNP and nucleotides located within the branch point (BP) sequence.

Once the recruitment of U1 and U2 snRNPs is complete, and SF1 is displaced, complex A is formed (as shown in Figure 3). Subsequently, the tri-snRNP, consisting of the U4, U5, and U6 snRNPs preassembled together, joins the complex, leading to the formation of the B complex. As U1 and U4 snRNPs become destabilized and exit the complex, the activated B complex (Bact) emerges and becomes catalytically

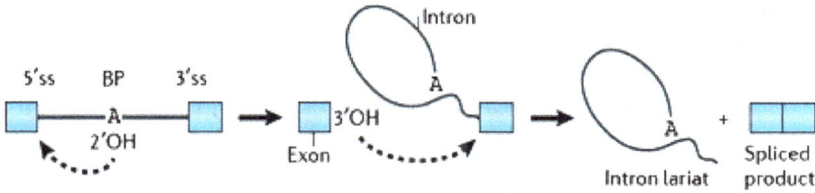

Figure 2: Illustrates the two stages of splicing catalysis. In the initial step, a 2'–5' nucleophilic attack links the 5'ss and the branch adenosine, resulting in the formation of a lariat structure and a free upstream exon. The subsequent second step involves a 3'–5' nucleophilic bond formation between the free exon and the 3'ss [16].

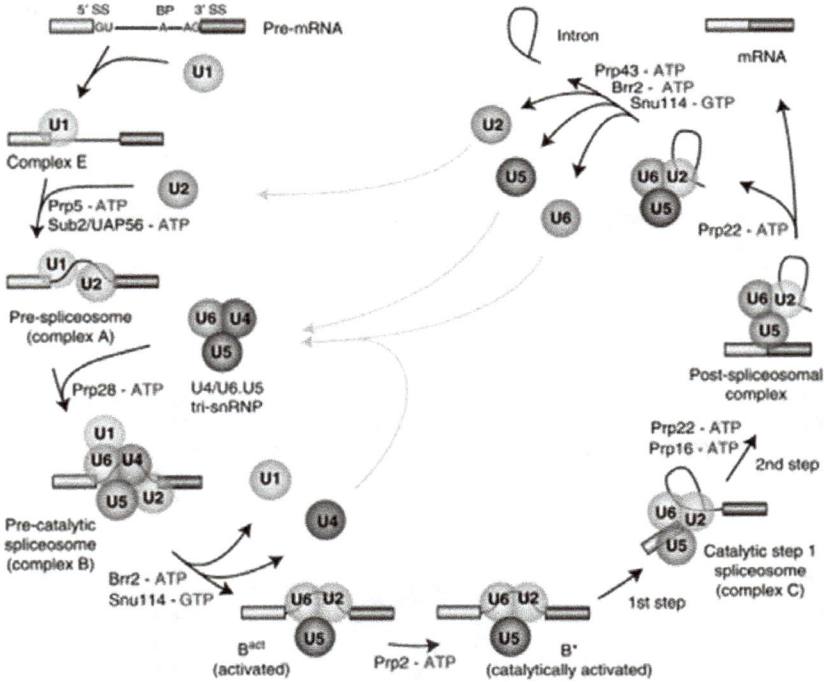

Figure 3: Illustrates the process of spliceosome assembly and the catalysis of intron removal. At the top, a representative precursor mRNA (pre-mRNA) is depicted, containing intronic sequences (represented by thin lines) and exonic sequences (depicted as colored boxes). The consensus sequences at the intron's 5' and 3' ends are indicated, with "Y" representing pyrimidines, "N" representing any nucleotide, and "(n)" signifying an undefined number of pyrimidines. The branch point (BP) adenosine is highlighted in bold. The splicing factors, shown within the box, sequentially assemble on the pre-mRNA, forming the indicated complexes (E, A, and B). Conformational changes within the assembled spliceosome (complexes Bact and C) lead to splicing catalysis, resulting in the production of a mature mRNA and the release of the intron in a lariat configuration. The binding of U2 small nuclear ribonucleoprotein (snRNP) involves base-pairing interactions between U2 small nuclear RNA (snRNA) and nucleotides flanking the BP, as well as interactions between the U2 snRNP proteins splicing factor 3B subunit 1 (SF3B1) and p14 with the pre-mRNA. Additionally, "BPRS" denotes the BP recognition sequence, "SF1" stands for splicing factor 1, "ss" represents splice site, and "U2AF" and "U2 snRNP" signify U2 snRNP auxiliary factor.

active (referred to as B*). After the initial step, the spliceosome adopts the C complex conformation (as depicted in Figure 3) and progresses to the post-splicing P complex at the conclusion of the entire process before ultimately disassembling [6].

These steps, with some potential variations, appear to be the general process for removing most eukaryotic introns. The complexity and dynamic nature of the spliceosome, which is among the most intricate cellular molecular machinery, offer multiple opportunities for regulation at various stages during its assembly.

Throughout spliceosome assembly, a multitude of interactions, including protein-protein, protein-RNA, and RNA-RNA interactions, are orchestrated to facilitate changes in composition and conformation. Notably, significant modifications occur in U2 snRNA (as depicted in Figure 4). The latter changes (as illustrated in Figure 5) are coordinated by essential helicases and are essential for bringing the 5'ss

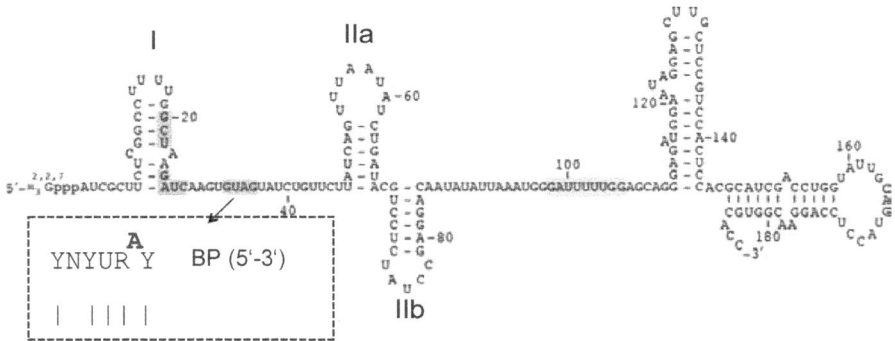

Figure 4: Displays the secondary structure of U2 snRNA, with stem-loop structures denoted by Roman numerals (adapted from [21]). Stem-loops IIa and IIb, along with the BP recognition sequence, undergo multiple remodeling steps during the assembly of the spliceosome. The interaction involving base-pairing between the pre-mRNA's branch point (BP) and the BP recognition sequence of the U2 snRNA (BPRS) results in the branch adenosine extending from the pre-mRNA (as indicated within the box).

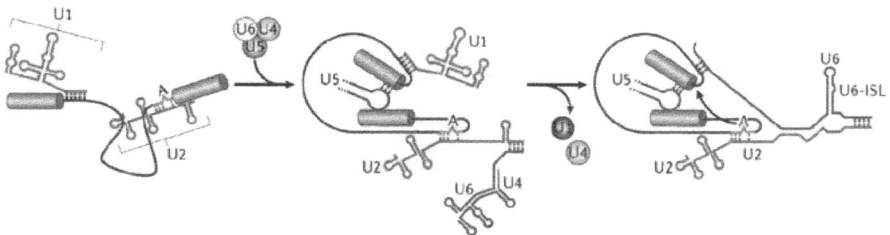

Figure 5: Illustrates the reconfiguration of RNA-RNA interactions throughout the progression of spliceosome assembly. As spliceosome assembly proceeds in a stepwise manner, RNA-RNA interactions are meticulously coordinated to establish the catalytic core. In the complex A stage (depicted in the left panel), U1 and U2 snRNAs engage in base-pairing interactions with the 5'ss and the BP region, respectively. In the B complex (shown in the middle panel), U5 snRNA binds to exonic sequences adjacent to the 5'ss, while U6 snRNA associates with U2 snRNA. U4 snRNA partially forms base-pairing interactions with U6 snRNA. Through precise helicase activities, the binding between U4 and U6 is unwound, resulting in U4 and U1 exiting the spliceosome. Simultaneously, U6 extends its base-pairing interactions with U2 snRNA and contacts the 5'ss. This conformation corresponds to the catalytically active Bact spliceosome (displayed in the right panel), where the branch adenosine and the 5'ss are brought into close proximity, poised for the nucleophilic attack (adapted from [13]).

and the branch sequence into proximity for splicing catalysis [19, 20]. The structure of the yeast catalytically activated spliceosome has recently been elucidated, providing detailed insights into these interactions at the heart of the catalytic process [9, 10].

3. Regulation of Alternative Splicing

Alternative splicing primarily presents itself in various forms, including exon skipping, intron retention, the utilization of alternative 5' or 3' splice sites, and the exclusion of exons (as illustrated in Figure 1). These decisions seem to be orchestrated through intricate control mechanisms operating during various stages of spliceosome assembly, particularly in the initial stages of intron recognition [6, 22].

Specific sequence elements within exons and introns exert regulatory effects, either promoting or inhibiting spliceosome assembly at particular intron boundaries. These elements are categorized as ESE (exonic splicing enhancers), ESS (exonic splicing silencers), ISE (intronic splicing enhancers), and ISS (intronic splicing silencers) (as depicted in Figure 6). Typically, their influence on the same transcript molecule, known as "cis", is mediated through the binding of trans-acting splicing factors, which operate on molecules other than their own. These proteins serve as supplementary splicing factors and are not strictly essential for spliceosome assembly and catalysis. As a result, they are not considered part of the core splicing machinery, which comprises the indispensable components of the spliceosome. However, these regulatory factors can impact the spliceosome's interaction with specific introns through synergistic or antagonistic interactions [6, 16]. Given the combinatorial effects they exert, the relative abundance of these regulators appears to be carefully controlled in various cell types. Interestingly, one of the ways in which splicing factors are regulated is through alternative splicing itself [13, 23].

Recent insights into the significance of primary sequences have enabled the reasonably accurate prediction of broad tissue-specific or disease-associated alternative splicing patterns based on transcript sequences and mutations that affect potential regulatory sequence motifs [12].

Apart from the presence of accessory splicing factors and regulatory sequences, the abundance of "core" spliceosomal proteins can also exert an impact on splicing and the choices made in alternative splicing [24]. Conversely, a crucial interplay between chromatin and transcription has been elucidated [13]. While splicing has historically been regarded, and often still is, as a "post-transcriptional" mechanism of RNA processing, it is now evident that a significant portion of splicing occurs concomitantly with transcription—referred to as cotranscriptional splicing— meaning that RNA polymerase II is still transcribing the DNA sequence while splicing takes place to form a pre-mRNA molecule [25]. This coupling of the two processes is so efficient that, in some instances, introns can undergo splicing as soon as the 3' splice site is transcribed [26]. Consequently, it becomes evident that a robust interplay exists among various layers of gene expression, with transcription kinetics, chromatin state, and splicing choices mutually influencing one another [13]. Notably, in yeast, it has been demonstrated that the recognition of the 3' splice site leads to RNA polymerase II stalling at intronic 3' ends, suggesting the existence of a checkpoint closely associated with cotranscriptional splicing [6].

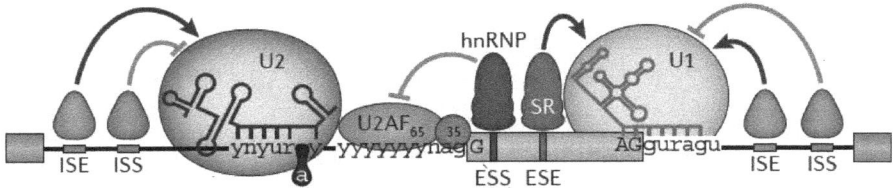

Figure 6: Presents an illustration of splicing silencers and enhancers. Intronic splicing silencers and enhancers (ISE and ISS) can exert both positive and negative effects on the binding of U2 or U1 small nuclear ribonucleoproteins (snRNPs) through their interactions with mediator proteins. Similarly, exonic silencers or enhancers (ESS and ESE) play analogous roles in this context. The hnRNP and SR protein families are widely recognized RNA-binding proteins that identify these regulatory elements, influencing the process of alternative splicing. Different regulatory sequences within the same region exhibit combinatory effects. Hence, the abundance of these regulators constitutes a critical parameter in shaping the splicing outcome (adapted from [16]).

4. The Spliceosome in Disease and Cancer

4.1 Splicing and Genetic Disease

Given the fundamental role of splicing in cellular processes, it is highly probable that any alterations to this mechanism would have pathological consequences. Indeed, numerous mutations associated with diseases are located within crucial regulatory sequences essential for splicing or within genes encoding splicing factors. In both scenarios, the resulting dysregulation of a subset of transcripts can lead to specific molecular defects and pathological manifestations [24].

One illustrative example of a disease related to splicing is Spinal Muscular Atrophy (SMA), which is the primary genetic factor contributing to infant mortality. This degenerative disorder arises from the loss of function of the SMN1 gene, responsible for encoding the SMN protein, a critical component in the assembly of snRNPs. The SMN2 gene, a counterpart of SMN1, is incapable of adequately compensating for pathological mutations in SMN1. This limitation arises from a single nucleotide difference within exon 7 of SMN2, resulting in the exclusion of this exon during the splicing process and consequently generating an abbreviated and unstable variant of the SMN protein. Although the precise reasons behind how a deficiency in this factor causes a distinct motor neuron-related issue are not yet fully understood, it's important to highlight that the characteristic of tissue-specificity is a shared feature with numerous other splicing-related disorders [24].

Remarkably, two primary therapeutic strategies are currently in development for treating SMA, and they serve as exemplary approaches to summarize the current techniques for modulating alternative splicing. On one front, antisense oligonucleotides designed to complement specific intronic silencer sequences within SMN2 have shown the capability to enhance the inclusion of exon 7, thereby restoring motor neuron function [16]. Conversely, small molecules have been identified to produce similar effects with notable specificity [9]. Among the array of conditions associated with splicing irregularities, Duchenne Muscular Dystrophy (DMD) stands out prominently. This condition arises due to frame-shift mutations occurring within the dystrophin gene. These mutations lead to the production of truncated proteins that are incapable of preserving the structural integrity of muscle fibers, ultimately leading to the development of the pathological state. Approaches employing antisense oligonucleotides have been employed to induce the skipping of specific exons, thereby restoring the reading frame. Due to the repetitive domains present in the extensive dystrophin protein, in-frame deletions generated through exon skipping yield proteins that can offer at least partial functionality, thereby imparting therapeutic effects [27]. Lastly, retinitis pigmentosa represents a genetic disorder characterized by progressive degeneration of the retina. It is initiated by mutations in splicing factor genes, primarily PRPF8 [16]. PRPF8, as the largest and exceptionally conserved spliceosomal protein, assumes a pivotal role in the process of splicing catalysis [24]. The tissue-specific consequences of mutations in splicing factors are notable and still not comprehensively understood. Even more intriguing is the fact that mutations within other spliceosomal components also lead to pathological conditions confined to distinct yet diverse organs. For example, mutations in SF3B4 have been associated with Nager syndrome, which is characterized by abnormalities in craniofacial features and limb development [20]. Conversely, mutations in one of the multiple copies of U2 snRNA genes in mice can lead to changes in splicing patterns that are connected to ataxia and neurodegenerative disorders [21]. A deeper comprehension of splicing-related diseases stands as one of the current challenges within this field.

4.2 Alternative Splicing and Cancer

Frequently, tumor cells display a disrupted equilibrium of alternative isoforms, which can significantly impact either the suppression of apoptosis or the stimulation of proliferation and invasion. One of the extensively studied alternative splicing events involves FAS exon 6 skipping. The full-length FAS (or CD95) is a membrane protein that triggers the apoptotic cascade upon binding to the FAS ligand, FASL. Skipping exon 6 leads to the generation of a truncated FAS isoform, this lacks the transmembrane domain and is subsequently released from the cell into the extracellular space. In the extracellular space, it can sequester FASL, thereby inhibiting the apoptotic process (as depicted in Figure 7). Several types of tumors shift FAS splicing towards the antiapoptotic isoform to evade apoptotic signals. Recent high-throughput screenings have revealed that this switch depends not only on the abundance of splicing factors but also on iron metabolism [24]. Likewise, the MCL1 (Myeloid Cell Leukemia 1) transcript encodes an apoptosis regulator whose role is dependent on alternative splicing decisions. MCL1 is a member of the Bcl-2 protein family, which modulates apoptosis by regulating the permeability of the mitochondrial outer membrane. Ordinarily, MCL1 carries out its antiapoptotic functions primarily through its full-length isoform, which is typically the most prevalent form. However, when exon 2 is skipped, a protein isoform is generated that lacks the transmembrane domain and the BH1 and BH2 Bcl-like domains, retaining only the BH3 domain. This particular isoform undertakes pro-apoptotic roles by disrupting its usual interactions with MCL1's typical associates and trapping the full-length protein, as depicted in Figure 7. Given MCL1's relevance in cancer and drug

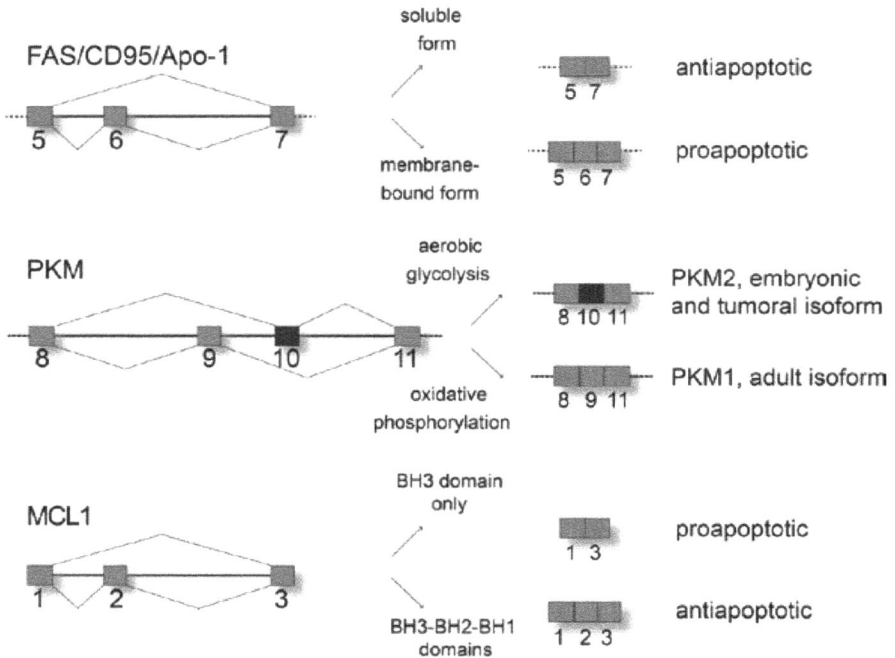

Figure 7. Showcases instances of controlled alternative splicing events with significance for the advancement of cancer. Each number corresponds to the exons participating in the alternative splicing event for the respective genes. These examples also underscore the various functional consequences resulting from the alternative protein products they generate.

resistance mechanisms, extensive high-throughput screenings have been conducted to explore its control over alternative splicing [12].

In specific situations, alternative splicing has the potential to alter the metabolic pathways within cancer cells. For instance, consider the PKM gene, responsible for encoding the metabolic enzyme known as pyruvate kinase M. This gene can produce two distinct PKM isoforms, each with mutually exclusive exons—one predominantly present in embryonic tissues and the other prevalent in adult cells. Many cancer cells opt to reintroduce the expression of the embryonic isoform, which promotes aerobic glycolysis, as illustrated in Figure 7. This modification leads to the Warburg effect, where cancer cells reactivate aerobic glycolysis, granting them a growth advantage [18].

Notably, recent analyses have revealed that synonymous mutations often function as driver mutations in cancer. In many instances, this phenomenon may be linked to differences in regulatory sequences that result in distinct splicing preferences in tumor cells. Consequently, oncogenes and tumor suppressors may experience increased or decreased activity due to splicing alterations [18].

Several additional cases have been recorded, underscoring the impact of activated signaling pathways within cancer cells on splicing irregularities and their role in governing diverse processes such as angiogenesis, metastasis, and the cell cycle [18].

4.3 Splicing Factors and Cancer

Although aberrant splicing in cancer has been recognized for a considerable duration, it's only recently, owing to extensive cancer sequencing initiatives that mutations in splicing factors are starting to emerge. Intriguingly, numerous mutated proteins are involved in the early stages of spliceosome assembly, particularly in 3'ss recognition. This suggests a potential link between this level of regulation and the control of cell proliferation or other aspects of tumor progression. Furthermore, it's important to highlight that the majority of these mutations are heterozygous and exhibit mutual exclusivity. This suggests that various factors may lead to similar outcomes, and the complete absence of wild-type versions of these factors can be deleterious [10].

Initially, mutations in splicing factors like SRSF2, SF3B1, ZRSR2 and U2AF35were linked with myelodysplastic disorders [10]. Significantly, SF3B1 mutations exhibit a robust association with the occurrence of ring sideroblasts (RS), which are erythroblasts characterized by perinuclear granules containing mitochondrial iron deposits. These mutations were later observed in chronic lymphocytic leukemia (CLL) [10] and solid tumors, including uveal melanoma. Subsequently, a flurry of clinical observations has emerged, associating these mutations with clear prognostic indicators [10]. Even though splicing is crucial for the expression of every gene, a relatively small number of transcripts have been identified as being changed in mutant samples [10]. Notably, alterations in the sequence of affected transcripts contribute partially to the splicing changes associated with SF3B1, U2AF1 and SRSF2 mutations. In the case of U2AF1, mutations have been identified in hematological malignancies, lung cancer, and various solid tumors [15]. This splicing factor is involved in recognizing 3' splice sites, as depicted in Figure 3. The S34 mutation leads to a preference for 3' splice sites where the AG is preceded by C or A rather than T, while the Q157 mutation favors introns that start with G rather than A [6].

In contrast, SRSF2 serves as a regulatory SR protein that interacts with exonic splicing enhancers (ESEs), as depicted in Figure 6. Pathogenic mutations lead to modifications in its RNA recognition motif (RRM), leading to changes in sequence preferences. Mutant SRSF2 exhibits stronger binding to CCNG motifs rather than GGNG motifs, whereas the unaltered, wild-type protein can effectively recognize both types [28]. In summary, these discoveries signify that mutated splicing factors undergo alterations in their RNA affinity, which, in turn, can result in sequence-specific changes in alternative splicing. Some of these modifications can provide partial insights into the pathological characteristics [15]. Nevertheless, it's worth noting that processes beyond alternative splicing may also be affected and contributes to tumor progression, as observed with the autophagy regulator ATG7, where 3' end processing is disrupted in the presence of U2AF1 mutations. While research in this field is progressing swiftly, numerous mechanistic questions remain unresolved as researchers strive to translate this knowledge into potential therapeutic applications.

4.4 Splicing Inhibitors: A Novel Class of Drugs

Many small molecules that can modulate splicing efficiency, either in a general or event-specific manner, have been discovered [19, 24]. Given the growing interest in these compounds, high-throughput systems have been established to screen drug libraries in the search for splicing modulators [5, 19, 24–26]. However, it's important to note that a direct interaction between small molecules and core spliceosome components has been confirmed for only a few of them. Interestingly, several of these molecules share a common target—the SF3B (splicing factor 3B) complex.

The discovery of splicing modulators emerged during screenings for potential antitumor candidates within bacterial Streptomyces plateniensis and Pseudomonas fermentation products. This effort led to the identification of compounds like Herboxidiene, FR901464, Thailanstatines, GEX1A and Pladienolides [21, 25, 27]. These compounds exhibit cytotoxic effects against various tumor cell lines, causing cell cycle arrest in the G1 and G2/M phases [7, 25]. Their distinct behavior compared to known antitumor drugs sparked significant interest, suggesting the involvement of a novel antitumor mechanism. Additionally, the detection of Cdc25a and Cdc2 short isoforms in GEX1A-treated cells hinted at the critical role of splicing in the drugs' effects on cell proliferation control.

The validation of this hypothesis occurred subsequently when a potent derivative of FR901464 was discovered, which hindered splicing by targeting the SF3B complex, and it was named Spliceostatin A. Similar splicing-inhibiting effects were also observed for Pladienolide, and more recently, for GEX1A. These biochemical findings were further supported by genetic evidence, demonstrating that a novel SF3B1 mutation could reverse both the splicing alterations and the inhibition of cell proliferation induced by Pladienolide B, effectively thwarting the drug's binding to the SF3B complex [9, 10, 28]. Interestingly, the drug-resistant mutation R1074H was found to be located in very close proximity to the branch adenosine [21]. Nevertheless, comprehensive structural elucidation of the interactions between these drugs and their respective targets has yet to be accomplished.

The extensive chirality and limited solubility of natural drugs have imposed restrictions on their utilization in both in vitro and in vivo studies. To tackle these challenges, scientists have developed derivatives of FR901464. Meayamycin and Meayamycin B exhibit increased stability, potency, and solubility [25], whereas Sudemycins are simpler to chemically synthesize and can potentially capture the presumed common pharmacophore shared by FR901464 and Pladienolide. As the interest in targeting the spliceosome for therapeutic purposes continues to grow, these molecules hold promise as powerful tools for therapy and for advancing our understanding of the complex and dynamic ribonucleoprotein machinery within cells, much like antibiotics have been invaluable in the study of the ribosome [18].

References

[1] Zhang, Y., Qian, J., Gu, C. and Yang, Y. (2021). Alternative splicing and cancer: A systematic review. Signal Transduction and Targeted Therapy 6(1): 78.

[2] Qi, F., Li, Y., Yang, X., Wu, Y.-P., Lin, L.-J. and Liu, X.-M. (2020). Significance of alternative splicing in cancer cells. Chinese Medical Journal 133(2).

[3] Farina, A.R., Cappabianca, L., Sebastiano, M., Zelli, V., Guadagni, S. and Mackay, A.R. (2020). Hypoxia-induced alternative splicing: the 11th Hallmark of Cancer. Journal of Experimental & Clinical Cancer Research 39(1): 110.

[4] Peng, Q., Zhou, Y., Oyang, L., Wu, N., Tang, Y., Su, M., Luo, X., Wang, Y., Sheng, X., Ma, J. et al. (2022). Impacts and mechanisms of alternative mRNA splicing in cancer metabolism, immune response, and therapeutics. Molecular Therapy 30(3): 1018–1035.

[5] Carrillo Oesterreich, F., Herzel, L., Straube, K., Hujer, K., Howard, J. and Neugebauer Karla, M. (2016). Splicing of nascent RNA coincides with intron exit from RNA polymerase II. Cell 165(2): 372–381.

[6] Braunschweig, U., Gueroussov, S., Plocik, A.M., Graveley Brenton, R. and Blencowe Benjamin, J. (2013). Dynamic integration of splicing within gene regulatory pathways. Cell 152(6): 1252–1269.

[7] David, C.J. and Manley, J.L. (2010). Alternative pre-mRNA splicing regulation in cancer: Pathways and programs unhinged. Genes & Development 24(21): 2343–2364.

[8] Nilsen, T.W. and Graveley, B.R. (2010). Expansion of the eukaryotic proteome by alternative splicing. Nature 463(7280): 457–463.

[9] Irimia, M. and Roy, S.W. (2014). Origin of spliceosomal introns and alternative splicing. Cold Spring Harbor Perspectives in Biology 6(6).

[10] Yoshida, K. and Ogawa, S. (2014). Splicing factor mutations and cancer. WIREs RNA 5(4): 445–459.

[11] Zhang, S., Mao, M., Lv, Y., Yang, Y., He, W., Song, Y., Wang, Y., Yang, Y., Al Abo, M., Freedman, J.A. et al. (2022). A widespread length-dependent splicing dysregulation in cancer. Science Advances 8(33): eabn9232.

[12] Buljan, M., Chalancon, G., Dunker, A.K., Bateman, A., Balaji, S., Fuxreiter, M. and Babu, M.M. (2013). Alternative splicing of intrinsically disordered regions and rewiring of protein interactions. Current Opinion in Structural Biology 23(3): 443–450.

[13] Iannone, C. and Valcárcel, J. (2013). Chromatin's thread to alternative splicing regulation. Chromosoma 122(6): 465–474.

[14] Yang, Q., Zhao, J., Zhang, W., Chen, D. and Wang, Y. (2019). Aberrant alternative splicing in breast cancer. Journal of Molecular Cell Biology 11(10): 920–929.

[15] Dvinge, H., Kim, E., Abdel-Wahab, O. and Bradley, R.K. (2016). RNA splicing factors as oncoproteins and tumour suppressors. Nature Reviews Cancer 16(7): 413–430.

[16] Scotti, M.M. and Swanson, M.S. (2016). RNA mis-splicing in disease. Nature Reviews Genetics 17(1): 19–32.

[17] Zhang, D., Hu, Q., Ji, Y., Chao, H.-P., Tracz, A., Kirk, J., Buonamici, S., Zhu, P., Wang, J., Liu, S. et al. (2019). Dysregulated alternative splicing landscape identifies intron retention as a hallmark and spliceosome as a therapeutic vulnerability in aggressive prostate cancer. BioRxiv 2019: 634402.

[18] Bonnal, S., Vigevani, L. and Valcárcel, J. (2012). The spliceosome as a target of novel antitumour drugs. Nature Reviews Drug Discovery 11(11): 847–859.

[19] Barash, Y., Calarco, J.A., Gao, W., Pan, Q., Wang, X., Shai, O., Blencowe, B.J. and Frey, B.J. (2010). Deciphering the splicing code. Nature 465(7294): 53–59.

[20] Bernier Francois, P., Caluseriu, O., Ng, S., Schwartzentruber, J., Buckingham Kati, J., Innes, A.M., Jabs Ethylin, W., Innis Jeffrey, W., Schuette Jane, L., Gorski Jerome, L. et al. (2012). Haploinsufficiency of SF3B4, a component of the pre-mRNA spliceosomal complex, causes nager syndrome. The American Journal of Human Genetics 90(5): 925–933.

[21] Hua, Y., Sahashi, K., Hung, G., Rigo, F., Passini, M.A., Bennett, C.F. and Krainer, A.R. (2010). Antisense correction of SMN2 splicing in the CNS rescues necrosis in a type III SMA mouse model. Genes & Development 24(15): 1634–1644.

[22] Sette, C., Ladomery, M. and Ghigna, C. (2013). Alternative splicing: Role in cancer development and progression. International Journal of Cell Biology 2013: 421606.

[23] Yamauchi, H., Nishimura, K. and Yoshimi, A. (2022). Aberrant RNA splicing and therapeutic opportunities in cancers. Cancer Science 113(2): 373–381.

[24] Daguenet, E., Dujardin, G. and Valcárcel, J. (2015). The pathogenicity of splicing defects: mechanistic insights into pre-mRNA processing inform novel therapeutic approaches. EMBO Reports 16(12): 1640–1655.

[25] Albert, B.J., McPherson, P.A., O'Brien, K., Czaicki, N.L., DeStefino, V., Osman, S., Li, M., Day, B.W., Grabowski, P.J., Moore, M.J. et al. (2009). Meayamycin inhibits pre–messenger RNA splicing and exhibits picomolar activity against multidrug-resistant cells. Molecular Cancer Therapeutics 8(8): 2308–2318.

[26] Chathoth Keerthi, T., Barrass, J.D., Webb, S. and Beggs Jean, D. (2014). A splicing-dependent transcriptional checkpoint associated with prespliceosome formation. Molecular Cell 53(5): 779–790.

[27] Kole, R., Krainer, A.R. and Altman, S. (2012). RNA therapeutics: Beyond RNA interference and antisense oligonucleotides. Nature Reviews Drug Discovery 11(2): 125–140.

[28] Zhang, J., Lieu, Y.K., Ali, A.M., Penson, A., Reggio, K.S., Rabadan, R., Raza, A., Mukherjee, S. and Manley, J.L. (2015). Disease-associated mutation in <i>SRSF2</i> misregulates splicing by altering RNA-binding affinities. Proceedings of the National Academy of Sciences 112(34): E4726–E4734.

4

Deregulated Spliceosome and Splice Variants in Cancers

Farheen Naz

1. Introduction

The deadliest disease in the whole world is still cancer. Day by day, research is increasing in the area of oncology. Bull still no specific reason has come out to unconditionally identify the growth of cells. Several therapies, such as radiation therapy, chemotherapy, immunotherapy, hormone therapy, targeted drug therapy, cryoablation, bone marrow transplant, surgery, etc., are used to kill cancer cells. Depending upon the situation and condition of the patient, particular therapies are given to kill the unwanted cells, shrink the growth of cancer, cure the disease, or it has been said that to stop the progression of uncontrolled growth of cells. Now it has been seen that spliceosomes can also be one of the therapeutic therapies to cure cancer patients. In this chapter, the role of the spliceosome, its deregulation, and splice variants in various types of cancer has been discussed. It has been reported that spliceosome therapy can lead to the control of cancer patients.

2. Splicing

The human genome consists of multiple exons and introns (Figure 1), which undergo the process known as splicing. This process is carried out by a spliceosome, a large complex consisting of RNA and proteins. It consists of five nuclear ribonucleoprotein particles, subunits of snRNPs named U1, U2, U4, U5, and U6 and they are associated with more than 200 proteins.

Department of Biosciences, Jamia Millia Islamia, New Delhi, 110025.
Email: farheenqh11@gmail.com

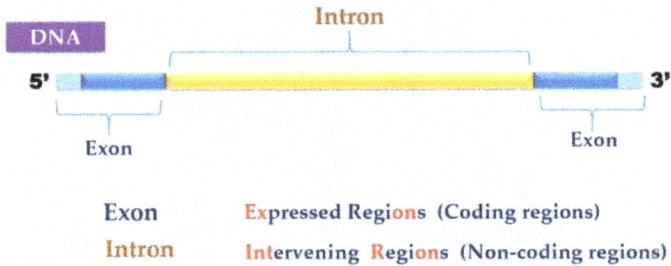

Figure 1: Structure of Intron.

Splicing, which can also be said as mRNA splicing, is the process of removing introns from the mRNA and joining of exons through ligation to form mature mRNAs. It is a post-transcriptional modification because of which a single gene can code for multiple proteins. The mRNAs splicing is processed in eukaryotes (Figure 2), prior to mRNA translation, by the differential cutting and joining of regions of pre-mRNA. The mRNAs splicing is the prime source of protein diversity. During the process of mRNAs/gene splicing event, the pre-mRNA transcribed from one gene can be processed into various mature mRNA molecules that form multiple functional proteins. Thus, gene/mRNAs splicing forms a single gene to increase its coding capacity, allowing the formation of protein isoforms that are functionally and structurally distinct. Gene/mRNAs splicing is supposed to be observed in a high proportion of genes. Human cells contain about 40–60% of the genes that are known to exhibit alternative splicing.

In the case of cancer, a number of splicing alterations have been noticed in the last few decades. There are various reasons for the alterations to cause cancer. These include mutations and abnormalities in the site of splicing. For example, mutations in splicing machinery or splice sites (such as small nuclear RNAs), in

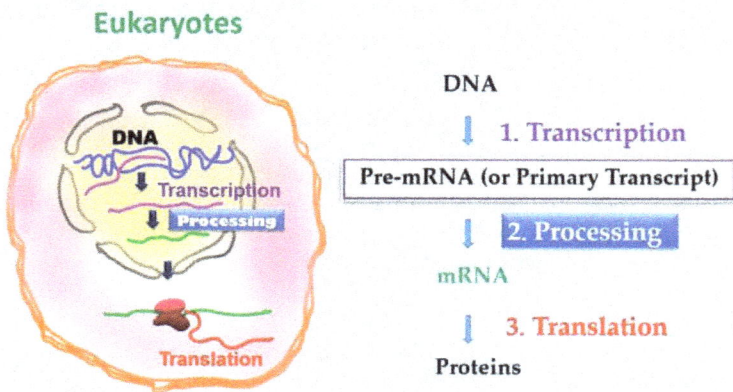

Figure 2: Gene expression in eukaryotes.

case of abnormalities in different splicing regulators, like deregulated expressions, deregulated spliceosome, post-translational modifications and somatic mutations, etc. Apart from these mutations and abnormalities, there are a large number of divergently or abnormal or deregulated-expressed splice variants that have been noticed in the case of cancer disease. Hence, the reason and functional defects of a large number of deregulated splice variants are still unknown, and some of the researchers, scientists, or authors declared that deregulated splicing from deregulated splice variants is a new type of cancer. It has been examined that targeting deregulated spliceosome machinery, i.e., spliceosome and splice variants, could evolve the latest anti-cancer therapies. Further, spliceosome machinery, its structure, function, and cause of cancer have been discussed.

Main points to define RNA splicing (Figure 3):

- The process of RNA splicing involves removing non-coding sequences or introns and joining coding sequences or exons.
- RNA splicing takes place during or immediately after transcription within the nucleus in the case of nucleus-encoded genes.
- In eukaryotic cells, RNA splicing is crucial as it ensures that an immature RNA molecule is converted into a mature molecule that can then be translated into proteins. The post-transcriptional modification is not necessary for prokaryotic cells.
- RNA splicing is a controlled process that is regulated by various ribonucleoproteins.
- RNA splicing facilitates the formation of multiple functional mRNAs from a single transcript, which codes for different proteins.
- It also helps regulate gene expression and protein content of the cell.
- It assists in the evolution process by forming different combinations of exons and thereby making new and improved proteins.
- New exons can be inserted into the introns to create new proteins without disrupting the functionality of the original gene.

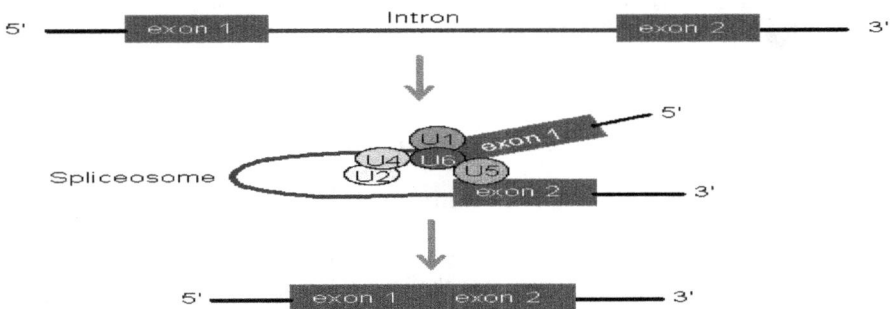

Figure 3: Structure of spliceosome and its mechanism.

3. Gene Splicing Mechanism

When it comes to common gene splicing events, there are several types. After the mRNA is formed from the transcription step of the central dogma of molecular biology, these are the events that can simultaneously occur in the genes.

3.1 Exon Skipping

The most commonly known gene splicing mechanism, in which exon(s) are included or excluded from the final gene transcript leading to extended or shortened mRNA variants, is known as exon skipping. The coding regions of a gene are known as exons, and these are responsible for formation of proteins. These proteins are utilised for a number of functions in different types of cells.

3.2 Intron Retention

Intron retention is nothing but an event in which an intron is retained in the last transcript. It has been reported that 2–5% of the genes retain introns in humans. The gene splicing mechanism retains (junk) portions, which are non-coding of the gene, leading to a demornity protein structure and functionality.

3.3 Alternative 3' Splice Site and 5' Splice Site

It is nothing but the joining of different 5' and 3' splice sites (Figure 4). Here two or more alternative 5' splice sites compete for joining two or more alternate 3' splice sites.

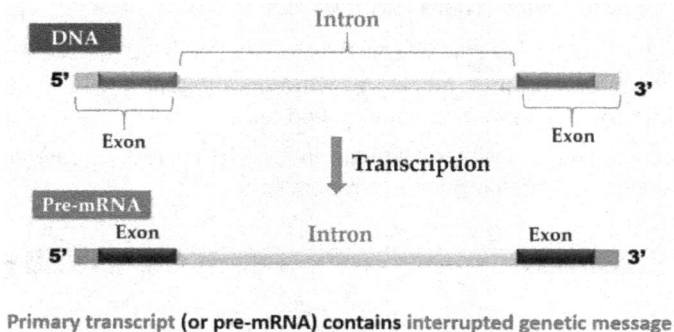

Figure 4: Primary transcript.

4. mRNA Splicing Phenomenon/Mechanism/Process

- The mechanism of mRNA splicing starts with the joining of the ribonucleoproteins, or it can be said that spliceosomes to the introns present on the area of the splice site.
- The joining of the spliceosome leads to a biochemical process known as transesterification between RNA nucleotides.

- During this particular reaction, which is called spliceosome assembly, the 3'OH group of specific start point nucleotide on the intron causes a nucleophilic attack on the prime nucleotide of the intron at the end of the 5' splice site.
- This forms the folding of the 5' and 3' ends, which results in a loop. Along with it, all the adjacent exons in mRNA are brought together.
- Finally, in this reaction, the looped introns are detached from the sequence by the machinery spliceosomes.
- After that, during the ligation process of adjacent segments of the exon, a second transesterification reaction takes place.
- In this reaction/process, the prime 3'OH group of the end released 5' exon then performs an electrophilic attack reaction on the first nucleotide, which is present just behind the last nucleotide of the intron that is at the 3' splice site.
- This gradually leads to the binding of the two exon segments and, along with it, the removal of the intron segment from mRNA.
- Previously, the intron, which is released during the process of splicing, is thought to be known as a junk unit. It has been recently examined that these introns are involved in other processes too, which are related to proteins after their removal from mRNA.
- Apart from the spliceosomes, another group of enzymes/proteins called 'ribozymes' are also involved in regulating and controlling the splicing mechanism.

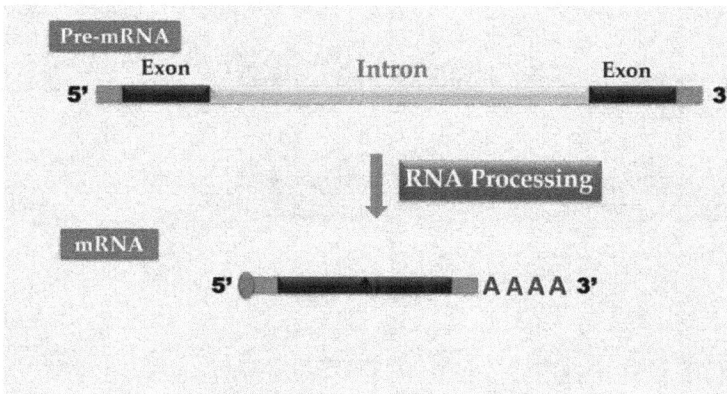

Figure 5: RNA processing.

5. Types of RNA Splicing

There are three types of splicing mechanisms involved in preparing mRNA, which are:

- Self-splicing
- Alternative Splicing
- tRNA Splicing

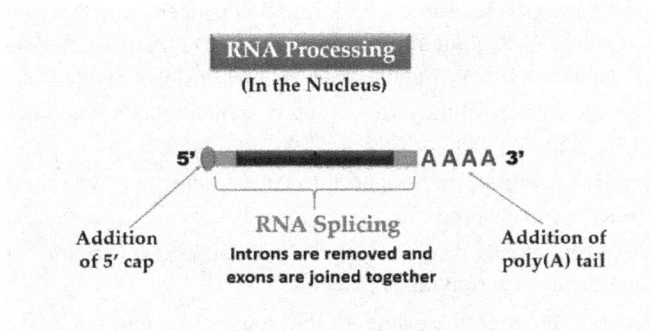

Figure 6: RNA splicing.

Self-Splicing

In self-splicing, the introns can catalyse their excision from their parent RNA. Many of the genes undergo self-splicing; some mitochondrial genes are also capable of self-splicing, whereas some of the genes undergo self-splicing. To name a few, protozoan ribosomal RNA genes, phage genes, etc.

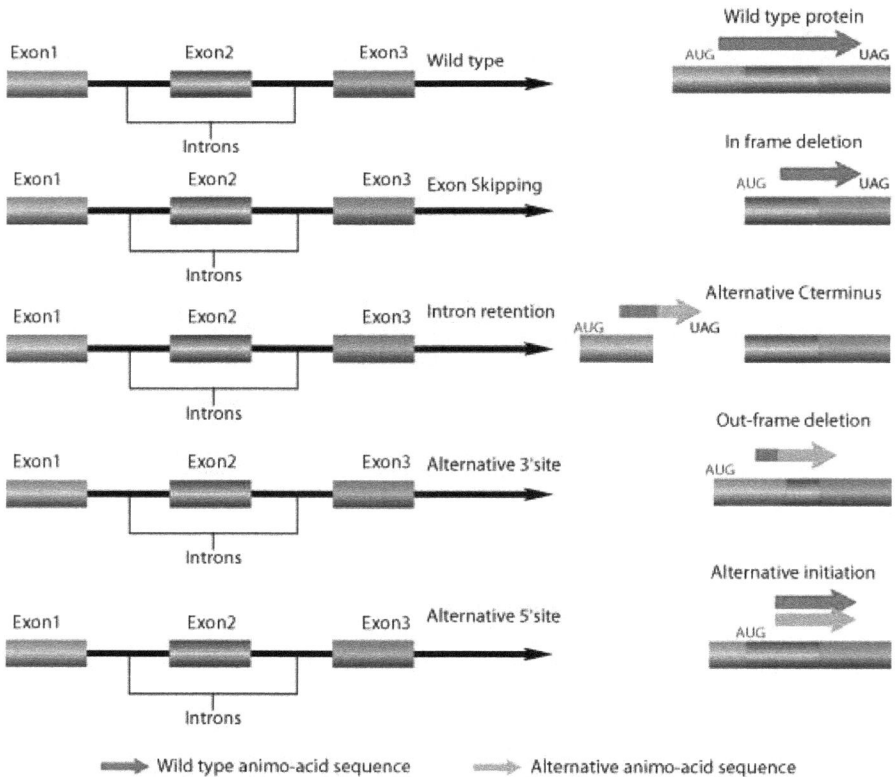

Figure 7: Gene splicing.

Self-Splicing Process is Listed Below

It is a type of RNA splicing which occurs in some scarce introns. They are capable of promoting phosphodiester bond cleavage and formation without the help of other proteins or spliceosomes.

- They can mediate their excision from precursor RNA and the subsequent ligation of the flanking exons in a simple salt buffer; hence, they are unique.
- This splicing reaction is stated by the tertiary structure of the intron. It provides the ability to recognise the splice sites of the precursor RNA and to perform the splicing and cutting reactions in a very short manner.
- The overall process is regulated by a ribozyme, which is nothing but a sequence present in such introns.
- Self-splicing introns are grouped three kinds, namely Group I, Group II, and Group III.
- The splicing process is performed by Group I and Group II introns, which is a similar mechanism to that type of spliceosome. Clearly indicating that these introns might be evolutionarily related to the spliceosomes.
- In the self-splicing process, the 5' splice site is recognised by a short sequence element in the intron, which is known as the internal guide sequence.
- Added to that, other strongly conserved sequences of the introns known as P, Q, R, and S are needed to 'catalyse' the splicing and ligation reactions.
- Self-splicing follows a similar phenomenon involving two transesterification reactions resulting in the removal of introns and ligation of exons.

5.1 Alternative Splicing

Alternative Splicing is nothing but splicing of RNAs, giving rise to different mRNA molecules that code for various proteins. In most eukaryotes, this process increases the diversity of proteins and occurs as a normal splicing mechanism.

5.1.1 Some Important Points to Remember Regarding Alternative Splicing

- It is a splicing process resulting in a different composition of exons in the same RNA and then forming a range of selected proteins.
- An essential mechanism to increase the complexity of gene expression in alternative splicing of pre-mRNA is necessary. It also plays an important role in cellular differentiation and organism development.
- This type of splicing enables exons to be arranged in various combinations, where various configuration results in varied proteins.
- The process of alternative splicing mechanism might occur either by leaving/ skipping or extending/increasing some exons or by remaining particular introns, resulting in various varieties of mRNA formed.
- Regulation of this type of splicing is a complex process that forms numerous components which interact with each other, including trans-acting factors and cis-acting elements.

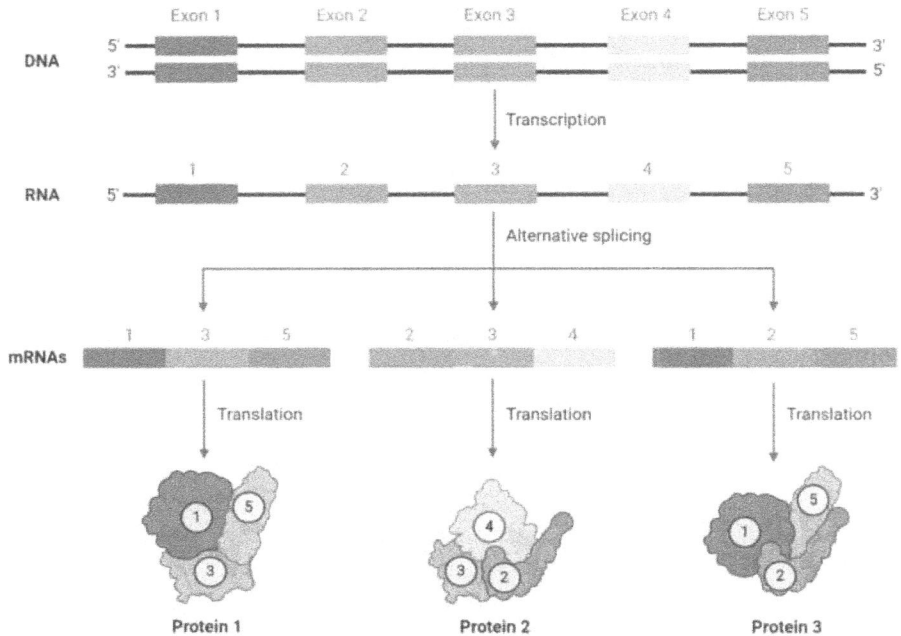

Figure 8: Alternative splicing.

- The alternative splicing process is further followed by the functional coupling between transcription and splicing.
- Additional molecular features, such as RNA structure, chromatin structure and alternative transcription initiation, or alternative transcription termination, collaborate with these basic components to contribute to the protein diversity because of this alternative splicing.
- This process is also essential for many other functions like identifying prognostic biomarkers, novel diagnostics, as well as new strategies for therapy in human cancer patients.
- Thus, this process has a great role in almost every aspect of protein function, including the interaction between proteins and ligands, localisation, nucleic acids or membranes, and enzymatic properties.

5.2 tRNA Splicing

tRNA is a process gene interrupted by introns. In this, the process of splicing is different from the other two splicing methods.

5.2.1 Some Important Points to Remember Regarding tRNA Splicing

In this method, splicing process is different; like, in mRNA, the genes in tRNA are also interrupted by introns.

- Splicing process in tRNA is catalysed by three enzymes with an intrinsic requirement for ATP hydrolysis.
- tRNA splicing process occurs in all three major lines of descent as follows:
  1. Bacteria show self-splicing.
  2. Archaea shows tRNA splicing.
  3. Eukarya shows tRNA splicing.

This mechanism might differ in bacteria as well as in higher organisms.

- In bacteria, the self-splicing process is used in the introns in the tRNA.
- In Archaea as well as Eukarya, however, the tRNA splicing reaction occurs in three steps; whereas each step is catalysed by a different enzyme, each of which can function interchangeably on all of the substrates.
- The first step of the pre-tRNA process is cleaved at the two splice sites by an endonuclease, resulting in two tRNA half molecules and a linear intron with 5'-OH and 3'-cyclic PO4 ends.
- Then cleavage is followed by the ligation of the two RNA half molecules in the presence of a tRNA ligase enzyme.
- Finally, the PO4 ends formed from splicing are transferred to NAD in a process catalysed by nicotinamide adenine dinucleotide (NAD)-dependent phosphotransferase.

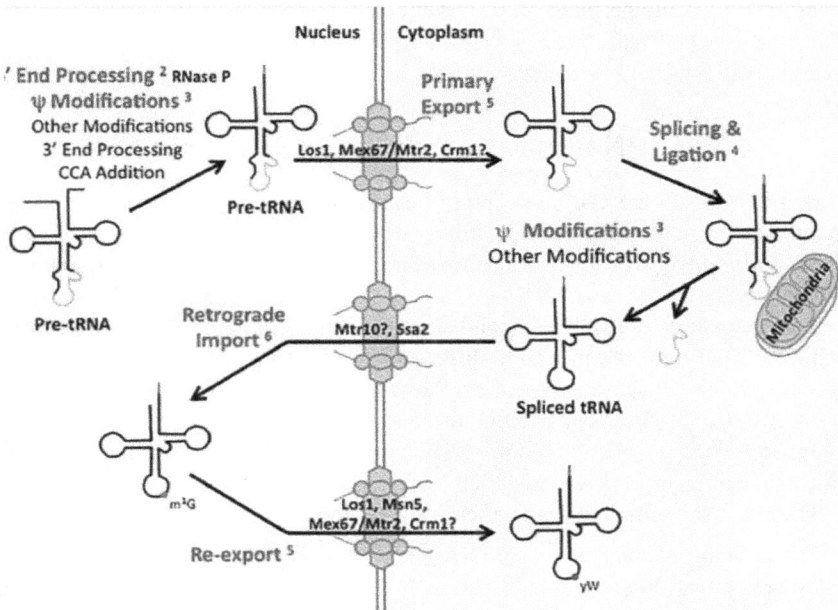

Figure 9: tRNA splicing.

Table 1: Types of splicing and examples.

S. No.	Types of Splicing	Few Examples
1.	Self-splicing	Protozoan ribosomal RNA genes; phage genes
2.	Alternative Splicing	Neurexin genes
3.	tRNA splicing	Bacteria, Archaea, and Eukarya

6. Significance of RNA Processing

RNA processing is required in eukaryotic gene expression. There is various biological and medical significance associated with premature RNA splicing, some of which are:

1. Pre-mRNA splicing is a fundamental process in cellular metabolism that plays an important role in generating protein diversity. The diversity is brought about by changes in the number and sequence of exons and introns present in the RNA sequence.
2. RNA splicing also provides the regulation of gene and protein content in the cell.
3. The splicing of RNA sequences assists the process of the evolution of new and improved proteins.
4. Various mutational splicing isoforms act as markers for cancer and as targets for cancer patient therapy.
5. Pre-mRNA splicing plays an important role in the pathology of cancer patients where it regulates the three functional aspects of cancer: proliferation, metastasis, and apoptosis.

7. RNA Splicing Errors Causes Cancer

Errors in RNA splicing can be occurred because of various reasons like mutations, structural and functional imbalance, impairment, etc. Errors can be integrated for any reason, but in mRNA splicing, it occurs because of the deregulation of spliceosome or the deregulation of splice variants. These were caused because of mutations at the displacement of the splice site and structural and functional loss of spliceosome machinery. These errors can lead to different harmful diseases, which will be related to splicing-related diseases. One of the causes of the deadly disease of cancer is the error in mRNA splicing. And mainly, these types of mutations and loss in structural and functional units were noticed in alternative splicing. Stress conditions might also be the cause of error in splicing, which leads to harmful effects.

These are some main points which were responsible for the error in slicing:

- The splicing of nuclear pre-mRNAs is a fundamental process required for the expression of most metazoan genes. However, errors in splicing might occur because of mutations that result in various splicing-related diseases.
- Mostly in alternative splicing, splicing results in biological products that are not functionally active.

- Mutation can cause errors during splicing at the splice site, which causes loss of exons or inclusion of an intron, disturbing the structure and function of the RNA sequence.

- Similarly, displacement of a splice site might also cause the formation of longer or shorter exons, resulting in faulty products.

- In the stress condition of various living organisms like plants, alternative splicing associated with different metabolic pathways might bring changes in the normal structure and functioning of the plant.

- The chances of unwanted splicing are more in eukaryotic cells with increased levels of alternative splicing at splice sites that have evolved to offer a weak interaction potential for components of the spliceosome.

- The chance of accuracy of splice site pairing is also limited by the accuracy of transcription as the transcription machinery makes a mistake once in every 10^3–10^5 nucleotide insertion step.

8. mRNA Splicing Alterations in Cancers

Cancer is a deadly disease which can be caused by various factors. Till now, no specific reason or factor was known. mRNA splicing alterations have been noticed to contribute to the initiation or maintenance of cancer in various terms [1]. Consistent with high rates of mis-splicing or error in splicing in cancer/tumour cells related to normal tissues, cancer cells contain a larger number of transcripts with premature termination codons (PTCs), which are deregulated by the nonsense mediated mRNA decay (NMD) system [2]. These PTCs were encoded by tumour suppressor genes and also appeared to be more frequently in transcripts than those encoded by onco-genes, which contributed to cancer caused by mis-splicing and is non-random. Recently, researchers examined on the basis of high-throughput results from complete exome sequencing and RNA-seq the data across 32 cancer/tumour, which varied from 8,705 cancer patients and 670 who matched normal controls; Kahles and colleagues validated that cancer harbour more non-/aberrant splicing events than paired normal tissue from the same patient/individual, which includes abnormalities in intron retention and high enrichment in novel splicing events that were not explained in reference databases such as GENCODE [3]. The authors also declared new exon-exon junctions (EJs). Polypeptides generated from these cancer specific to EEJs might have the ability to interact with the major histocompatibility complex (MHC-I) and to serve as neo-antigens. Though the immunogenicity of neo-antigens which was derived from deregulated splicing was still to be validated, these data stated that deregulated/non-splicing could contribute to various tumour-specific, A-derived neo-antigens. All these results concluded by authors propose that alternative splicing might be one of the new paths to immunogenicity in cancers [4]. Splicing alterations associated with cancer or tumour can result from cis-alterations; for example, mutations that disturb the structure and function of splicing regulatory motifs of critically cancer-associated genes. Such as, it has been declared that single-nucleotide variants (SNVs), which affect the splice site, are correlated with intron retention [5]. Interestingly, these SNV-induced intron retention events occur more frequently in tumour suppressor genes than in oncogenes. These are the main cause of tumour suppressor inaction, which leads to cancer [5]. Apart from this, cancer-associated synonymous mutations, which were close to exon boundaries, result in a gain of ESE motifs and a loss of ESS motifs, thereby favouring inclusion rather than skipping alternative exons [6]. These synonymous mutations lead to acting as a driver mutation causing human cancer. Cis-alterations, as well as trans-alterations splicing, have been studied and highlighted a number of times in human cancer cases, including various somatic mutations of different components of the spliceosome machinery and deregulated expression (either overexpression or downregulation) [7]. Further, the expression of RBP encoding genes in alterations has been shown to constitute multiple alternative splicing changes in oncogenic pathways and cancer drivers [8]. Such as, different SR phosphorylating kinases (including SRSF1, SRSF6, SRSF3, and SRPK1) and SR proteins exhibit abnormal/carcinogenic properties that are upregulated in tumour cells, such as lung cancers [9]. In addition to this, different components of the exon recognition machinery are overexpressed in hepatocellular cancer [10]. In contrast to the above abnormal behaviours, many other splicing factors, including QKI, RBM5, RBM6, and RBM 1 0, act as tumour suppressors and are preferably found to be mostly downregulated in some various tumour types [7].

9. Deregulated Expression or Mutations of Splicing Factors Explained

Deregulated expression or mutations of splicing factors seriously leads to cause cancer. It has been examined that mutually exclusive heterozygous hotspot mutations in the SF3B1, SRSF2 [11], and U2AF1 genes have been coming across in lymphoma, leukaemia and several solid cancer/tumours, including uveal melanoma and lung, breast and pancreatic carcinomas. These above-mentioned mutations are said to be confirmed the status of spliceosome 'sickness' in human cancer cells, which could provide mutated cells more dependent on the remaining wild-type splicing factor for survival and, consequently, more vulnerable to spliceosome inhibition than unmutated cells. Consistently, E7107 treatment with or without mutant SRSF2 of isogenic murine myeloid leukaemias, preferential cell death of leukaemic cells having mutated SRSF2 [12]. Similar results were seen in patient-xenograft models and cell lines that derived from epithelial and haematological tumours, in which H3B-8800 was equipped to preferentially kill spliceosome-mutant cells at well-tolerated doses [13]. These types of effects depended on a direct binding between the SF3B and H3B-8800 complex, although H3B-8800 modulated alternative splicing in the same range in wild-type and mutant cells; the preferential killing of spliceosome-mutant cells was due to retention of short, GC-rich introns enriched in genes encoding spliceosome components [13]. It has been reported that similar synthetic-lethal bindings between the expression of a mutated U2AF1 protein and exposure to sudemycins [14]. Hence, a human being with spliceosome mutations in their cancer cells could better benefit from spliceosome inhibitors than those with non-mutated ones.

10. Spliceosome

10.1 Structure of Spliceosome

10.1.1 What is Spliceosome?

A spliceosome consists of a large and complex molecule, which is formed of RNAs and proteins that are used to regulate the process of mRNA splicing.

- The spliceosome is composed of five small nuclear RNAs (snRNA) and about 80 protein molecules.
- It is formed by the combination of RNAs, along with the proteins resulting in the formation of an RNA-protein complex, which is termed small nuclear ribonucleoproteins (snRNPs).
- These are mainly present within the nucleus, where they remain associated with immature pre-RNA transcripts.
- These formed spliceosomes, in addition to working on RNA-RNA binding, also interact in RNA-protein interactions.
- These spliceosome are significant as an important component that selectively ligates out unnecessary and unwanted materials (introns) to make a functional final cut.
- Spliceosome machinery helped in both the removal of introns as well as ligation of remaining exons.
- 'Minor spliceosomes' are another set of spliceosomes. These are also found in eukaryotic cells, which have abundant RNAs and help in the splicing of a special class of pre-mRNA introns.

11. Main Classes of Small Molecules that Triggers the Spliceosome Machinery and their Mechanisms of Action Involved

There are some small molecules which can control the spliceosome machinery. Specific action of mechanism is involved is this process. Within the postulate the deregulated splicing and its mechanism represents an Achilles' s heel for human cancer cells/tumour cells. This could be suitable for potential therapeutic purposes. Many researchers and several of laboratories have taken an action to design and develop small molecule inhibitors that can target different components of the spliceosome machinery. Several potential molecules have been now examined as splicing inhibitors, either in cell- based assays or cell-free splicing systems, such as those using cells transfected with minigene splicing reporters [15].

11.1 Main Classes of Spliceosome Inhibitor

As discussed above spliceosome inhibitors which include a variety of small compounds, natural or synthetic, which prevent different steps of the splicing mechanism (Table 1). A number of splicing inhibitors are involved in such process to prevent the machinery. These inhibitor classes are given as below:

1. Spliceostatins
2. Pladienolides
3. Herboxidienes
4. Isoginkgetin
5. Madrasin
6. CLK inhibitors
7. SRPK inhibitors
8. CLK and SRPK inhibitors

These classes of inhibitors interrupt [16] the early stage of spliceosome assembly so that no splicing complex is formed during the process [12]. In a similar way, protein arginine-methyltranferase 5 (PRMT5) inhibitors interrupt the assembly of snRNPs [17]. Many splicing inhibitors identified to date cause spliceosome assembly to stall at the A complex (i.e., the stage at which U2 snRNP binds to the BP sequence), including isoginkgetin, madrasin, tetrocarcin A, and members of the pladienolide, spliceostatin and herboxidiene families [18]. Compounds from these last these three families target SF3Bl, the core spliceosome component of U2 snRNP. Sulfonamides, such as E7820, tasisulam, indisulam and chloroquinoxaline sulfonamide, interfere with splicing by targeting the U2AF-related splicing factors, namely RNA binding motif protein 39 (RBM39) and coactivator of activating protein 1 and oestrogen receptor (CAPERalpha), for ubiquitin-mediated degradation [16]. Degradation of RBM39 leads to exon skipping and intron retention and causes cell death in a subset of cancer cells [16]. Other spliceosome inhibitors target kinases that regulate splicing factor activities and subcellular localisation, including the ATP-competitive inhibitors of the SR-phosphorylating kinases SRPKl/SRPK.2 and CLK. Finally, different splicing inhibitors, such as quinones, have been shown to interfere with later assembly stages, but their effects are often concentration-dependent [19].

11.2 Aberrant Spliceosomal Machinery

Due to the faulty components of spliceosome there could be abnormal structure as well as abnormal function of spliceosome. Hence, it has been said as aberrant spliceosomal machinery. Canonical spliceosomal splicing mechanism and back-splicing mechanism are involved in the biogenesis of circRNA [20]. It has been reported that aberrant/abnormal RNA splicing has been linked to cancer disease [21]. In the case of cancer, the spliceosomal machinery may subscribe to circRNA deregulation. Mutations in spliceosome components as well as splice sites and, also including five small nuclear RNAs (snRNA), affect the steady state levels of circular RNAs [22].

11.3 Factors Dependent on Spliceosome in Cancer

1. Primarily in cancer, recurrent mutations in spliceosomal genes, such as SF3B1, SRSF2 and U2AF1, are responsible for vulnerabilities and mis-splicing [1].

2. Secondly, splicing factors (SR proteins, hnRNPs) increase Laccase2 circular RNA levels in conjunction with intronic repeats [23].

3. Thirdly, dozens of splicing factor genes are variably expressed in cancer cells [24]. Alternative RNA splicing events, which are diverse in the biogenesis of circRNA, also characterise cancer [25].

11.4 Function of Spliceosome

The main function of spliceosome machinery is the removal of introns and the joining of the exons to form mRNA. It includes three cycles to complete this process:

1. Assembly
2. Function
3. Disassembly

By controlling the assembly of the spliceosome, a cell can regulate the quality and quantity of splicing and so regulate gene expression.

11.5 Splice Variants

Different types of cancer have been caused by splice variants. These are caused by splicing errors due to mutation. Structure and function have been disrupted due to mutations.

12. Splice Variant Detection Methods

Gene splicing leads to the synthesis of alternate proteins that play an important role in human physiology and deadly disease such as cancer. Currently, the most efficient methods for large-scale detection of splice variants include microarray analysis and computational prediction methods. Microarray is based on splice variant detection, which is the most popular method currently in use. The highly parallel and sensitive nature of microarrays makes them ideal for monitoring gene expression on a tissue-specific, genome-wide level. Microarray-based methods for detecting splice variants provide a robust, scalable platform for high-throughput discovery of alternative gene splicing. A number of novel gene transcripts were detected by using microarray-based methods that were not detected by ESTs using computational methods. Another commonly used method for discovering of novel gene isoforms is RT-PCR followed by sequencing. This is a powerful approach and can be effectively used for analysing a small number of genes. However, it only provides only a limited view of the gene structure, is labour-intensive, and does not easily scale to thousands of genes or hundreds of tissues.

13. Challenges in Microarray Design for Splice Variant Detection

Microarray is based on gene splicing detection and poses some unique challenges in designing probes for isoforms that show a high degree of homology. In order to differentiate between these isoforms, a microarray that uses a combination of probes for exons and exon-exon junctions is used. Exon skipping events or other deletions can be monitored by using junction probes. For example, a probe spanning the exon 1 and exon 3 of the gene will detect the skipping of exon 2 from the gene that is translated into a protein.

14. Software for Designing Splice Variant Microarrays

AlleleID® automatically designs junction probes as well as intra-exon probes. With AlleleID® you can design any combination of probes to detect any alternative splicing event.

15. Future Directions: Key Issues and Challenges

Defects caused in RNA splicing are hallmarks of the deadly disease cancer, which can be compiled from various meta-analyses of transcriptomic or proteomic datasets. It has been demonstrated that various factors, including irradiation, stresses, hypoxia or any one of ten different chemotherapeutic/targeted drugs, induced an overall decreased splicing efficiency and global intron retention in more than 101 cancer cell lines of various types [26]. These data show that therapy-induced stress response is associated with the downregulation of pre-mRNA splicing in cancer cells, further supporting the idea that the development of small molecules acting as spliceosome inhibitors constitutes a promising therapeutic area in cancer. However, many questions and several issues are still remaining to be answered.

The prime issue is to improve our basic knowledge regarding the effects of these compounds on human cancer cells. Notably, we need to characterise further which RNA splicing patterns are specifically affected by these compounds and to what extent reprogramming of RNA splicing contributes to distinct cellular phenotypes upon treatment. Based on previous studies, it appears that splicing modulators do not broadly impact splicing but rather modulate the splicing of specific genes. However, it is likely that these effects vary according to the structure and mechanisms of action of spliceosome inhibitors and cellular contexts. Also important is the impact of spliceosome inhibition on the general landscape of non-coding RNAs (ncRNA), which needs to be investigated, as this is still largely unclear. For example, it was recently found that pharmacologically depleting or inhibiting spliceosome components in drosophila increases the steady-state levels of circular. RNAs (circRNAs), while concomitantly decreasing the expression of their associated linear mRNAs [27]. CircRNAs are emerging as a new category of ncRNAs, with a role in tumour progression and response to therapies. Therefore, deciphering the impact of splicing modulators on circRNA expression is of critical interest. Moreover, as splicing-independent functions of various components of the spliceosome machinery have been reported, and especially in the maintenance of genomic stability, a better understanding of the consequences of spliceosome inhibitors on these splicing independent functions is essential.

A second issue is our need to identify predictive biomarkers of response to these therapies in order to select patients who will most benefit from these treatments. Mutations or loss of copy number of spliceosome components, oncogene activation, and specific dependency on splicing switches of BCL2 family members are among the first biomarkers of response discovered. We need more in tumour types. Besides MYC, numerous other oncogenic signalling pathways are activated in tumours, sometimes with specificity for cancer types. Deciphering the contribution of deregulated splicing in response to oncogene hyperactivation should allow us to identify additional oncogene-dependent vulnerabilities to splicing inhibition and to help us define innovative therapeutic strategies, as most of these oncogenic alterations and/or their downstream signalling pathways are drug-targetable.

A third issue is related to the toxic side effects of spliceosome inhibitors. Different ways exist to bypass toxicity. As mentioned above, lowering the doses of these compounds and/or having a way to select patients are alternatives. Another alternative is to find strategies that directly target the splicing regulators that are aberrantly expressed in cancer, which could be less toxic than global targeting of the spliceosome machinery. Decoy oligonucleotides comprising several repeats of an RNA motif recognised by a specific splicing factor could be used to preclude the binding of these splicing factors to their cognate sequence. Proof-of concept of such approaches recently showed that decoy oligonucleotides targeting the splicing factors RBFOX 1/2, SRSF1 and polypyrimidine tract binding protein 1 (PTBP1) specifically bind to their respective splicing factors and inhibit their splicing and biological activities both *in vitro* and *in vivo* [28]. Another possibility is to use spliceosome inhibitors in combination with either chemotherapeutic drugs or targeted therapies to try to obtain additive or synergistic cytotoxic effects. Concerning inhibitors of SR-phosphorylating kinases, synergistic anti-AML effects, without noticeable toxicity in mice, were found by combining SPHINX3 1 and i-BET- 151, an epigenetic drug targeting BRD4 [29]. These studies highly support the need to identify potent therapeutic combinations using spliceosome inhibitors. In this setting, another promising lead deals with immune checkpoint therapies, such as anti- PD-1 or anti-PD-LI. These therapies have demonstrated remarkable and persistent anti-cancer efficacy, particularly for advanced SCLC. As discussed above, SF3B1 inhibitors increase intron retention, which is a source of neoepitopes in cancer [30] and leads to genomic instability through the accumulation of highly mutagenic R-loops and inhibition of D A repair processes. Although this has yet to be tested, spliceosome inhibition could thus contribute to increasing both the tumour mutational burden

(TMB) and tumour immunogenicity. As these are two determinants of tumour response to immune checkpoint inhibitors, it is tempting to speculate that combining SF3B1 inhibitors with immune therapies could provide a therapeutic benefit in patients.

One of the main problems in the management of cancer patients remains the occurrence of secondary resistance to therapies. Deciphering whether spliceosome inhibition could provide therapeutic advantages in this setting is another issue to be examined. Various splice variants have been highlighted as determinants of primary or secondary resistance to therapies [31]. Nonetheless, almost nothing is known in regard to the effects of spliceosome inhibitors in this context. As proofs-of-concept, it was demonstrated that E7 1 07 counteracts resistance to venetoclax in CLL cells [32] and that spliceostatin A can restore sensitivity to melanoma cells with secondary resistance to vemurafenib [33]. More preclinical studies are now needed to identify additional models of resistance in which spliceosome inhibitors might offer therapeutic benefits. Importantly, the acquisition of secondary mutations in SF3B1 associated with resistance to F3B1 inhibitors has been shown in cancer cells, indicating that cancer cells are also able to escape from these therapies. This point should therefore be considered in the future.

Conclusion

RNA splicing modulators are attractive anti-cancer agents. However, much work remains to be done in order to determine their clinical potential, safety, and usefulness. In particular, it will be important to characterise the landscape of splicing changes mediated by these compounds, as well as to identify cancer patients for whom they are more prone to act. Similarly, more systematic efforts are required to understand the effects of these splicing-inhibitory compounds on the proteome and the expression of non-coding RNAs.

References

[1] Chabot, B. and Shkreta, L. (2016). Defective control of pre–messenger RNA splicing in human disease. Journal of Cell Biology 212(1): 13–27.

[2] Chen, L., Tovar-Corona, J.M. and Urrutia, A.O. (2011). Increased levels of noisy splicing in cancers, but not for oncogene-derived transcripts. Human Molecular Genetics 20(22): 4422–4429.

[3] Kahles, A. et al. (2018). Comprehensive analysis of alternative splicing across tumors from 8,705 patients. Cancer Cell 34(2): 211–224.e6.

[4] Slansky, J.E. and Spellman, P.T. (2019). Alternative splicing in tumors—A path to immunogenicity? New England Journal of Medicine 380(9): 877–880.

[5] Jung, H. et al. (2015). Intron retention is a widespread mechanism of tumor-suppressor inactivation. Nature Genetics 47(11): 1242–1248.

[6] Supek, F. et al. (2014). Synonymous mutations frequently act as driver mutations in human cancers. Cell 156(6): 1324–1335.

[7] Anczuków, O. and Krainer, A.R. (2016). Splicing-factor alterations in cancers. RNA 22(9): 1285–1301.

[8] Sebestyén, E. et al. (2016). Large-scale analysis of genome and transcriptome alterations in multiple tumors unveils novel cancer-relevant splicing networks. Genome Research 26(6): 732–744.

[9] Gout, S. et al. (2012). Abnormal expression of the pre-mRNA splicing regulators SRSF1, SRSF2, SRPK1 and SRPK2 in non small cell lung carcinoma. 2012.

[10] Soto, M. et al. (2020). Relationship between changes in the exon-recognition machinery and SLC22A1 alternative splicing in hepatocellular carcinoma. Biochimica et Biophysica Acta (BBA)-Molecular Basis of Disease 1866(5): 165687.

[11] Lee, S.C.-W. et al. (2016). Modulation of splicing catalysis for therapeutic targeting of leukemia with mutations in genes encoding spliceosomal proteins. Nature Medicine 22(6): 672–678.

[12] Soret, J. et al. (2005). Selective modification of alternative splicing by indole derivatives that target serine-arginine-rich protein splicing factors. Proceedings of the National Academy of Sciences 102(24): 8764–8769.

[13] Seiler, M. et al. (2018). H3B-8800, an orally available small-molecule splicing modulator, induces lethality in spliceosome-mutant cancers. Nature Medicine 24(4): 497–504.

[14] Shirai, C.L. et al. (2017). Mutant U2AF1-expressing cells are sensitive to pharmacological modulation of the spliceosome. Nature Communications 8(1): 1–10.

[15] Effenberger, K.A., Urabe, V.K. and Jurica, M.S. (2017). Modulating splicing with small molecular inhibitors of the spliceosome. Wiley Interdisciplinary Reviews: RNA 8(2): e1381.

[16] Han, T. et al. (2017). Anticancer sulfonamides target splicing by inducing RBM39 degradation via recruitment to DCAF15. Science 356(6336): eaal3755.

[17] Fong, J.Y. et al. (2019). Therapeutic targeting of RNA splicing catalysis through inhibition of protein arginine methylation. Cancer Cell 36(2): 194–209.e9.

[18] Lee, S.C. and Abdel-Wahab, O. (2016). Therapeutic targeting of splicing in cancer. Nat. Med. 22(9): 976–86.

[19] Berg, M.G. et al. (2012). A quantitative high-throughput *in vitro* splicing assay identifies inhibitors of spliceosome catalysis. Molecular and Cellular Biology 32(7): 1271–1283.

[20] Quan, G. and Li, J. (2018). Circular RNAs: Biogenesis, expression and their potential roles in reproduction. Journal of Ovarian Research 11(1): 1–12.

[21] Scotti, M.M. and Swanson, M.S. (2016). RNA mis-splicing in disease. Nature Reviews Genetics 17(1): 19–32.

[22] Liang, D. et al. (2017). The output of protein-coding genes shifts to circular RNAs when the pre-mRNA processing machinery is limiting. Mol. Cell 68(5): 940–954.e3.

[23] Kramer, M.C. et al. (2015). Combinatorial control of Drosophila circular RNA expression by intronic repeats, hnRNPs, and SR proteins. Genes & Development 29(20): 2168–2182.

[24] Sveen, A. et al. (2016). Aberrant RNA splicing in cancer; expression changes and driver mutations of splicing factor genes. Oncogene 35(19): 2413–2427.

[25] Tremblay, M.-P. et al. (2016). Global profiling of alternative RNA splicing events provides insights into molecular differences between various types of hepatocellular carcinoma. BMC Genomics 17(1): 1–16.

[26] Anufrieva, K.S. et al. (2018). Therapy-induced stress response is associated with downregulation of pre-mRNA splicing in cancer cells. Genome Medicine 10(1): 1–18.

[27] Liang, D. et al. (2017). The output of protein-coding genes shifts to circular RNAs when the pre-mRNA processing machinery is limiting. Molecular Cell 68(5): 940–954.e3.

[28] Denichenko, P. et al. (2019). Specific inhibition of splicing factor activity by decoy RNA oligonucleotides. Nature Communications 10(1): 1–15.

[29] Tzelepis, K. et al. (2018). SRPK1 maintains acute myeloid leukemia through effects on isoform usage of epigenetic regulators including BRD4. Nature Communications 9(1): 1–13.

[30] Smart, A.C. et al. (2018). Intron retention is a source of neoepitopes in cancer. Nature Biotechnology 36(11): 1056–1058.

[31] Wang, B.-D. and Lee, N.H. (2018). Aberrant RNA splicing in cancer and drug resistance. Cancers 10(11): 458.

[32] Ten Hacken, E. et al. (2018). Splicing modulation sensitizes chronic lymphocytic leukemia cells to venetoclax by remodeling mitochondrial apoptotic dependencies. JCI Insight 3(19).

[33] Salton, M. et al. (2015). Inhibition of vemurafenib-resistant melanoma by interference with pre-mRNA splicing. Nature Communications 6(1): 1–7.

5

Deregulated Oncogenic Signaling Pathway(s) and Alternative Splicing

*Bhawana** and *Pratyusha Vavilala**

1. Introduction

There are more than 100 million living species on Earth today. The parent organism transmits information about the features that the offspring will have. The hereditary information of all living cells on Earth is stored in the form of double-stranded DNA molecules, which is ultimately responsible for RNA and protein synthesis. This expression is mediated by a mechanism known as the central dogma, which is present in all living creatures and leads to the development of two additional important polymers: RNAs and proteins. DNA molecules are often enormous, containing the blueprints for thousands of proteins. Individual parts of the lengthy DNA sequence are translated into individual mRNA molecules, each of which codes for a different protein. Each of these DNA segments corresponds to a single gene. Individual gene expression is regulated in all cells rather than making its whole repertoire of possible proteins. The cell regulates the pace of transcription and translation of distinct genes independently. Furthermore, cells have the means to correct mistakes made during DNA replication and repair DNA damage caused by external agents such as radiation. A failure of proper replication and preservation of genomic DNA results in abnormalities, which can have disastrous effects such as the development of cancer [1, 2].

Cancer is the second biggest cause of death worldwide, accounting for over 9.6 million fatalities to date. Lung, prostate, colorectal, stomach, and liver cancers are more prevalent in males, whereas women's cancers include breast, colorectal, lung, cervical, and thyroid cancer [3–5]. There is widespread agreement that cancer

Shaheed Rajguru College of Applied Sciences for Women, University of Delhi, Delhi-110096, India.
* Corresponding authors: bhawana.sharma@rajguru.du.ac.in, bhawanasharma85@gmail.com; pratyusha@rajguru.du.ac.in

is essentially a genetic illness, and its growth is based on the accumulation of molecular changes in the genome of somatic cells [6]. Carcinogenesis is the result of increased cell proliferation, evasion of growth suppressors, resistance to cell death, enabling replicative immortality, inducing angiogenesis, activating invasion and metastasis, reprogramming energy metabolism, evading immune destruction, genome instability, and immune inflammation [4, 7–11]. The disruption of cellular signal transduction pathways is the most common cause of these alterations [9]. Cell surface receptors that attach to growth factors, proteins that interact with DNA to initiate the replication, and signaling molecules that connect the receptors to the replication initiators via multiple pathways are examples of crucial functions that, if altered, can lead to tumorigenesis [9, 12]. Proto-oncogenes are genes that code for the normal proteins that control these crucial functions in their normal condition. When they are mutated to become oncogenes, their aberrant protein products have greater activity, which aids in tumor growth [13–15]. Furthermore, oncogenes have the ability to save cells from planned cell death [16]. Ras proteins, for example, act as molecular switches that turn on and off depending on the nucleotide form (di-phosphate or tri-phosphate) to which they are attached. When a proto-oncogene becomes an oncogene because of mutations, Ras becomes permanently activated regardless of the signals the cell gets. Chromosomal translocation is the second form of genetic change that turns a proto-oncogene into an oncogene. This happens when fragmented chromosome segments reattach randomly, resulting in the production of a fusion protein with the N-terminus of one protein and the C-terminus of another or altered protein expression regulation. BCR/ABL is an oncogene that is produced by this sort of chromosomal translocation. Tumor suppressor genes, on the other hand, encode proteins that have the ability to inhibit carcinogenesis [17]. Inhibitors of cell cycle advancement, components involved in cell cycle checkpoint maintenance, and proteins essential for apoptosis induction are only a few examples.

Alternative splicing of precursor mRNA is an essential mechanism to increase the complexity of gene expression, and it plays an important role in cellular differentiation and organism development. Precursor mRNA or pre-mRNA is the initial RNA produced during transcription. Processing of pre-mRNA occurs to remove the introns through a process known as alternative splicing. Regulation of alternative splicing is a complicated process; any alteration of the process may lead to disruption of normal cellular function and eventually lead to disease [18, 19]. Cancer is one of those diseases where alternative splicing may be the basis for identifying novel diagnostic and prognostic biomarkers as well as new strategies for therapy [20, 21]. Thus, an in-depth understanding of alternative splicing regulation, oncogenic pathways, and the relationship between the two has the potential not only to elucidate fundamental biological principles but to provide solutions for various diseases [22–24].

Advances in next-generation sequencing technologies and genome-wide analyses have revealed that the vast majority of multi-exon genes under normal physiology engage in alternative splicing in tissue-specific and developmental-specific manner. Cancer cells exhibit remarkable transcriptome alterations, and it is evident that the pathways that regulate alternative splicing in cancer and splicing factors that bind to pre-mRNAs are of great importance [21, 22]. Changes in cancer-related genes that directly affect pre-mRNA splicing, mutations in genes encoding splicing factors or core spliceosomal subunits, and seemingly mutation-free disruptions in the balance of the expression of RNA-binding proteins, including components of both the major (U2-dependent) and minor (U12-dependent) spliceosomes, are among the splicing aberrations highlighted [13, 25–27]. Given that the latter two classes generate global splicing changes that influence a wide variety of genes, identifying the ones contributing to cancer progression remains difficult [23, 28]. A methodical approach is required to interpret these abnormalities and their impact on cancer. This chapter focuses on the deregulation of oncogenic pathways and the role of alternative splicing, explaining a few oncogenes and the change in their functions because of alternative splicing.

2. Splicing: An Essential Step in Gene Expression

In eukaryotes, most genes have intervening regions (introns) that cause the expressed sequences (exons) to disrupt. In comparison to exons (median size 120 bp), introns in eukaryotes are substantially longer (median size 1,000 bp but can be > 100,000 bp), making introns the largest contributors to gene sequence. After transcription, the expressed transcript (pre-mRNA) has to undergo substantial processing to remove the introns by a molecular machine—i.e., the spliceosome to become a suitable message for downstream processes like translation [6].

Due to the large size of introns, the transcription process of genes sometimes takes hours to complete. Introns pose a conundrum because their transcription, only to be spliced out and degraded, appears to be a wasteful operation in terms of time and energy required in their transcription as well as their removal and degradation. Furthermore, the splicing process must be both highly efficient and precise. To ensure that all introns removes from the pre-mRNA on time and in a coordinated manner, efficiency is essential. Because connecting exons with even a single base error might have disastrous implications on the reading frame, fidelity is crucial [29].

Moreover, outside the actual splice sites, the cis-sequences or splice sites at the boundaries of each intron are too basic, occasionally degenerate, and too redundant to serve as efficient landmarks for spliceosome assembly. These characteristics of introns, as well as those of splicing in general, make the presence of introns in eukaryotes seem paradoxical. Introns, on the other hand, are not simply superfluous sequences that delete by splicing; they have a number of benefits, including connecting various RNA processing steps for increased gene expression efficiency and regulation and providing a checkpoint for mRNA quality control. Through the process of alternative splicing, they also allow any gene that contains them to have a vast capacity for diversification. Because the presence of introns in eukaryotic genomes and their position in genes are highly conserved, it is likely that the benefits of having introns outweigh the disadvantages [30].

3. Alternative Splicing: Reason for Transcriptome Diversification

Most multi-exon genes in eukaryotes experience at least one event of alternative splicing, resulting in two or more different mRNAs from the same gene, with the number of alternatively spliced transcripts for some genes even higher. Many of these transcripts, interestingly, are expressed in tissue-specific ways, at specific developmental stages, or in disease-specific ways. While the role of some of these alternative transcripts is not always clear or even recognized, a growing body of evidence suggests that alternatively spliced exons are translated and tend to encode essential polypeptide domains Figure 1.

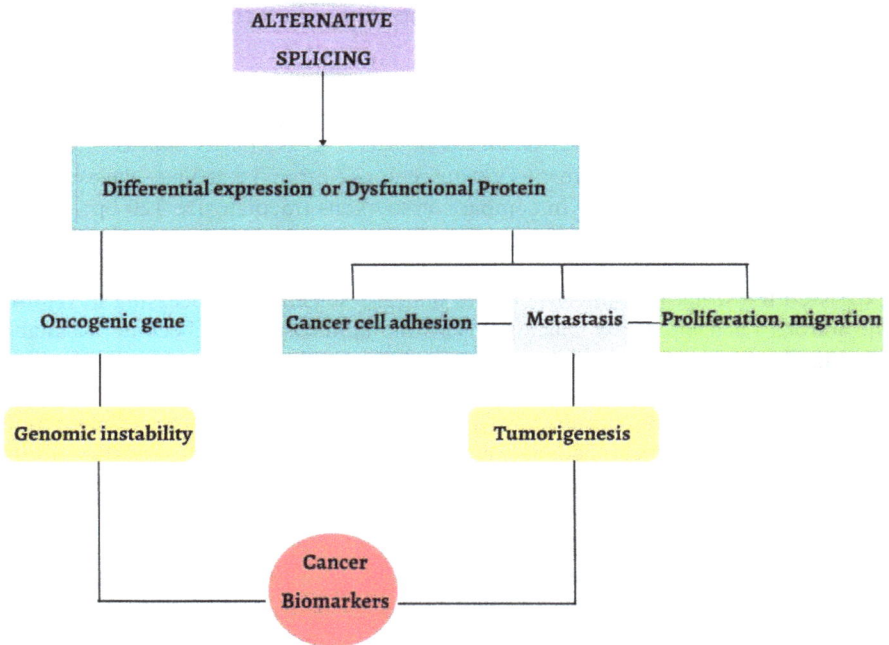

Figure 1: Alternative splicing in cancer-related genes (oncogenic and tumor suppressor genes) produces dysfunctional proteins, increasing genomic instability. These abnormal proteins aid malignant cells in their continual growth, proliferation, and metastasis (tumorigenesis). Variants of cancer-related genes that are expressed differently or specifically in relation to their critical functions in carcinogenesis could be used as biomarkers to distinguish between normal and malignant cells.

This proposes an evolutionary conserved molecular architecture for transcriptome diversification that avoids the need to enlarge the genome, which would necessitate the creation of homologous genes that perform comparable but unique tasks. Alternative splicing is a term that refers to a variety of splicing processes. Alternative exons (cassette exons: skipped/included complete exons), retained introns, and alternative 5' and 3' splice sites (5' ss and 3' ss) are all examples of alternative splicing, as seen in Figure 2. Alternative first and last exons are two less visible alternative splicing

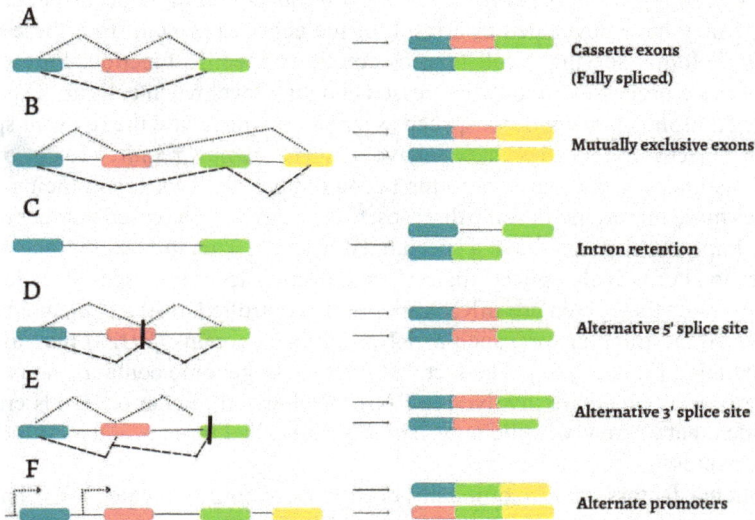

Figure 2: Various types of alternative splicing events potentially exist in cells. Alternative splicing acts as a source of diversification. The colored rectangles represent the exons, and the introns are presented by solid horizontal lines. The dashed and dotted lines indicate the various splicing events.

events that are intimately connected to and could result from transcription regulation. Nonetheless, all of these phenomena have been well described in eukaryotes and have significant consequences for transcriptome diversification. Intron retention is a type of alternative splicing frequently neglected because it mistakenly interprets as a splicing error resulting in an intron not being spliced out. While this may be true in some circumstances, a growing body of evidence suggests that intron retention is regulated to govern gene expression after transcription. In fact, cancer cells of all types have a higher level of retained intron than normal cells, resulting in a greater diversity of transcriptomes [30].

Intron retention and regulation are clearly seen in a family of introns known as minor or U12 introns. In the human genome, some 800 minor introns are found in genes involved in signal transduction and information relay, cell cycle control, and DNA damage repair. U12 introns are highly conserved and are used as molecular switches to control gene expression without requiring the transcription of new pre-mRNA, which is especially important when the gene product is required immediately, such as when cells are stressed. Given the activities of minor intron-containing genes, it is plausible that they are controlled similarly in cancer [30].

4. Regulation of Alternative Splicing and its Mechanism

Several documented alternative splicing events account only for a small percentage of the processed mRNA expressed at any given moment. While this may seem like alternative splicing events are just biological noise in a highly busy cell process, it is shown that these events are carefully regulated and play important functions in a variety of cell types and tissues [31]. The low frequency of these occurrences in one cell type may have developed as a result of the encoded protein from these unique splicing isoforms serving a cell type or condition-specific function. Furthermore, some of these processes are only expressed at a high level when cells are exposed to specific environmental conditions, such as stress, in which case the relevant splicing isoform is necessary. As a result, extensive searches are now required to discover the conditions under which these isoforms become more numerous, and their function becomes more important. In some diseases, tissues have an increased number of these events, implying that they have a specialized purpose. This implies that the number of specific alternatively spliced transcripts and the selection of specific alternative splicing events for a given pre-mRNA are tightly controlled. Both cis-elements in the pre-mRNA and trans-factors, such as RNA-binding proteins (RBPs) play a role in this regulation [7, 8, 32, 33]. The fact that the human genome contains thousands of RBPs, many of which are involved in RNA splicing and regulation, lends credence to the idea that alternative splicing is a highly controlled process and a critical stage in gene expression regulation [4, 5].

Splicing factors have traditionally been divided into two categories: hnRNPs, which repress splicing, and SR proteins, which promote splicing control. On the other hand, a closer examination of the function of any individual hnRNP or SR protein soon reveals that they do not necessarily fit into these categories [4, 34–36]. Multiple factors influence an RBP's eventual involvement in splicing control. The strength and context of its pre-mRNA binding sites and competing or cooperative binding of several RBPs on or around the regulated exon or intron are all factors to consider. Reduced or increased binding of a single splicing component in a normal or diseased state has a difficult time predicting the splicing outcome because of combinatorial regulation. Another issue is that a splicing factor will likely affect the pre-mRNA splicing of other splicing factors. To better understand how alternative splicing controls in a given state, a systems biology approach is required in which the expression status and targets of numerous, if not all, RBPs are examined to begin constructing complex networks of co-regulated pathways [37].

5. Deregulation of Oncogenic Signaling Pathways and Alternative Splicing

Individual cells or populations of cells expand in response to contextual cues that regulate their ability to enter and continue through the cell cycle, migrate, and survive within temporary microenvironments throughout development and tissue repair. Multiple signaling systems, which often function in a tissue and cell type specific manner to manage the cell cycle, anti-apoptotic, and pro-migratory machineries, have been discovered by cell biology investigations [38]. Cancer cells genetic alterations influence signaling pathways, resulting in tumorigenesis-related control process disturbance. Oncogenic mutations disturb the normal operation of these pathways, resulting in unregulated mitogenesis, resistance to pro-apoptotic insults, and an increase in motility, but the extent, mechanisms, and co- occurrence of changes in these pathways vary by tissue and tumor type [32, 39].

Oncogenic signaling changes in cancer not only cause aberrant cell differentiation and proliferation but also play a key role in tumor immune evasion. Cancer-related signaling influences the immunological components of tumors through the production of cytokines, chemokines, and growth factors [29]. To date, it has been shown that genetic changes in various signaling pathways found in malignancies impact the tumor immune microenvironment. Normal stem cell homeostasis is governed by highly controlled molecular signaling networks. Many of these pathways are inappropriately active or suppressed in human malignancies and experimental carcinogenesis models, which is not surprising. CSCs' self-renewal, proliferation, survival, and differentiation capabilities are all influenced by such aberrations. These pathways are generally complex, containing numerous external and intrinsic molecular signals and regulatory factors. Many of these 'pathways' are interwoven networks of signaling mediators that feed into one another, allowing for cross-pathway communication [40]. A few of the oncogenic pathways affected by deregulated alternative splicing are explained below.

The binding of several ligands to their corresponding receptors, including interleukins, interferons, hormones, and growth factors, activates this pathway [41]. JAK/STAT signaling has been implicated in maintaining embryonic stem cell self-renewal capabilities, hematopoiesis, and neurogenesis. Cells isolated from tumors of the breast, prostate, blood, and glia showed abnormal activation of this pathway. In solid tumor model systems, modulation of the JAK/STAT pathway in cancer stem cells (CSCs) demonstrated acceleration or repression of their proliferation. Chemical suppression of STAT3 reduced CSC number, proliferation, and clonogenicity in breast CSCs, indicating that JAK/STAT signaling promotes tumorigenesis. Alternative splicing of STAT3 results in dominant negative regulation of transcription because of splicing of intron 22 by alternative 3' splice site mechanism [42–44]. It has been observed that various splicing activators and repressors are direct transcriptional targets of MYC [21]. In the presence of hyperactive MYC, downregulation of some spliceosomal components or mutations of a few proteins can result in lower cell viability and increased apoptosis, and the accumulation of transcripts with one or more retained introns. Furthermore, oncogenic MYC reprograms the spliceosome in B lymphocytes, causing the insertion of alternative exons with weak 5' splice sites. The mechanisms by which MYC appears to alter splicing in lymphocytes differ from those in breast cancer [35, 45]. Alternative splicing of MYC results in reduced tumor suppression and apoptosis regulation via cassette alternative splicing of exon 12a and exon 13.

Apoptosis-related genes, including the Bcl2 family and various caspases, can produce oncogenic isoforms after alternative splicing events. Intron 2 of the Bcl2L1 gene, coding for the Bcl-X protein, has two 5' splice sites. Consequently, the mRNA generated can be large (Bcl-XL), which encodes a Bcl-X protein with anti-apoptotic action, or small (Bcl-XS), which encodes a Bcl-X protein that lacks the BH domain and is pro-apoptotic, depending on which 5'ss is chosen [23]. Another example is caspase 2 pre-mRNA splicing, wherein the inclusion of exon 9 produces the pro-apoptotic Casp2L protein. On the other hand, the anti-apoptotic Casp2S protein is encoded by the short isoform. Studies have shown that cancer cells reprogram the splicing machinery and/or splicing factors that bind to these pre-mRNAs to ensure that the cancer-specific isoforms are enriched [46, 47].

Additionally, the loss of function of some tumor suppressor genes can result in various cancers [4]. p53 can cause cell cycle arrest in either the G1, G2, or G2/M phase of the cell cycle once activated, for example, by DNA damage. Multiple cell cycle checkpoints that allow cells to repair DNA damage or commit to apoptosis need the presence of a functional p53. Because p53 is a tetramer, the synthesis of dominant negative subunits by alternative splicing, for instance, even at low levels, can have catastrophic consequences. The full-length and functional p53 protein are encoded by a fully spliced mRNA comprising the 11 canonical exons. For p53 pre-mRNA, several isoforms with different initial exons can be produced [48, 49].

Splicing variations in tumor suppressor genes like p53 and PTEN (Phosphatidylinositol 3,4,5-trisphosphate 3-phosphatase and dual-specificity protein phosphatase) have been linked to cancer. Alternative splicing of the p53 protein results in isoforms that play key roles in a variety of biological processes, implying that dysregulation affects carcinogenesis. Varied tissue types have different levels of PTEN expression and alternatively spliced transcripts. PTEN controls the stability of p53, which in turn controls its transcriptional activity. Breast cancer has been shown to have PTEN splice variants retained in the intron 3 and intron 5 regions [29, 48, 49].

Not only mutations in the splicing factor, but also an alteration in their expression can lead to many disorders, including cancer. Because splicing factors bind to and regulate the splicing of hundreds of pre-mRNAs, changes in splicing factors, whether due to mutations or altered expression, have a significant impact on cell phenotype [26, 50]. By deregulating a few splicing factors, cancer cells can in turn change the splicing of a huge number of genes. There are various other oncogenic pathways/genes that are also affected by alternative splicing, such as VEGFA, MCL-1, MDM2, Caspase 8, CD44, etc. [11, 12, 51–54]. However, further research is required to gain a better perspective regarding these genes as well as those which are still undiscovered (Figure 3).

6. Methods to Identify Splicing Aberrations Linked to Cancer

When a splicing factor or a key component of the spliceosome is altered, the expression of a wide variety of genes can be altered [15]. Two major challenges that should be focused on during future research are identifying those splicing alterations that contribute significantly to cancer progression among the thousands of splicing alterations and therapeutically targeting the splicing factors without having massive side effects. It is critical to building a systematic and standardized approach to get a sufficient understanding of splicing dysregulation in cancer and its impact on cancer in order to adequately address these aspects [55].

Transcriptome profiling to uncover global splicing alterations depended mostly on gene expression exon microarrays, which contain probes for almost all exons and many introns before the introduction of next-generation RNA sequencing (RNA-seq). While these arrays were a significant improvement over standard microarrays

Alternative splicing related changes in oncogenic functions	Gene name protein product	Alternative splicing events
• Reduced anti-proliferative response to stress. • Reduced tumor suppression	TP53/ P53	• Alternative 1st exon • Intron 9 retention and cryptic exon inclusion • Intron 2 retention • Cassette Exon 8 • Intron 7 alternative 5' ss
• Apoptosis regulator: • Anti-apoptotic (BCL-XL) • Pro-apoptotic (BCL-XS)	Bcl2l1 BclX	• Intron 2 alternative 5' ss
• Apoptosis regulator: • Anti-apoptotic (Casp2S) • Pro-apoptotic (Casp2L)	CASP2/ Caspase 2	• Cassette Exon 9
• Reduced tumor suppression and • pro-apoptotic activity	CASP8/ Caspase-8	• Cassette Exon 8 • Alternative Exon 8 and 9 splicing (136 bp insertion between exon 8 and 9)
• Dominant negative regulation of transcription (STAT3b)	STAT3	• Intron 22 alternative 3' ss
• Promotes EMT and metastasis	FGFR2	• Mutually exclusive exon 8 (IIIb) or 9 (IIIc)
• Enhanced pro-angiogenic function.	VEGFA	• Cassette Exons 6 and 6b • Cassette Exon 7 and 7b • Alternative Intron 6 5' ss • Alternative Intron 6 3' ss • Cassette Exon 8
• Acts as an oncogene	EGFR	• Alternativer splicing of a combination of Exons 9a, 10,16, or 17 • Skipping of Exon 2-7 • Skipping of Exons 2-22
• Reduces tumor suppression	PTEN	• Intron 3 retention and Intron 5 retention • Inclusion of partial Intron 5 • Inclusion of 5' end of intron H between Exons 8 and 9
• Anti-apoptotic	MCL1	• Cassette Exon 2

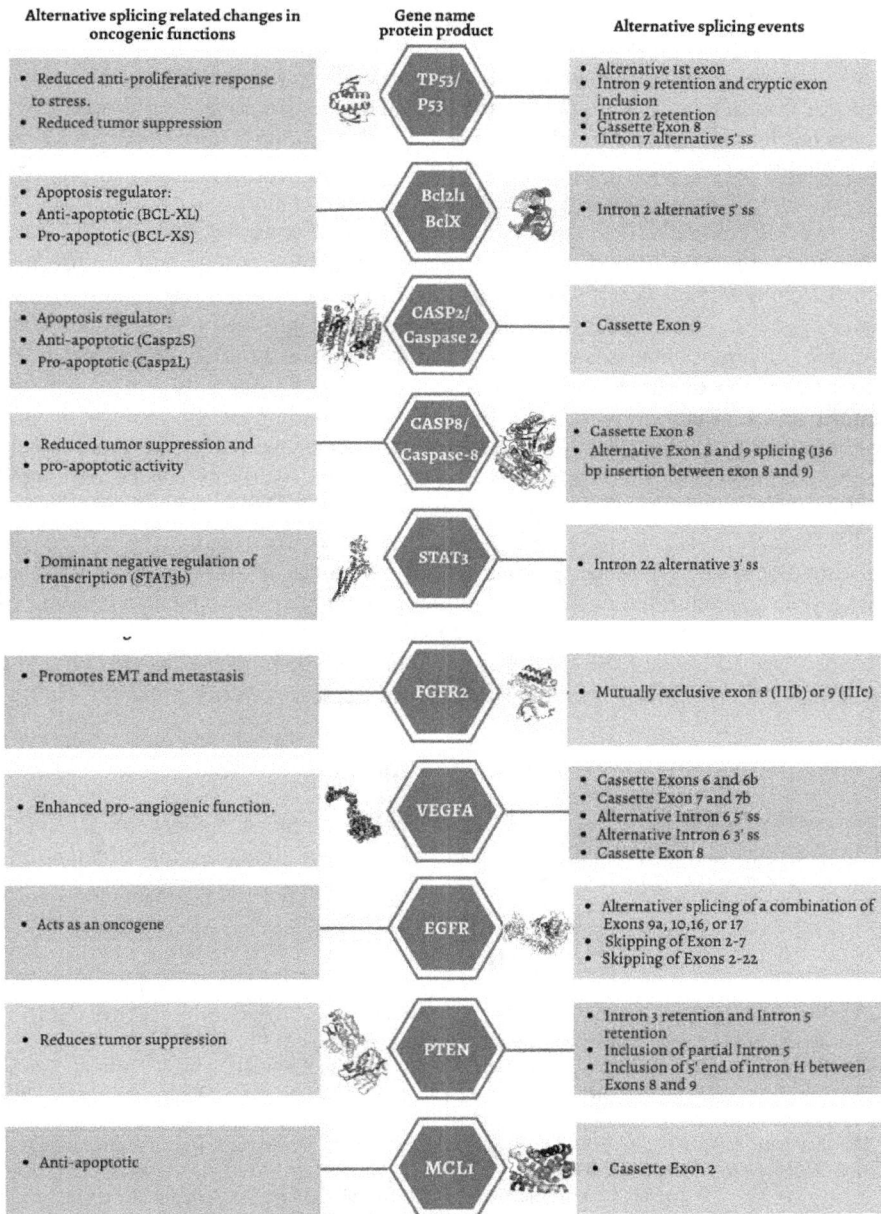

Figure 3: A few examples of genes with cancer-related alternatively spliced isoforms. TP53: Tumor Protein P53; Bcl-2 B-cell lymphoma 2; STAT3: Signal transducer and activator of transcription 3; FGFR2: Fibroblast growth factor receptor 2; VEGFA: vascular endothelial growth factor A; EGFR: epidermal growth factor receptor; PTEN: Phosphatidylinositol 3,4,5-trisphosphate 3-phosphatase and dual-specificity protein phosphatase; MCL1: Myeloid Cell Leukemia-1.

with a restricted number of probes per gene, they are difficult to read and can result in a large number of false positives. RNA-seq is now the gold standard for transcriptome profiling, which includes data on both genome-wide expression level changes and splicing disruptions. While this method is quantitative and qualitative, it also has its own set of difficulties. To begin with, creating RNA-seq libraries from high-quality RNA is costly, time-consuming, and requires highly skilled workers. The amount of data generated (several gigabytes per sample) necessitates the use of powerful computer devices for both storage and analysis. As a result, many laboratories have developed their own in-house pipelines that are suitable for their specific analyses but are not suitable for worldwide use [56]. The fact that RNA-seq data analysis is difficult does not change the fact that it has been widely employed to produce important databases of splicing abnormalities in a variety of malignancies. As our capacity to standardize the analysis process improves, the list of splicing abnormalities in cancer will grow, and our understanding of the molecular basis of these changes, as well as their relevance to cancer, will improve dramatically in the coming years. Finding the appropriate splicing isoforms can be extremely useful and they should be employed as new biomarkers for a variety of cancer types and subtypes [57]. Innovative RNA-based therapies are required to correct the splicing alterations or induce splicing changes in cancer cells that make them more susceptible to traditional chemotherapy; once enough molecular understanding of the splicing aberrations has been gained and their impact on cancer has been proven. Recent approval by the FDA for the use of antisense oligonucleotides for correcting the splicing of exons has paved the way for future research in this field.

Acknowledgments

The authors would like to thank Ms. Anamika Sinha, a student of B.Sc (H) Biochemistry, Shaheed Rajguru College of Applied Sciences for Women, the University of Delhi, for her help and contributions in preparing the figures.

References

[1] Roy, B.M., Haupt, L.R. and Griffiths L. (2013, Apr). Review: Alternative splicing (As) of genes as an approach for generating protein complexity. Curr. Genomics 14(3): 182–94.

[2] Sciarrillo, R., Wojtuszkiewicz, A., Assaraf, Y.G., Jansen, G., Kaspers, G.J.L., Giovannetti, E. et al. (2020, Dec). The role of alternative splicing in cancer: From oncogenesis to drug resistance. Drug Resist. Updat. 53: 100728.

[3] Jeanteur, P. (ed.). (2006). Alternative Splicing and Disease. Berlin: Springer, 257 p. (Progress in Molecular and Subcellular Biology).

[4] Urbanski, L.M., Leclair, N. and Anczuków, O. (2018, Jul). Alternative-splicing defects in cancer: Splicing regulators and their downstream targets, guiding the way to novel cancer therapeutics. WIREs RNA [Internet] [cited 2022 Sep 6] 9(4). Available from: https://onlinelibrary.wiley.com/doi/10.1002/wrna.1476.

[5] Zong, F.Y., Fu, X., Wei, W.J., Luo, Y.G., Heiner, M., Cao, L.J. et al. (2014, Apr). The RNA-binding protein QKI suppresses cancer-associated aberrant splicing. In: Cheung, V.G. (ed.). PLoS Genet. 10(4): e1004289.

[6] Adler, A.S., McCleland, M.L., Yee, S., Yaylaoglu, M., Hussain, S., Cosino, E. et al. (2014, May). An integrative analysis of colon cancer identifies an essential function for PRPF6 in tumor growth. Genes Dev. 28(10): 1068–84.

[7] Bonnal, S.C., López-Oreja, I. and Valcárcel, J. (2020, Aug). Roles and mechanisms of alternative splicing in cancer-implications for care. Nat. Rev. Clin. Oncol. 17(8): 457–74.

[8] Bonomi, S., Gallo, S., Catillo, M., Pignataro, D., Biamonti, G. and Ghigna, C. (2013). Oncogenic alternative splicing switches: Role in cancer progression and prospects for therapy. Int. J. Cell Biol. 2013: 1–17.

[9] Derakhshani, A., Rostami, Z., Taefehshokr, S., Safarpour, H., Astamal, R.V., Taefehshokr, N. et al. (2020, Mar). An overview of the oncogenic signaling pathways in different types of cancers [Internet]. Medicine & Pharmacology [cited 2022 Sep 6]. Available from: https://www.preprints.org/manuscript/202003.0110/v1.

[10] Oltean, S. and Bates, D.O. (2014, Nov). Hallmarks of alternative splicing in cancer. Oncogene 33(46): 5311–8.

[11] Zhao, Q., Caballero, O.L., Davis, I.D., Jonasch, E., Tamboli, P., Yung, W.K.A. et al. (2013, May). Tumor-specific isoform switch of the fibroblast growth factor receptor 2 underlies the mesenchymal and malignant phenotypes of clear cell renal cell carcinomas. Clin. Cancer Res. 19(9): 2460–72.

[12] Abou-Fayçal, C., Hatat, A.S., Gazzeri, S. and Eymin, B. (2017, Feb). Splice variants of the RTK family: Their role in tumour progression and response to targeted therapy. Int. J. Mol. Sci. 18(2): 383.

[13] Ghigna, C., Riva, S. and Biamonti, G. (2013). Alternative splicing of tumor suppressors and oncogenes. pp. 95–117. *In*: Wu, J.Y. (ed.). RNA and Cancer [Internet]. Berlin, Heidelberg: Springer Berlin Heidelberg [cited 2022 Sep 6]. (Cancer Treatment and Research; vol. 158). Available from: http://link.springer.com/10.1007/978-3-642-31659-3_4.

[14] Gonçalves, V., Pereira, J. and Jordan, P. (2017, Dec). Signaling pathways driving aberrant splicing in cancer cells. Genes 9(1): 9.

[15] Venables, J.P. (2004, Nov). Aberrant and alternative splicing in cancer. Cancer Res. 64(21): 7647–54.

[16] Karni, R., de Stanchina, E., Lowe, S.W., Sinha, R., Mu, D. and Krainer, A.R. (2007, Mar). The gene encoding the splicing factor SF2/ASF is a proto-oncogene. Nat. Struct. Mol. Biol. 14(3): 185–93.

[17] Dvinge, H., Kim, E., Abdel-Wahab, O. and Bradley, R.K. (2016, Jul). RNA splicing factors as oncoproteins and tumour suppressors. Nat. Rev. Cancer 16(7): 413–30.

[18] Zhang, J. and Manley, J.L. (2013, Nov). Misregulation of Pre-mRNA alternative splicing in cancer. Cancer Discov. 3(11): 1228–37.

[19] Zhang, Y., Qian, J., Gu, C. and Yang, Y. (2021, Dec). Alternative splicing and cancer: A systematic review. Signal Transduct Target Ther. 6(1): 78.

[20] El Marabti, E. and Younis, I. (2018, Sep). The cancer spliceome: Reprograming of alternative splicing in cancer. Front. Mol. Biosci. 5: 80.

[21] Singh, B. and Eyras, E. (2017, Mar). The role of alternative splicing in cancer. Transcription 8(2): 91–8.

[22] Climente-González, H., Porta-Pardo, E., Godzik, A. and Eyras, E. (2017, Aug). The functional impact of alternative splicing in cancer. Cell Rep. 20(9): 2215–26.

[23] David, C.J. and Manley, J.L. (2010, Nov). Alternative pre-mRNA splicing regulation in cancer: Pathways and programs unhinged. Genes Dev. 24(21): 2343–64.

[24] Li, F., Wu, T., Xu, Y., Dong, Q., Xiao, J., Xu, Y. et al. (2020, May). A comprehensive overview of oncogenic pathways in human cancer. Brief Bioinform. 21(3): 957–69.

[25] Ladomery, M. (2013). Aberrant alternative splicing is another hallmark of cancer. Int. J. Cell Biol. 2013: 1–6.

[26] Sveen, A., Kilpinen, S., Ruusulehto, A., Lothe, R.A. and Skotheim, R.I. (2016, May). Aberrant RNA splicing in cancer; expression changes and driver mutations of splicing factor genes. Oncogene 35(19): 2413–27.

[27] Turunen, J.J., Niemelä, E.H., Verma, B. and Frilander, M.J. (2013, Jan). The significant other: Splicing by the minor spliceosome: Splicing by the minor spliceosome. Wiley Interdiscip. Rev. RNA 4(1): 61–76.

[28] Chabot, B. and Shkreta, L. (2016, Jan). Defective control of pre–messenger RNA splicing in human disease. J. Cell Biol. 212(1): 13–27.

[29] Younis, I., Dittmar, K., Wang, W., Foley, S.W., Berg, M.G., Hu, K.Y. et al. (2013, Jul). Minor introns are embedded molecular switches regulated by highly unstable U6atac snRNA. eLife 2: e00780.

[30] Dvinge, H. and Bradley, R.K. (2015, Dec). Widespread intron retention diversifies most cancer transcriptomes. Genome Med. 7(1): 45.

[31] Qi, F., Li, Y., Yang, X., Wu, Y.P., Lin, L.J. and Liu, X.M. (2020, Jan). Significance of alternative splicing in cancer cells. Chin. Med. J. (Engl.) 133(2): 221–8.

[32] Schwerk, C. and Schulze-Osthoff, K. (2005, Jul). Regulation of apoptosis by alternative pre-mRNA splicing. Mol. Cell 19(1): 1–13.

[33] Scotti, M.M. and Swanson, M.S. (2016, Jan). RNA mis-splicing in disease. Nat. Rev. Genet. 17(1): 19–32.

[34] Akerman, M., Fregoso, O.I., Das, S., Ruse, C., Jensen, M.A., Pappin, D.J. et al. (2015, Jun). Differential connectivity of splicing activators and repressors to the human spliceosome. Genome Biol. 16: 119.

[35] Babic, I., Anderson, E.S., Tanaka, K., Guo, D., Masui, K., Li, B. et al. (2013, Jun). EGFR mutation-induced alternative splicing of max contributes to growth of glycolytic tumors in brain cancer. Cell Metab. 17(6): 1000–8.

[36] Graham, S.V. and Faizo, A.A.A. (2017, Mar). Control of human papillomavirus gene expression by alternative splicing. Virus Res. 231: 83–95.

[37] Shkreta, L., Bell, B., Revil, T., Venables, J.P., Prinos, P., Elela, S.A. et al. (2013). Cancer-Associated perturbations in alternative pre-messenger RNA splicing. pp. 41–94. *In*: Wu, J.Y. (ed.). RNA and Cancer [Internet]. Berlin, Heidelberg: Springer Berlin Heidelberg [cited 2022 Sep 6]. (Cancer Treatment and Research; vol. 158). Available from: http://link.springer.com/10.1007/978-3-642-31659-3_3.

[38] Sumithra, B., Saxena, U. and Das, A.B. (2016, Feb). Alternative splicing within the WNT signaling pathway: Role in cancer development. Cell Oncol. 39(1): 1–13.

[39] Shilo, A., Siegfried, Z. and Karni, R. (2015, Jan). The role of splicing factors in deregulation of alternative splicing during oncogenesis and tumor progression. Mol. Cell Oncol. 2(1): e970955.

[40] Paronetto, M.P., Passacantilli, I. and Sette, C. (2016, Dec). Alternative splicing and cell survival: From tissue homeostasis to disease. Cell Death Differ. 23(12): 1919–29.

[41] Guillaudeau, A., Durand, K., Bessette, B., Chaunavel, A., Pommepuy, I., Projetti, F. et al. (2012, May). EGFR soluble isoforms and their transcripts are expressed in meningiomas. *In*: Monleon, D. (ed.). PLoS ONE 7(5): e37204.

[42] Caldenhoven, E., van Dijk, T.B., Solari, R., Armstrong, J., Raaijmakers, J.A.M., Lammers, J.W.J. et al. (1996, May). STAT3β, a splice variant of transcription factor STAT3, is a dominant negative regulator of transcription. J. Biol. Chem. 271(22): 13221–7.

[43] Pencik, J., Pham, H.T.T., Schmoellerl, J., Javaheri, T., Schlederer, M., Culig, Z. et al. (2016, Nov). JAK-STAT signaling in cancer: From cytokines to non-coding genome. Cytokine 87: 26–36.

[44] Zammarchi, F., de Stanchina, E., Bournazou, E., Supakorndej, T., Martires, K., Riedel, E. et al. (2011, Oct). Antitumorigenic potential of STAT3 alternative splicing modulation. Proc. Natl. Acad. Sci. 108(43): 17779–84.

[45] Anczuków, O., Rosenberg, A.Z., Akerman, M., Das, S., Zhan, L., Karni, R. et al. (2012, Feb). The splicing factor SRSF1 regulates apoptosis and proliferation to promote mammary epithelial cell transformation. Nat. Struct. Mol. Biol. 19(2): 220–8.

[46] Jang, H.N., Lee, M., Loh, T.J., Choi, S.W., Ohm H.K., Moon, H. et al. (2014, Jan). Exon 9 skipping of apoptotic caspase-2 pre-mRNA is promoted by SRSF3 through interaction with exon 8. Biochim. Biophys. Acta BBA - Gene Regul. Mech. 1839(1): 25–32.

[47] Olsson, M. and Zhivotovsky, B. (2011, Sep). Caspases and cancer. Cell Death Differ. 18(9): 1441–9.

[48] Bourdon, J.C., Surget, S. and Khoury, M.P. (2013, Dec). Uncovering the role of p53 splice variants in human malignancy: A clinical perspective. OncoTargets Ther. 57.

[49] Tang, Y., Horikawa, I., Ajiro, M., Robles, A.I., Fujita, K., Mondal, A.M. et al. (2013, May). Downregulation of splicing factor SRSF3 induces p53β, an alternatively spliced isoform of p53 that promotes cellular senescence. Oncogene 32(22): 2792–8.

[50] Chen, L., Chen, J.Y., Huang, Y.J., Gu, Y., Qiu, J., Qian, H. et al. (2018, Feb). The augmented R-Loop is a unifying mechanism for myelodysplastic syndromes induced by high-risk splicing factor mutations. Mol. Cell. 69(3): 412–425.e6.

[51] Harper, S.J. and Bates, D.O. (2008, Nov). VEGF-A splicing: The key to anti-angiogenic therapeutics? Nat. Rev. Cancer 8(11): 880–7.

[52] Pritchard-Jones, R.O., Dunn, D.B.A., Qiu, Y., Varey, A.H.R., Orlando, A., Rigby, H. et al. (2007, Jul). Expression of VEGFxxxb, the inhibitory isoforms of VEGF, in malignant melanoma. Br. J. Cancer 97(2): 223–30.

[53] Prochazka, L., Tesarik, R. and Turanek, J. (2014, Oct). Regulation of alternative splicing of CD44 in cancer. Cell Signal. 26(10): 2234–9.

[54] Shieh, J.J., Liu, K.T., Huang, S.W., Chen, Y.J. and Hsieh, T.Y. (2009, Oct). Modification of alternative splicing of Mcl-1 Pre-mRNA using antisense morpholino oligonucleotides induces apoptosis in basal cell carcinoma cells. J. Invest. Dermatol. 129(10): 2497–506.

[55] Martinez-Montiel, N., Rosas-Murrieta, N., Anaya Ruiz, M., Monjaraz-Guzman, E. and Martinez-Contreras, R. (2018, Feb). Alternative splicing as a target for cancer treatment. Int. J. Mol. Sci. 19(2): 545.

[56] Weatheritt, R.J., Sterne-Weiler, T. and Blencowe, B.J. (2016, Dec). The ribosome-engaged landscape of alternative splicing. Nat. Struct. Mol. Biol. 23(12): 1117–23.

[57] Tapial, J., Ha, K.C.H., Sterne-Weiler, T., Gohr, A., Braunschweig, U., Hermoso-Pulido, A. et al. (2017, Oct). An atlas of alternative splicing profiles and functional associations reveals new regulatory programs and genes that simultaneously express multiple major isoforms. Genome Res. 27(10): 1759–68.

6

Alternative Splicing and Metastasis

Bilal A. Naikoo[1,]* *and Muzafar A. Macha*[2]

1. Introduction

One of the prominent hallmarks of cancer is the enhanced ability to migrate from its site of origin to secondary sites inside the body, a complex process called metastasis [1]. A tumor cell becomes deadly when it turns invasive and acquires morphological and functional features that render it free to detach from its surrounding niche to travel through the body to conquer a new target site. Tumor metastasis is marked by epithelial-mesenchymal transition (EMT), a multistep process that promotes the cell state switching from an epithelial state to a mesenchymal state, which can easily undergo migration. EMT is a conserved critical part of metastasis that affects each step of metastasis, including tumor invasion, tumor migration, intravasation/ extravasation, dissemination, and finally angiogenesis to promote tumor cell plasticity [2]. EMT is a critical developmental process responsible for proper embryogenesis [3], responsible for the migration of cells, such as neural crest cells, to new sites that mark the formation of an organ. The migration of undifferentiated cells or stem cells to a secondary site is followed by the reversal of the EMT program to the mesenchymal-epithelial transition (MET) program, resulting in the differentiation of these cells to form a new organ. Similarly, the process of EMT/MET plays a critical role in wound healing and regeneration and therefore helps in maintaining tissue integrity [3]. The processes of EMT and MET are modulated as per the developmental needs for providing a switch between epithelial and mesenchymal states. In tumor cells, EMT is often a reason for the lack of sensitivity toward chemotherapy and anti-apoptotic behavior. These obstacles lead to drug resistance in tumors and increase the likelihood of cancer recurrence [4]. Cancer patients who responded poorly to chemotherapy were found to have more mesenchymal circulating tumor cells

[1] Tata Institute of Fundamental Research, Hyderabad.
[2] Watson Crick Center for Molecular Medicine, IUST Awantipora.
* Corresponding author: bilalnaik100@gmail.com

(CTCs) than epithelial ones in the blood [5]. The process of EMT also imparts stem cell-like features to cancer cells to make cancer stem cells (CSCs) by overexpressing stemness-associated transcription factors (TFs) such as SOX2 and OCT4 [6]. These TFs are, in turn, regulated by EMT-specific TFs, including SNAIL1/2 and ZEB1.

During EMT, the epithelial cell shows a loss of intercellular interactions and apical-basal polarity, which basically renders them physically free to detach from the primary site [7]. Moreover, the epithelial marker proteins like E-cadherin are found to decrease in expression, whereas the mesenchymal marker genes like N-cadherin and vimentin have clearly an increased expression during EMT. A tumor cell undergoes massive cytoskeletal remodeling to become motile and crawl through tissues to enter into the bloodstream and then invade a new site. It is well established that there is a shift in the transcriptomic and proteomic profiles of the cell from epithelial cell-type specific to the mesenchymal state, which subsequently leads to phenotypic changes for achieving an effective tumor spread, colonization, and therapeutic resistance.

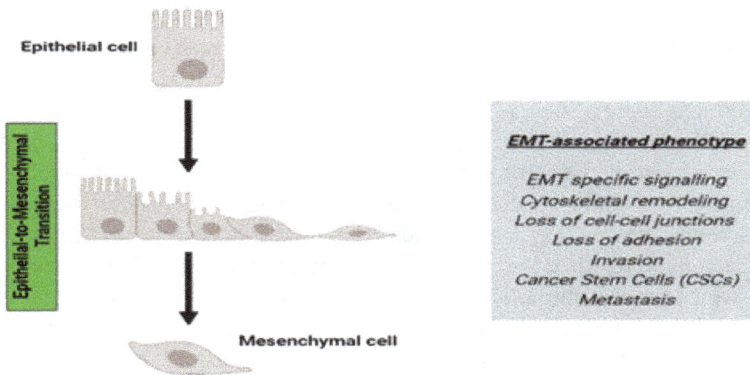

Figure 1: Schema of Epithelial-to-Mesenchymal Transition (EMT).
EMT is a process in which cells acquire features which make them motile. The most visible changes occur with cell morphology, cell-cell interactions and cell adhesion. Intercellular junctions and adhesion are broken down and cells remodel their cytoskeletal network dynamically to be able to migrate in the body. EMT is essential during early development and also for cancer cell metastasis. During disease associated EMT, cancer epithelial cells acquire a mesenchymal cell state associated phenotypes to migrate and invade secondary sites the body.

2. Alternative Splicing and EMT

EMT is a multistep process whose functioning is determined by multiple levels of regulation, from transcriptional to post-translational control of gene expression. The alterations in any of the layers of regulation lead to changes in dynamic processes, such as cell shape, cell adhesion, and intercellular connections. EMT-specific TFs, including ZEB1/2, TWIST1, SNAI1/2, etc., have been studied extensively in this regard. Majority of the research on EMT is focused on studying signaling pathways that determine cellular morphology and intercellular connections. Notch signaling, Wnt/β-catenin, and TGF-β are some of the prominent signaling pathways that regulate EMT/MET processes [8]. There are also other lesser-studied mechanisms of gene regulation, such as epigenetic modifications and chromatin remodeling, which also play an important role in EMT [9]. These regulations are important in generating desired structural and functional variants of structural, functional, and regulatory proteins involved in different steps leading to EMT.

Apart from the transcriptional regulation, EMT is also regulated post-transcriptionally through mRNA splicing and post-translationally by specific modifications [10]. Alternative splicing is one of the critical components of gene regulation that contributes to mRNA and proteome diversity in our cells. The different steps in metastatic progression are positively promoted as a direct result of the reprogramming of EMT-related splicing events. In this way, a cell produces diverse proteins that possess altered structure and function, such as enzymatic activity, binding affinity, solubility, etc. Instead of conventional intron-exon combination, cells process a variant combination of exons and introns from pre-mRNA that ultimately make these EMT-specific protein variants. In cancer, the splicing switch from epithelial to mesenchymal type splicing program promotes tumor malignancy, invasion, and metastasis by altering the transcriptomic and proteomic repertoire of the cell in favor of the mesenchymal state. Following reprogrammed alternative splicing machinery, there is a cell type specific production of protein variants that play an important role in critical events, such as cytoskeletal organization, cell interactions, and cell motility. The dysregulation of alternative splicing ultimately confers adaptability and evasion against different obstacles presented by surroundings against tumor cell motility. The whole process of switching from epithelial state type splicing to mesenchymal state splicing is modulated by altered spliceosome machinery. During EMT, the cells upregulate the levels of RNA-binding proteins (RBPs) and splicing factors, which process the splicing events in favor of mesenchymal-type transcripts. Here we will discuss some of the most important splicing events contributing to EMT progression during development and tumor metastasis.

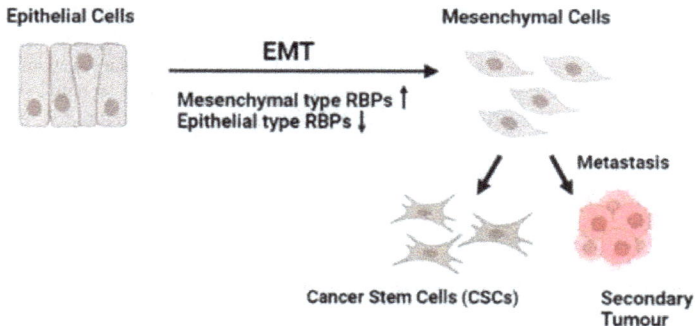

Figure 2: EMT related RBPs promote tumor metastasis.
Epithelial state of cells is governed by numerous splicing events that are promoted by epithelial specific RNA Binding Proteins (RBPs). A primary tumor EMT for generation of mesenchymal cells. The mesenchymal state is promoted when epithelial RBPs are downregulated, whereas mesenchymal specific RBPs are upregulated that process EMT specific isoforms of numerous proteins. The switch in splicing pattern promotes cancer metastasis and generation of cancer stem cells. These cells now possess the potential for recurrence and therapeutic resistance.

3. Alternative Splicing Alters Cell Signaling in Favor of Mesenchymal Cell State

The growth signals mostly initiate and transduce from the cell surface as a result of receptor-ligand interaction. Other times the cell interacts with the extracellular matrix and surrounding niche of cells to determine its morphology and growth. In tumor metastasis, cells produce protein isoforms, such as receptors that promote invasiveness. These isoforms are a direct consequence of the aberrant alternative splicing that results from a lack of tight control over splicing regulation machinery. The EMT-specific alternative splicing of important receptor proteins either renders them constitutively active or heavily enhances the ligand binding affinity of these proteins. As a result of the incessant signal transduction from these EMT-specific isoforms, the tumor cell attains a mesenchymal-type phenotype and becomes motile. Multiple splicing occurrences of alternative splicing have been studied that have been shown to induce metastasis in tumor cells. We discuss some prominent cases of alternative splicing that generate isoforms that promote EMT signaling.

3.1 Recepteur d'Origine Naintas (RON)

RON (Recepteur d'origine naintas) proto-oncogene is a receptor tyrosine kinase (RTK) that gets activated by binding with its ligand macrophage stimulating protein (MSP). RON protein possesses docking sites for many important adapters that transduce signals, which are responsible for controlling cell adhesion, proliferation, and apoptosis. RON is an important player in metastasis; it plays an important role in cell motility and helps invade extracellular matrixes [11]. It drives EMT during embryonic development and is important for the metastasis of epithelial cancers. RON is a 180 kDa heterodimeric transmembrane protein composed of an extracellular a-chain and a transmembrane b-chain belonging to a single precursor. The extracellular a-domain is responsible for ligand binding to transmit growth signals to the cell's internal machinery. The skipping of the variable exon 11 from the mature transcript as a result of alternative splicing produces a tumor-specific protein variant (Figure 3(A)) that remains constitutively active even in the absence of an MSP ligand [12]. This constitutively active isoform, ΔRON, is upregulated in cancer cells and during development. The upregulation of ΔRON induces an EMT-specific morphology and promotes cell motility and invasion. The mechanistic details underlying the production of ΔRON reveal that it is controlled by an enhancer regulator element on exon 12. In a tumor, the splicing factor SF2/ASF gets upregulated and binds the enhancer element in the constitutive exon 12, thereby inducing the exclusion of exon 11. The resulting ΔRON isoform promotes oncogenic signals and confers motility to these cells [13]. Targeting these signals may provide another option for intervening with novel anti-cancer therapeutic strategies. There have been attempts already for stopping the oncogenic signals by targeting the oncogenic variant isoform with antibody [14]. However, the activation mechanism independent of the MSP ligand can still make it difficult to stop the tumor cells from eliciting incessant transformative growth signals. People have also targeted SF2/ASF splicing factors by providing inhibitors against them that were shown to affect the invasive phenotype of the cells [15].

3.2 CD44

CD44 receptor is a complex transmembrane glycoprotein that plays an important role in numerous signaling pathways. It is a multifunctional protein that is involved in cell-cell interactions, cell adhesion, and migration. It undergoes an extensive post-translational modification (PTM), such as glycosylation, because of which it shows an extensive size heterogeneity. Post-translational modifications of CD44 affect its interaction with the cytoskeleton and therefore determine its altered functioning across its variants. Besides PTMs, CD44 undergoes extensive alternative splicing that leads to changes in its extracellular stem region. The stem region is located in the N-terminal globular region of the protein Figure 3(B). CD44 is present on chromosome 11, containing 19 exons (20 in the case of mice). Alternative splicing gives rise to two major isoforms of CD44 that reflect a cell's potential to behave as an epithelial or a mesenchymal cell. The two isoforms are the standard isoform CD44s (or CD44h) and the variant isoform CD44v.

Generally speaking, a CD44 transcript contains two variable regions; one is located on the extracellular side, and the other is present in the cytoplasmic tail region, of which the former is implicated more in EMT-related alternative splicing. The standard isoform (CD44v) is formed when all the possible variable exons are excluded following the splicing process. CD44s isoform is therefore devoid of any variable exons; in contrast, the variable form (CD44v) retains at least one of the variable exons [6–15] in the mature transcript, apart from standard exons (1–5 and 16–20). These isoforms differ in ligand interaction affinity and alter cell state phenotype by directly influencing processes, such as signal transduction, interaction with extracellular matrix (ECM), and interaction with CSCs niche. The variable isoform of CD44 is normally present on the epithelial cells and is associated with a normal proliferation, differentiation, and adhesion of cells. The standard isoform, on the other side, is mesenchymal and is responsible for inducing EMT and leading to metastasis in cancer cells [16]. CD44s is a crucial player in the formation of CSCs where its levels are found elevated [17]. CD44s binds its ligand hyaluronic acid (HA) and leads to uncontrolled proliferation by augmenting the activation of important growth signaling pathways, including PI3K, Ras-MAPK, etc. The production of CD44s also leads to increased drug resistance through the activation of invasion and EMT. The stemness marker aldehyde dehydrogenase (ALDH1) in the case of breast cancer is elevated as a direct consequence of the upregulation of CD44s.

In lung cancer, CD44 leads to an upregulation of matrix metalloproteinases (MMPs) and promotes invasion [18]. A key splicing factor, ESRP1 promotes the inclusion of variable exons in the mature CD44 transcript by interacting with GU-rich elements in the pre-mRNA. Downregulation of ESRP1 is associated with the induction of metastasis and invasiveness. In this regard, ESRP1 is also seen to be downregulated by EMT-promoting TFs, including ZEB1 and Snail. The absence of ESRP1 is responsible for switching in the alternative splicing of the CD44 transcript.

Likewise, its upregulation of ESRPs prevents the promotion of mesenchymal type phenotype and blocks the induction of EMT by TGF-β, Twist, Snail, and Cadherins. CD44 isoforms are not strict markers of either of the epithelial and mesenchymal cell types, but there is definitely a switch in the specific isoform production during EMT from CD44v to CD44s, which is an important event in tumor progression and metastasis.

3.3 Fibroblast Growth Factors

Fibroblast growth factors (FGFs) are a 23-member family of ligand molecules that play a crucial role in embryogenesis, tissue repair, and cancer. These molecules transduce their signals via the corresponding receptor fibroblast growth factor receptor (FGFR2) belonging to the RTK family of growth receptors. FGF ligand molecules play an important role in several pathological conditions, including cancer. FGFR2, with a total of 20 exons, is predominantly found in two isoforms resulting from alternative splicing. FGFR2 protein contains three immunoglobulin-like domains (Ig-loops I–III) in the extracellular portion, of which Ig-III loop C-terminus is the target for EMT-related alternative splicing. Alternative splicing (Figure 3(C)) leads to the production of an epithelial-specific isoform (FGFR2 IIIb) and a mesenchymal-specific isoform (FGFR2 IIIc). The IIIb isoform is produced as a result of the alternative inclusion of exon 8, whereas the inclusion of exon 9 leads to the formation of the IIIc isoform. This splicing alteration is responsible for altered ligand binding specificities of the two isoforms [19]. The mesenchymal isoform IIIc has a greater affinity for binding the FGF ligand [20].

The strong receptor-ligand binding enhances the downstream signaling that ultimately induces cancer cell aggression [21]. The isoform switch from IIIb to IIIc is considered an important hallmark of tumor progression and EMT. The presence of IIIc isoform is listed among a set of criteria for EMT of several carcinomas where it is regarded as a strong oncogene. However, very little is known about the role of FGFR2 led signaling in inducing pathology in soft tissue sarcomas (STSs). The upregulation of FGFR2 is caused by several factors, such as N-cadherin, which results in EMT progression, stemness, and increased levels of TFs-like Snail and Slug. Apart from TFs like Zeb1/2, the isoform switching in FGFR2 is mediated by splicing factors, including ESRP, hnRNPA1, PTBP1, and RBFOX2, all of which act as endogenous regulators. The epithelial isoform IIIb is supported by the binding of ESRP1 to the intronic segment present between exon 8 (isoform IIIb) and exon 9 (isoform IIIc). Apart from endogenous regulators, the mesenchymal-type alternative splicing event is also regulated by exogenous regulators like TGF-β, Wnt signaling, and the activation of RTKs. FGFR and its splice isoforms produced by alternative splicing govern drug resistance in cancer cells. In esophageal squamous cell carcinoma (ESCC), fibroblast supernatant induced cancer cell proliferation and blocked the action of the drug lapatinib [22]. Based on the understanding of splice events pertaining to FGFR, there are numerous drugs available to target tumor cells. However, the problem of drug resistance is still hindering the prospects of curing cancer disease.

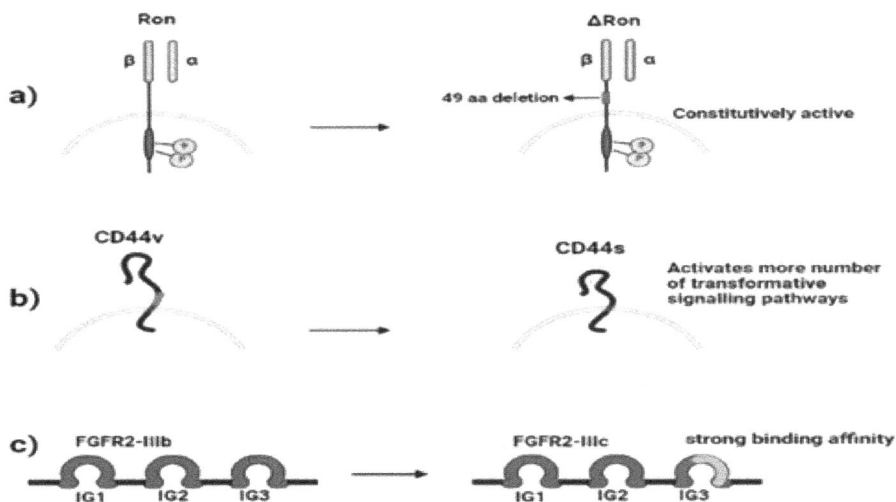

Figure 3: Crucial splicing events that alter cell signaling to promote EMT. (A) Isoform switching of FGFR2 protein from III-b to II-b results in structural changes at immunoglobin-like domain III (IG-III), which enhances ligand binding. (B) CD44 mesenchymal isoform CD44s has a shorter extracellular stem than its epithelial isoform CD44v. This switch to differential activation of signaling pathways from CD44 protein. (C) Alternative splicing of RON forms a mesenchymal isoform ΔRON that lacks a 49 amino acid stretch in the extracellular portion, which makes it a constitutively active protein isoform.

Table 1: Showing list of splice variants involved in EMT.

Alternatively Spliced Gene	EMT-specific Isoform Type	Outcome in the Splicing Event	Role in Cancer Progression and Metastasis
RON	ΔRON	Exclusion of exon 11, mediated by ASF/SF2	ΔRON is a constitutively active isoform that transmits constant signals leading to changes in tumor cell morphology and interaction with ECM. It also plays an important role during embryogenesis by facilitating migration and invasion of cells.
CD44	CD44s	Exclusion of all variable exons	The particular isoform differs in ligand binding affinity compared to the epithelial type isoform. It binds with HA and promotes cancer sternness, EMT, chemoresistance, and inhibition of apoptosis. CD44 facilitates invasion and migration by activating STAT3, PI3K, and Ras-MAPK pathways.
FGFR2	FGFR2 IIIc	Inclusion of exon 9	FGFR IIIs has a greater affinity for binding its ligand FGF to activate signaling cascades within the cell. It is found to be upregulated in colorectal cancer tissues and correlates with increased invasiveness. It is listed as a criterion for EMT of several carcinomas.

4. Alternative Splicing Leads to Cytoskeletal Remodeling and Breakdown of Cell Adhesion

In the process of EMT, cells undergo massive morphological reprogramming through modulation of splicing events associated with cell shape and cell motility dynamics. During the process of EMT, there is a disruption in junctions that interconnect cells, reorganization of the cytoskeletal network, and formation of invadopodia. Aberrant alternative splicing is one the ways to make sure the cells produce protein isoforms that contribute to establishing the mesenchymal type cell morphology that ultimately results in the cobblestone-shaped cells which are migratory. The associated proteins may be part of the structural framework of the cells or act as regulatory factors in deciding the cell-cell interactions, cell adhesion, cell shape, and formation of an invasive front. Destabilization of the splicing machinery in tumor cells produces specific isoforms of these proteins that lead to cell proliferation, cell motility, and evasion of apoptosis. Multiple cases of differential alternative splicing have been studied that are shown to affect these characteristics.

Rac1 is a Rho GTPase responsible for actin cytoskeleton organization in the cell. The Rho GTPases RhoA, Rac1, and molecular switch cdc42 regulate cell migration as well as adhesion and cell-cell junctions by organizing and reorganizing the actin cytoskeleton. In addition to its role in governing the cytoskeleton, Rac1 plays a crucial role in regulating gene expression in favor of cell cycle progression and division. Rac1 activates signal transduction pathways, including JNK and MAPK, that upregulate the levels of the cell cycle regulators leading to malignancy in these cells. Like other GTPase molecules, Rac1 switches its GDP and GTP-bound forms by interacting with GAPs and GEFs. The inclusion of a variable exon 3b in the final transcript of this protein results in the 19 amino acid insertion in its GDP/GTP binding domain which transforms the protein into a constitutively active GTP bound form [23]. The constitutively active isoform of Rac1b, by activating NF-kB and Akt signaling, promotes cell growth and division and enhances the anti-apoptotic response in cells. This tumor-associated mesenchymal isoform, Rac1b, causes growth transformation of the cells. The cells that express the Rac1b isoform exhibit typical cancer cell characteristics, such as foci formation and loss of anchorage-dependent and density-dependent growths. Rac1b is also responsible for developing resistance to anti-cancer therapies and preventing apoptosis in cancer cells. The splicing regulator ESRP1 is known to negatively regulate the production of EMT-specific Rac1b isoform, which is why this splicing factor is absent in highly metastatic cell lines. Moreover, Rac1b leads to increased ROS production from mitochondria which results in the activation of SNAI1. The expression of SNAI1 promotes EMT in the cells that express the Rac1b isoform [24].

MENA (ENAH) is another regulatory protein that is involved in the actin regulation and formation of invadopodia. In the context of metastasis, MENA is found in two isoforms; an epithelial isoform hMENA[11a] and a mesenchymal isoform hMENAΔv6. The epithelial isoform hMENA[11a] reduces the amount of filopodia,

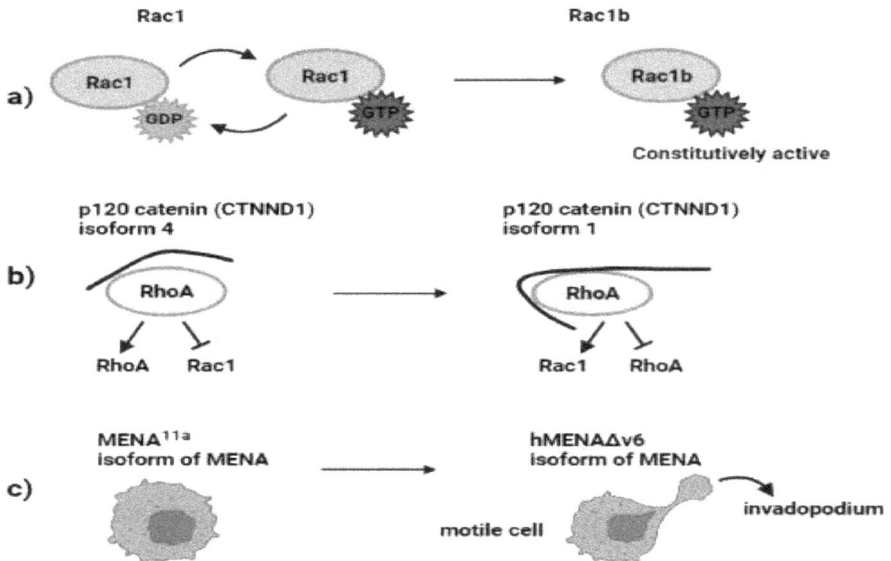

Figure 4: Critical alternative splicing events that promote EMT by altering morphology. (A) Alternative splicing of Rac1 results in the formation of Rac1b, which promotes EMT. It is a constitutively active GTP-bound form that leads to the generation of excess ROS from mitochondria, thereby activating SNAI1 expression. (B) Alternative splicing of p120 catenin (CTNND1) produces the mesenchymal type isoform 1, which is the longest of all other isoforms. It retains regulatory and coiled-coil domains, both of which are absent in epithelial isoform 4. The mesenchymal isoform activates Rac1 and inhibits RhoA activation. (C) EMT-specific alternative splicing of MENA leads to formation of the mesenchymal isoform hMENAΔv6, which enhances the formation of invadopodia and invasive front on tumor cells.

thereby decreasing cell migration. In contrast, the expression of hMENAΔv6 promotes invasiveness in cancer cells [25]. MENA acts by antagonizing actin-capping proteins in the cell. Alternative splicing of protein regulators associated with cytoskeleton and cell adhesion is crucial for determining the integrations of a cell with other cells and with the ECM.

Similarly, p-120 catenin (CTNND1), being a regulator of adherens junctions, plays an important role in tumor metastasis. This protein normally acts by binding the E-Cadherin from the cytosolic side, thereby stabilizing them on adherens junctions. Additional functions of CTNND1 include transcriptional regulation via Wnt signaling. The mesenchymal type isoforms of CTNND1 contain exons 2 and 3, which are lacking in epithelial isoforms. The full-length isoform 1, which is the mesenchymal type isoform of p-120 catenin, is responsible for inhibiting Rho GTPase activity and consequently leading to invasiveness in cells and tumors [26]. Moreover, adherens junctions are not stable on cells that express this isoform. The epithelial-specific isoforms 3 and 4 prevent invasiveness by enhanced binding to E-Cadherin and stabilizing it at adherens junctions.

Table 2: Showing list of splice variants involved in cytoskeleton reorganization.

Alternatively Spliced Gene	EMT-Specific Isoform Type	Outcome in the Splicing Event	Role in Cancer Progression and Metastasis
Rac1	Rac1b	Inclusion of exon 3b	The Rac1b isoform is a constitutively active GTP-bound form. It activates NFkB and Akt pathways and blocks apoptosis. Rac1b expressing cells exhibit typical cancer cell characteristics such as foci formation and loss of anchorage-dependent and density-dependent growths. It elevates ROS production from mitochondria to induce apoptosis.
MENA (ENAH)	hMENAΔv6	Exclusion of internal exon 6	hMENAΔv6 isoform promotes mesenchymal-specific cytoskeletal changes. It promotes the formation of invasive front on tumor cells by increasing the formation of invadopodia. Cancer cell migration is associated with elevated levels of hMENAΔv6.
CTNND1 (p-120 catenin)	isoform 1	Inclusion of exons 2 and 3	The full-length isoform 1 of p-120 catenin is responsible for inhibiting Rho GTPase activity, consequently leading to invasiveness in cells and tumors. The expression of this isoform is linked with a lack of E-cadherin stabilization on adherens junctions, thereby facilitating cell motility.

5. Regulation of EMT-Specific Alternative Splicing

Alternative splicing is controlled by trans-acting splicing factors that bind the splice sites at the exon-intron boundaries. The splice sites are prioritized and recognized by spliceosome machinery based on the degree of strength or weakness in their sequence. Stronger splices sites are more frequently recognized, leading to exon inclusion, whereas the weaker sites are more dependent on the cis-elements that facilitate the recognition of these sites by the spliceosome. The cis-acting elements that are found inside the introns and exons of transcript act as silencers or enhancers and are bound by RBPs, which determine the inclusion and exclusion of exons and introns. During EMT, these RBPs are dynamically regulated for tight control over the splicing patterns that produce protein isoforms essential for promoting a mesenchymal phenotype, including enhanced signaling of growth and division, disruption of cell-cell junctions, breakdown of adhesion, changes in cytoskeleton, and prevention of apoptosis. Broadly speaking, epithelial-type specific and mesenchymal-type specific RBPs regulate EMT-related alternative splicing in cells.

SRPs are the most discussed epithelial-specific RBPs that are indispensable for EMT during early embryogenesis. In mice, knockout of ESRP1 led to neonatal lethality [27]. It is worth noting that epithelial-type RBPs have not yet been studied extensively in the context of development-related EMT. In cancer cells, the downregulation of ESRPs is a crucial event in the progression of EMT and invasiveness. Its re-expression in mesenchymal cells can partially restore the epithelial alternative splicing program. ESRP1 is critical for ensuring EMT by regulating the mesenchymal type alternative splicing of proteins such as p-120 catenin, CD44, MENA, etc. Other important epithelial types of RBPs AKAP8 and RBM47 have been identified whose downregulation is associated with tumor metastasis, particularly in breast cancer. AKAP8 acts by preventing the EMT and promoting alternative splicing of important transcripts, such as CD44. Likewise, mesenchymal type RBPs RBFOX1/2, hnRNPM, etc., are crucial for producing EMT-inducing protein variants. The general trend is that contrary to epithelial-type RBPs, the upregulation of mesenchymal-type RBPs leads to EMT-specific alternative splicing. Another RBP hnRNPM leads to the production of CD44s isoform by inhibiting the variable exon inclusion.

Recent studies suggest that RBPs have the ability to regulate splicing events in both cell states, i.e., epithelial and mesenchymal. They act in coordination and coregulate splicing events during EMT. RFFOX2 is majorly known for regulating mesenchymal-type splicing patterns but can also induce some epithelial splicing events too [28]. These RBPs act in a combinatorial fashion to draw an outcome to follow a particular splicing type. Competitive or cooperative regulation of RBPs on the EMT-associated splicing events determines the type of protein isoform that will be the final outcome. Therefore, one RBP can reflect in many regulation networks connected to other RBPs responsible for the production of different protein isoforms. In this way, the presence of a single RBP may not necessarily reflect only a particular splicing event; it is a cumulative balance of different RBPs that seems to establish a particular cell state.

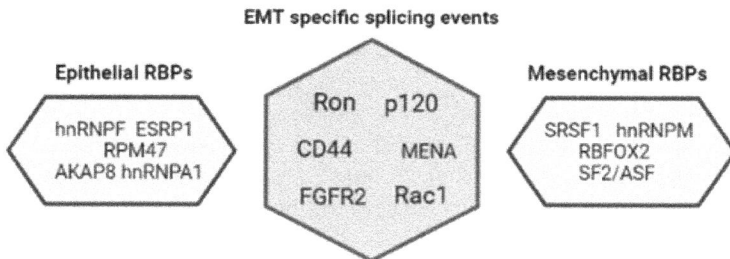

Figure 5: Critical RBPS that govern EMT.

EMT-related splicing events are regulated by RNA Binding Proteins (RBPs) that either promote an epithelial state or a mesenchymal state of the cell. These RBPs play a critical role in determining EMT during development and cancer metastasis. The RBPs bind cis-regulatory elements on pre-mRNAs to promote or block the inclusion of specific introns and exons that are crucial for determining the structural and functional properties of a protein isoform.

6. Therapeutic Strategies Targeting EMT-Related Splicing Events

The discovery of EMT-specific protein isoform expression as a result of alternative splicing has opened doors to novel strategies for treating cancer. The EMT-specific transcript isoforms are expressed in tumor cells and during development. It has also been seen that the metastatic-specific isoforms can also be expressed in a tumor-type specific manner. Some protein isoforms may only be expressed in certain tumors and not others. The pharmacological prospects of targeting these aberrant disease-causing splicing events and their corresponding products have led to discovery of numerous drugs that decrease invasiveness in cancerous cells. Some of the recent approaches to target EMT-specific splicing events include the use of Antisense-Oligos (ASOs) that target RNAs, inhibitor molecules targeting EMT-specific splicing regulators and isoforms, and treatment with novel antibodies against [29]. The use ASOs, which target coding and non-coding RNAs, is one of the prominent strategies to prevent the splicing events that produce mesenchymal isoforms. In melanoma cells, ASO resulted in the exclusion of exon 6 of MDM4, reduced its expression level, enhanced drug sensitivity, and induced programmed cell death. In mice transfected with melanoma cells, CD44-HA interaction gets blocked after administering soluble ectodomain of the CD44 protein that binds HA, blocking its binding to a CD44 protein present on the surface of cancer cells and preventing the subsequent activation. This interference in CD44-HA binding resulted in a visible impairment of tumor growth and metastasis. Numerous inhibitors, collectively called tyrosine kinase inhibitors (TKIs), have been tested against FGFR to prevent tumor cell proliferation and angiogenesis. One good example of such inhibitor molecules is PKC412, which is used to treat hematopoietic and solid tumors. The splicing factors SF2/ASF responsible for splicing events that produce ΔRON have also been targeted by such inhibitors, and the results showed decreased invasiveness [14]. Moreover, numerous antibodies targeting EMT-promoting proteins and splicing factors have shown promising effects in decreasing the invasive phenotype in cancer cells from different various tissue origins. The discovery of EMT-inducing alternative splicing and splice isoforms of several genes provides an exciting opportunity for designing new therapeutic strategies in the future as well.

Conclusion

EMT is an essential step during embryonic development, wound healing, and regeneration to maintain tissue integrity. In tumor cells, dysregulation of EMT leads to the promotion of invasion, metastasis, and therapeutic resistance. EMT is marked by a phenotypic switch from epithelial cell state to mesenchymal cell state, thereby conferring cell motility. Alternative splicing is one of the important modes for contributing to the transcriptomic and proteomic diversity inside our cells. Alternative splicing produces mesenchymal-specific isoforms of important proteins and regulators that function in the maintenance of cell morphology, intercellular interactions, and cell adhesion. During EMT, critical proteins such as RON, CD44, and FGFR undergo alternative splicing to produce their mesenchymal isoforms which are implicated in aberrant cell signaling alteration because of alterations in ligand binding affinity. Moreover, cells produce mesenchymal isoforms of morphology-associated proteins like Rac1, p120 catenin (CTNND1), and MENA, which are considered crucial events in cancer progression and metastasis.

We have outlined how different protein isoforms modulate the metastatic potential and lead to cancer aggressiveness by affecting important aspects, such as cell signaling and cytoskeletal remodeling. The mesenchymal isoforms of the proteins contain a different intron-exon combination than their epithelial counterparts, which leads to altered structure and function of these isoforms. RBPs are essential trans-acting regulators of alternative splicing that determine the epithelial and mesenchymal-specific splicing events. The regulation of alternative splicing by RBPs is combinatorial, which means that different RBP networks interact to determine the global alternative splicing. EMT-associated splicing events, their products, and regulators have been used as therapeutic targets for developing new therapeutics. The use of ASOs, ligand-receptor interaction inhibitors and antibodies have been extensively used as anti-cancer strategies and found to affect metastasis to a great deal.

To sum up, alternative splicing leads to isoform switching in tumor cells, which govern invasiveness and metastasis. The EMT-associated splice isoforms and their regulators provide a useful option for designing novel therapies against cancer metastasis. However, we still lack a clear understanding of EMT-associated alternative splicing and its regulation, though many critical splicing events have been identified. There is still obscurity related to the implications of structural alterations in EMT-associated protein isoforms.

Funding

This study was supported by Ramalingaswami Fellowship (Grant number: DO NO.BT/HRD/35/02/2006) from the Department of Biotechnology, & Core Research Grant (CRG/2021/003805) from the Science and Engineering Research Board (SERB), Govt. of India, New Delhi to Dr. Muzafar A. Macha.

References

[1] Hanahan, D. and Weinberg, R.A. (2011). Hallmarks of cancer: The next generation. Cell 144: 646–674.

[2] Yeung, K.T. and Yang, J. (2017). Epithelial mesenchymal transition in tumor metastasis. Mol. Oncol. 11: 28–39.

[3] Kim, D.H., Xing, T., Yang, Z., Dudek, R., Lu, Q. and Chen, Y.H. (2017). Epithelial mesenchymal transition in embryonic development, tissue repair and cancer: A comprehensive overview. J. Clin. Med. 7(1): 1.

[4] Chaffer, C.L., San Juan, B.P., Lim, E. and Weinberg, R.A. (2016). EMT, cell plasticity and metastasis. Cancer Metastasis Rev. 35(4): 645–54.

[5] Aktas, B., Tewes, M., Fehm, T., Hauch, S., Kimmig, R. and Kasimir-Bauer, S. (2009). Stem cell and epithelial mesenchymal transition markers are frequently overexpressed in the circulating tumor cells of metastatic breast cancer patients. Breast Cancer Res. 11(4): R46.

[6] Zhou, P., Li, B., Liu, F., Zhang, M., Wang, Q., Liu, Y., Yao, Y. and Li, D. (2017). The epithelial to mesenchymal transition (or EMT) and cancer stem cells: Implication for treatment resistance in pancreatic cancer. Mol Cancer. 16(1): 52.

[7] Kalluri, R. and Weinberg, R.A. (2009). The basics of epithelial-mesenchymal transition. J. Clin. Invest. 119(6): 1420–8. doi: 10.1172/JCI39104. Erratum in: J. Clin. Invest. 2010 May 3; 120(5): 1786.

[8] Lamouille, S., Xu, J. and Derynck, R. (2014). Molecular mechanisms of epithelial-mesenchymal transition. Nature Rev. Mol. Cell Biol. 15(3): 178–96.

[9] Lee, J.Y. and Kong, G. (2016). Roles and epigenetic regulation of epithelial-mesenchymal transition and its transcription factors in cancer initiation and progression. Cell Mol. Life Sci. 73(24): 4643-4660.

[10] Pradella, D., Naro, C., Sette, C. and Ghigna, C. (2017). EMT and stemness: Flexible processes tuned by alternative splicing in development and cancer progression. Mol. Cancer 16(1): 8.

[11] Camp, E.R., Liu, W., Fan, F., Yang, A., Somcio, R. and Ellis, L.M. (2005). RON, a tyrosine kinase receptor involved in tumor progression and metastasis. Ann. Surg. Oncol. 12(4): 273–81.

[12] Ghigna, C., Giordano, S., Shen, H., Benvenuto, F., Castiglioni, F., Comoglio, P.M., Green, M.R., Riva, S. and Biamonti, G. (2005). Cell motility is controlled by SF2/ASF through alternative splicing of the RON protooncogene. Mol. Cell 20: 881–90.

[13] Zhou, Y.Q., He, C., Chen, Y.Q., Wang, D. and Wang, M.H. (2003). Altered expression of RON receptor tyrosine kinase in primary human colorectal adenocarcinomas: Generation of different splicing RON variants and their oncogenic potential. 22: 186–97.

[14] O'Toole, J.M., Rabenau, K.E., Burns, K., Lu, D., Mangalampalli, V., Balderes, P., Covino, N., Bassi, R., Prewett, M., Gottfredsen, K.J., Thobe, M.N., Cheng, Y., Li, Y., Hicklin, D.J., Zhu, Z., Waltz, S.E., Hayman, M.J., Ludwig, D.L. and Pereira, D.S. (2006). Therapeutic implications of a human neutralizing antibody to the macrophage-stimulating protein receptor tyrosine kinase (RON), a c-MET family member. Cancer Res. 66(18): 9162–70.

[15] Ghigna, C., De Toledo, M., Bonomi, S., Valacca, C., Gallo, S., Apicella, M., Eperon, I., Tazi, J. and Biamonti, G. (2010). Pro-metastatic splicing of Ron proto-oncogene mRNA can be reversed: Therapeutic potential of bifunctional oligonucleotides and indole derivatives. RNA Biology 7:4, 495–503.

[16] Brown, R.L., Reinke, L.M., Damerow, M.S., Perez, D., Chodosh, L.A., Yang, J. and Cheng, C. (2011). CD44 slice isoform switching in human and mouse epithelium is essential for epithelial-mesenchymal transition and breast cancer progression. J. Clin. Invest. 121(3): 1064–74.

[17] Biddle, A., Gammon, L., Fazil, B. and Mackenzie, I.C. (2013). CD44 staining of cancer stem-like cells is influenced by downregulation of CD44 variant isoforms and upregulation of the standard CD44 isoform in the population of cells that have undergone epithelial-to-mesenchymal transition. PLoS One 8: e57314.

[18] Li, L., Qi, L., Qu, T., Liu, C., Cao, L., Huang, Q., Song, W., Yang, L., Qi, H., Wang, Y., Gao, B., Guo, Y., Sun, B., Meng, B., Zhang, B. and Cao, W. (2018). Epithelial splicing regulatory protein 1 inhibits the invasion and metastasis of lung adenocarcinoma. Am. J. Pathol. 188: 1882–1894.

[19] Yeh, B.K., Igarashi, M., Eliseenkova, A.V., Plotnikov, A.N., Sher, I., Ron, D., Aaronson, S.A. and Mohammadi, M. (2003). Structural basis by which alternatives splicing confers specificity in fibroblast growth factor receptors. Proc. Natl. Acad. Sci. USA 100(5): 2266–71.

[20] Shirakihara, T., Horiguchi, K., Miyazawa, K., Ehata, S., Shibata, T., Morita, I., Miyazono, K. and Saitoh, M. (2011). TGF-b regulates isoform switching of FGF receptors and epithelial-mesenchymal transition. EMBO J. 30(4): 783–95.

[21] Zhang, X., Ibrahimi, O.A., Olsen, S.K., Umemori, H., Mohammadi, M. and Ornitz, D.M. (2006). Receptor specificity of the fibroblast growth factor family: The complete mammalian FGF family. J. Biol. Chem. 281(23): 15694–700.

[22] Saito, S., Morishima, K., Ui, T., Hoshino, H., Matsubara, D., Ishikawa, S., Aburatani, H., Fukayama, M., Hosoya, Y., Sata, N., Lefor, A.K., Yasuda, Y. and Niki, T. (2015). The role of HGF/MET and FGF/FGFR in fibroblast derived growth stimulation and lapatinib-resistance of esophageal squamous cell carcinoma. BMC Cancer 15: 82.

[23] Singh, A., Karnoub, A.E., Palmby, T.R., Lengyel, E., Sondek, J. and Der, C.J. (2004). Rac1b, a tumor associated constitutively active Rac1 splice variant promotes cellular transformation. Oncogene 23: 9369–80.

[24] Radisky, D.C., Levy, D.D., Littlepage, L.E., Liu, H., Nelson, C.M., Fata, J.E., Leake, D., Godden, E.L., Albertson, D.G., Nieto, M.A., Werb, Z. and Bissell, M.J. (2005). Rac1b and reactive oxygen species mediate MMP-3 induced EMT and genomic instability. Nature 436(7047): 123–7.

[25] Di Modugno, F., Iapicca, P., Boudreau, A., Mottolese, M., Terrenato, I., Perracchio, L., Carstens, R.P., Santoni, A., Bissell, M.J. and Nisticò, P. (2012). Splicing program of human MENA produces previously undescribed isoform associated with invasive, mesenchymal-like breast tumors. PNAS 109(47): 19280–5.

[26] Yanagisawa, M., Huveldt, D., Kreinest, P., Lohse, C.M., Cheville, J.C., Parker, A.S., Copland, J.A. and Anastasiadis, P.Z. (2008). A p120 catenin isoform switch affects Rho activity, induces tumor cell invasion, and predicts metastatic disease. J. Biol. Chem. 283(26): 18344–54.

[27] Bebee, T.W., Park, J.W., Sheridan, K.I., Warzecha, C.C., Cieply, B.W., Rohacek, A.M., Xing, Y., Carstens and R.P. (2015). The splicing regulators Esrp1 and Esrp2 direct an epithelial splicing program essential for mammalian development. eLife 4: e08954.

[28] Braeutigam, C., Rago, L., Rolke, A., Waldmeier, L., Christofori, G. and Winter, J. (2014). The RNA-binding protein Rbfox2: An essential regulator of EMT driven alternative splicing and a mediator of cellular invasion. Oncogene 33(9): 1082–92.

[29] Bielli, P., Pagliarini, V., Pieraccioli, M., Caggiano, C. and Sette, C. (2019). Splicing dysregulation as oncogenic driver and passenger factor in brain tumors. Cells 9(1): 10.

7

Role of Splice Variant of Pyruvate Kinase M2 (PKM2) in Cancer Metabolism

Mohd Amir,[1] *Ata Abbas,*[2] *Saleem Javed,*[1,]* *Saleh Alfuraih*[3] *and Rais Ansari*[3,]*

1. Introduction

1.1 Cancer and Pyruvate Kinase Isoform Expression

Metabolic reprogramming is a hallmark of cancer cells. Cancer cells are different from normal cells. Cancer cells rely on glycolysis in the presence of oxygen (aerobic glycolysis) to produce energy and provide biomolecules for the synthetic process [1, 2]. Preferential utilization of aerobic glycolysis was discovered by Warburg as the Warburg effect. By using aerobic glycolysis at a higher rate, cancer cells produce increased lactate and energy [3]. Pyruvate kinase is a rate-limiting enzyme that catalyzes the conversion of phosphoenolpyruvate (PEP) to pyruvate and ATP in the last phase of glycolysis [4]. Mammalian pyruvate kinase (PK) has four isoforms: PKM1, PKM2, PKR, and PKL [3, 5, 6].

Cells usually express one isoform of PK at higher levels, while tissues may have a complex expression pattern—i.e., expression of more than one isoform. The isoforms of PK and its expression are summarized in Table 1. PKM2 is expressed in most adult tissues, while other forms are restricted to specialized cells, such as

[1] Department of Biochemistry, Faculty of Life Sciences, Aligarh Muslim University, Aligarh-202002 (UP), India.
[2] Division of Hematology and Oncology, Department of Medicine, Case Western Reserve University, Cleveland, OH 44106, USA.
[3] Department of Pharmaceutical Sciences, College of Pharmacy, Health Professions Division, Nova Southeastern University, 3200 S University Drive, Fort Lauderdale, FL 33200, USA.
* Corresponding authors: saleemjaved70@gmail.com; ra557@nova.edu

Table 1: The isoforms of PK and their expression.

Isoform of PK	Expression	Reference
PKR	Red blood cells	[9]
PKL	Liver (major), kidney (minor), and intestine	[10]
PKM1	Tissues with a high demand for energy, such as skeletal, brain, and heart	[11]
PKM2	Cancer cells and proliferating cells, such as lymphocytes and intestinal epithelial cells	[12]

red blood cells (PKR), liver, and kidney (PKL). PKM1 is expressed in tissue with a demand for higher energy, such as skeletal muscles, brain tissue, and heart tissue [4, 7, 8]. PKM2 is found in cells that proliferate quickly, particularly cancer cells, and is an embryonic isoform. PKM2 is restricted to expression in cancer cells but is found in normal proliferating cells, such as lymphocyte, lung, adipose tissues, and intestinal epithelial cells [8].

PKM2 is allosterically modulated between its dimeric and tetrameric forms, unlike the other PKM isoforms, which exist as stable tetramers. The molecular flexibility of PKM2 permits cancer cells to retain their limitless replicative potential by switching between tetramer and dimer forms to maintain a constant flow of ATP and biomolecules for synthesis, which is needed for growth. The PKM2 tetramer is catalytically active, while the dimeric form has been shown to restrict itself for pyruvate generation and enhance the accumulation of metabolic pathway precursor biomolecules. Alternative splicing of an RNA transcript of the PKM gene, including sequences encoded by exons 9 and 10 for PKM1 and PKM2, respectively, produces the muscle isoforms of PKM1 and PKM2 [13]. A 56-amino acid region is encoded via alternative splicing of mutually exclusive exons in the PKM gene, which varies by 22 residues between PKM1 and PKM2. In terms of catalysis, both PKM isoforms are equivalent. Unlike PKM1, which is able to produce an active tetramer on its own because of its 22 amino acid changes, PKM2 requires the allosteric binding of fructose 1,6-bis phosphate (FBP) in order to form an active tetramer [14]. PKM2 expression is influenced by a number of variables, including heterogeneous ribonucleoproteins (hnRNPA1 and hnRNPA2) and polypyrimidine-tract binding protein (PTB), bind to the PKM gene and initiate alternate splicing of the transcript, resulting in the exclusion of exon 9 and simultaneous inclusion of exon 10 [15]. In addition, the PKM2 dimer/tetramer ratio is dependent on the availability of substrate regulators that allosterically stimulate or inhibit PKM2. Succinyl 5'-aminoimidazole-4-carboxamide-1-ribonucleotide-5'-phosphate (SAICAR), fructose-1,6-bisphosphate (FBP), and serine are known activators of the PKM2 tetramer [6, 16]. On the contrary, metabolites, such as phenylalanine and oxalate, deactivate PKM2 as depicted in Figure 1 [3, 17–19].

Mammalian PK isoforms possess a similar affinity with ADP; however, they differ in kinetics by allosteric regulation by PEP [7, 14, 20–25]. While ADP binds non-cooperatively with high affinity, PEP binding is affected cooperatively as part of allosteric regulation; however, PEP binding can be increased by FBP also

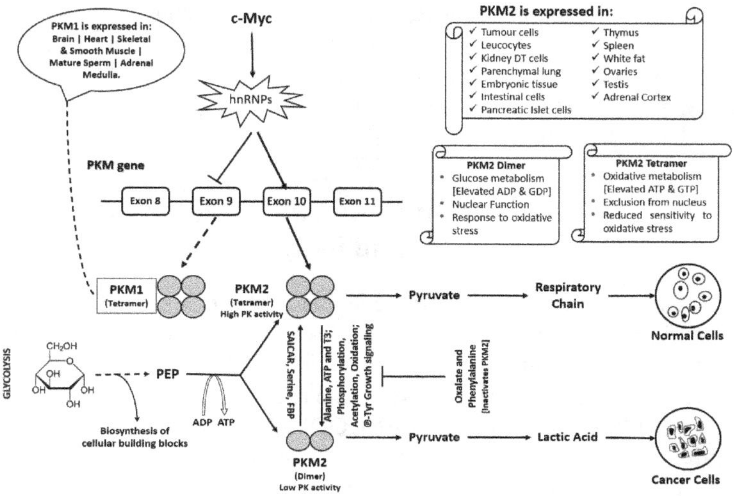

Figure 1: A schematic depiction illustrating glucose metabolism in healthy and malignant cells, including alternative splicing of the PKM gene. By activating hnRNP proteins, MYC expression in cancer cells promotes alternative splicing of mutually exclusive exons 9 and 10 of the PKM gene that encodes PKM1 and PKM2 isoforms, both of which have different tissue expression patterns. To enhance PKM2 expression, different hnRNPs suppress exon nine while activating exon 10.

[26, 27]. FBP also stabilizes the tetrameric active form of PKM2, PKL, and PKR [26, 27]. Due to structural differences, PKM1 is unable to bind FBP, hence no further activation by FB; however, PKM1 isoform is tetrameric and constitutively active [25]. Phenylalanine reduces the affinity of PKM1 and PKM2 to PEP [28, 29]. The phenylalanine, FBP, and active binding sites are different [30]. The allosteric activator, FBP, can partially activate the phenylalanine-bound PKM2 [31, 32]. The allosteric regulation of PK isozymes is summarized in Table 2.

Table 2: The allosteric regulation of PK.

Allosteric Regulator	PK Isozyme
PEP (Positive)	PKM2, PKR, and PKL
FBP (Positive)	PKM2, PKR, and PKL
SAICAR (Positive)	PKM2, PKR, and PKL
Serine (Positive)	PKM2
Phenylalanine (Negative)	PKM1 and PKM2

2. Structure and Mechanism of Action of PKM2

PKM2 is made up of 531 amino acids. A fully functioning PKM2 protein is made up of four monomeric subunits, each of which has three domains, namely domains A, B, and C. There is an active site between domains A and B where two substrates, PEP and ADP, along with K^+ and Mg^{2+}, bind [33]. In a two-step process, PKM2 transfers the phosphate group from PEP to ADP and generates pyruvate and ATP. The phosphate group is transferred from PEP to ADP in the first phase, resulting in the formation of an enolate and ATP. The proton is added to enolate in the second step, resulting in the formation of pyruvate. Tyr 105, Lys 305, and Cys 358 are found in the A-domain of the PKM2 monomer, while the C-domain contains the FBP binding site. Phosphorylation, acetylation, and oxidation of Tyr 105, Lys 305, and Cys 358, respectively, during post-translational modification; on the other hand, it may prevent FBP from binding, resulting in PKM2 remaining dimeric [34].

3. Genomic Organization and Regulation of Mammalian PK and its Link to Cancer

Human pyruvate kinase M (PKM) gene was first characterized by Takenaka et al. in 1991 [35]. As shown in Figure 2, the PKM gene is present on the long arm of chromosome 15 and is about 32 kb long (hg38: chr15:72,199,029 to chr15:72,231,190). PKM gene transcribes from the negative strand and may have multiple transcriptions start sites (TSS). ChIP-seq data from ENCODE (Encyclopedia of DNA Elements) shows enrichments of histone H3K4me1, H3K4me3, and H3K27ac around TSS along with DNaseI hypersensitivity and transcription factor clusters around TSS indicate tight regulation of the PKM gene at both epigenetic and transcriptional levels (Figure 2). A CpG island is also around TSS, covering the proximal promoter region (hg38: chr15:72229791 to chr15:72231897). It is interesting to see the enrichment of H3K4me1 and H3K27ac marks around exon 9 and 10 in UCSC browser tracks (Figure 2), which undergo alternative splicing to generate PKM1 and PKM2. Various histone marks, as splicing-associated chromatin signatures, have been reported

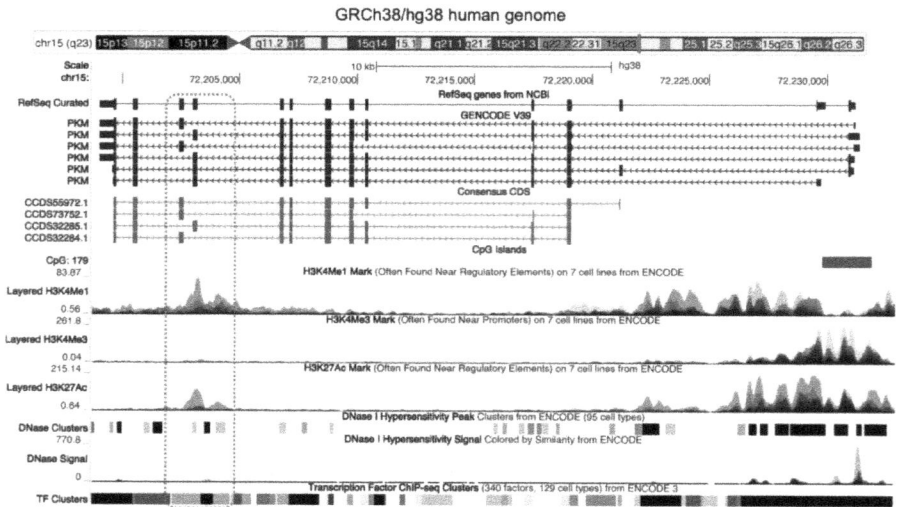

Figure 2: UCSC browser tracks showing PKM gene (RefSeq) and transcript (GENCODE) annotations, consensus CDS, and the presence of CpG island around transcription start site (TSS). Enrichments of H3K4me1, H3K4me3, and H3K27ac (ENCODE ChIP-seq data) are shown around TSS. DNaseI hypersensitivity and transcription factor ChIP-seq tracks from ENCODE indicate tight regulation of the PKM gene. Box with a dotted line indicates inclusion of one of the two exons results in PKM1/M2 isoforms. Differential enrichments of H3K4me1 and H3K27ac histone marks around these exons may play a role in PKM1/M2 alternative splicing.

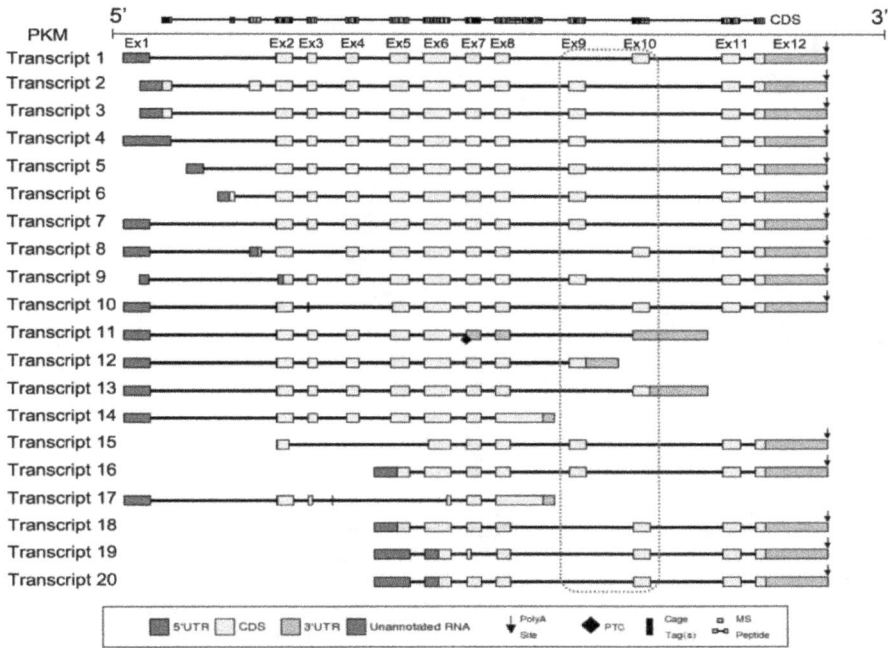

Figure 3: Schematic representation showing predicted transcripts (top 20) of PKM gene obtained from ASPicDB (Alternative Splicing Prediction Data Base) database. The overall exon-intron scheme along with the annotation for 5'UTR, 3'UTR, CDS, PTC (premature stop codon), and polyA site are shown. The PKM transcript 1 corresponds to the NCBI RefSeq transcript. The box with a dotted line indicates inclusion of either exon 9 (PKM1) or exon 10 (PKM2) of PKM gene.

[36, 37]. The presence of H3K4me1 and H3K27ac marks around exon 9/10, suggesting that they may directly or indirectly regulate PKM1/2 splicing (Figure 2).

The PK gene is alternatively spliced to produce an encoding transcripts for either PKM1 or PKM2. Human, mice, and rat PKM gene contains a total of twelve exons and eleven introns. Exons 9 and 10 get mutually spliced to produce either PKM1 or PKM2 [35, 38, 39]. If exon nine becomes part of the transcript, it produces PKM1, and when exon 10 becomes part of the transcript, it produces PKM2. Besides these transcripts, other transcripts are also produced whose significance is not well understood (Figure 3). A recent study by Li et al., characterized four different isoforms, including PKM2 and reported that two of them exhibited opposite prognosis in various cancers [40]. However, it is not precisely known how many PKM transcripts produce functional proteins and their respective functions in the physiopathology of cancer.

4. Role of PKM2 in Cancer

The PKM gene is highly expressed in various cancers. Looking into the TCGA (The Cancer Genome Atlas) dataset that contains 33 cohorts, PKM gene expression was significantly increased in 16 different cancer cohorts, including breast, cervix, colon, kidney, liver, lung, ovary, pancreas, skin, etc. (Figure 4).

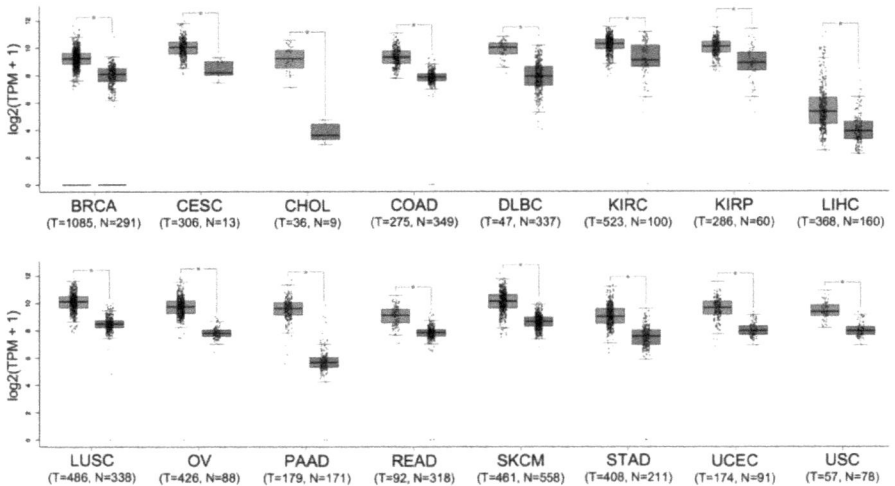

Figure 4: Box plots showing increased expression of PKM gene in various TCGA cohorts. Only 16 TCGA cohorts that showed a significant difference in PKM expression are shown. Red box indicates tumor samples. BRCA, breast invasive carcinoma; CESC, cervical squamous cell carcinoma and endocervical adenocarcinoma; CHOL, cholangiocarcinoma; COAD, colon adenocarcinoma; DLBC, diffuse large B-cell lymphoma; KIRC, kidney renal clear cell carcinoma; KIRP, kidney renal papillary cell carcinoma; LIHC, liver hepatocellular carcinoma; LUSC, lung squamous cell carcinoma; OV, ovarian serous cystadenocarcinoma; PAAD, pancreatic adenocarcinoma; READ, rectum adenocarcinoma; SKCM, skin cutaneous melanoma; STAD, stomach adenocarcinoma; UCEC, uterine corpus endometrial carcinoma; UCS, uterine carcinosarcoma; T, tumor; N, normal.

5. PKM2 and Cancer Cell Proliferation

In cancer cells, PKM2 is known to control a wide range of cellular processes that are critical to their survival. To efficiently absorb resources such as glucose into biomass in the growth of cancer cells, their metabolism must be changed. PKM2 expression is upregulated in a variety of human cancer cells, including lung, breast, blood, prostate, cervix, bladder, kidney, colon, and papillary thyroid carcinoma, as compared to corresponding normal tissues [41]. The PK, PKM2, regulates glycolytic flow as well as the transcription of numerous oncogenes by translocating to the nucleus [42]. Glycolytic activity mediated by PKM2 is also regulated by protein association, as shown in Figure 5 [1, 43–50].

PKM2 in the nucleus of cancer cells works as a kinase and has been shown to cause phosphorylation of histone and signal transducer and activator of transcription 3 (STAT3) in many studies [51]. Cancerous cells overexpress PKM2 [52]. Several aspects of cancer are influenced by the abnormal overexpression of PKM2. Increased expression of FEZF1-AS1 (a long non-coding RNA and lncRNA) in colorectal cancer has been observed. FEZF1-AS1 plays an important function in tumor cell proliferation and metastasis and stabilizes cytosolic and nuclear PKM2 to regulate these processes. The cytosolic PKM2 stimulates aerobic glycolysis, whereas the dimeric PKM2 enhances tumor growth by activating STAT3 [53]. Supporting evidence shows that microRNA (miR-124) inhibits polypyrimidine tract-binding

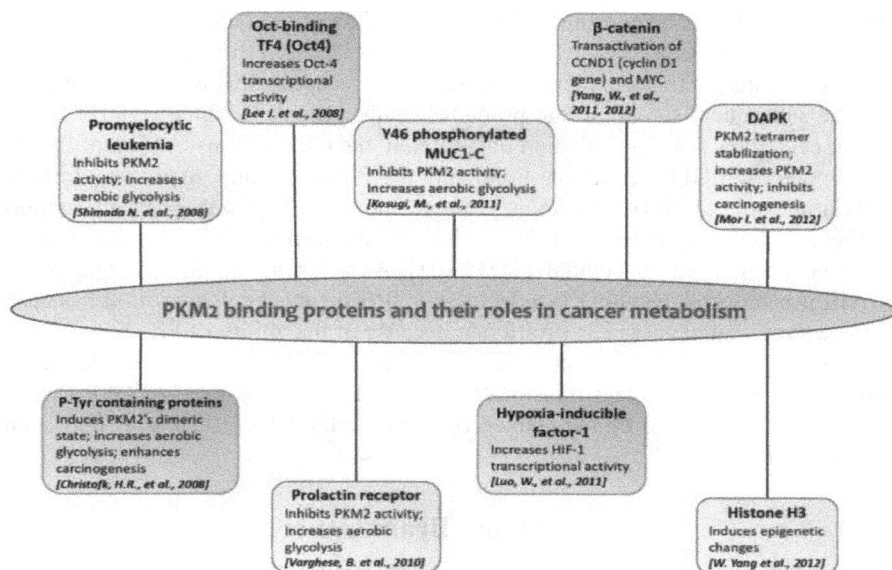

Figure 5: PKM2 binding proteins and their roles in cancer metabolism. In figure, nine proteins have been mentioned along with their functions that are involved in carcinogenesis [1, 43–50].

protein 1 (PTB1), which is a splicer of PKM1 and PKM2, in colorectal cancer cells [54].

The role of PKM2 in cell proliferation has been linked to another miRNA, miR-let-7a. MiR-let-7a has been shown to be downregulated in gastric cancer patients. In gastric cancer cells, overexpression of miR-let-7a led to a reduction in proliferation because of PKM2 inhibition [55]. Tumor suppressor maternally expressed gene 3 (MEG3) is a chromatin-interacting lncRNA that suppresses cancer growth by downregulating PKM2 and inhibiting cyclin D1 (CCND1), C-Myc, and β-catenin. According to recent research, ERK2 specifically binds to PKM2 Ile 429/Leu 431 and phosphorylates PKM2 at Ser 37, causing peptidylprolyl cis/trans isomerase (PIN1) to recruit PKM2 for cis-trans isomerization and PKM2 translocation to the nucleus. Then, nuclear PKM2 serves as a β-catenin coactivator, inducing the c-Myc expression and promoting tumor growth [50]. The interaction between metformin and nuclear PKM2 enhances renal cell carcinoma growth in glucose-deprived conditions as well [56]. Furthermore, in colon cancer cells, activation of the epithelial-mesenchymal transition (EMT) triggers nuclear translocation of PKM2, resulting in direct interaction of PKM2 in the nucleus with TGF-induced factor homeobox 2 (TGIF2). Histone deacetylase 3 is recruited to the E-cadherin promoter region by PKM2 binding to TGIF2, resulting in the deacetylation of histone H3 and inhibition of E-cadherin transcription, which is crucial in inducing EMT [57].

In cancer cells, PKM2 also moves to the mitochondria under conditions of significant oxidative stress. PKM2 phosphorylates and binds to Bcl2 after translocation, preventing its destruction. Bcl2 binds to BAX and inactivates it in addition to the inactivation of pro-apoptotic proteins, thus inhibiting apoptosis [58]. Apoptosis is prevented in cancer cells by phosphorylation of Bcl2. Different cell types release PKM2 as exosomes. The bloodstreams of cancer patients are flooded with PKM2, which is released by tumor cells [59]. It has been shown that PKM2 in the bloodstream may activate epidermal growth factor receptor (EGFR) phosphorylation and the subsequent signaling that promotes tumor growth in triple-negative breast cancer cells [60]. For the activation of β-catenin, the nuclear translocation of PKM2 is promoted by EGFR activation. Following the PKM2-catenin interaction, histone acetylation and CCND1 activation promote tumor cell growth [50]. In support of these results, lapatinib, an EGFR inhibitor, induced a significant decrease in the expression levels of cellular PKM2 [61]. As a result, the phosphorylation of STAT3 was blocked, resulting in a decrease in breast cancer cell growth. It has been hypothesized that cancer cells proliferate because of the ability of PKM2 to trigger an EGFR activation loop, which results in the activation of numerous proliferative genes following the nuclear translocation of the EGFR. The PKM2/EGFR/PKM2 signaling axis seems to be maintained by tumor cells, which secrete a significant quantity of PKM2 [13].

6. PKM2 and Brain Cancers

PKM2 expression is reported in a variety of different forms of brain tumors, including meningiomas, oligodendrogliomas, and ependymomas [62]. Meningiomas are related to elevated phosphatidylserine (PS) levels, which are required for PKM2 kinase activation. Thus, by lowering PS, the resources necessary to support a malignant phenotype are depleted [33]. In oligodendrogliomas, a mutation in the chromosomal arm of the PKM2 gene was discovered. This mutation may benefit brain cancer cells by supporting energy metabolism or promoting cell movement (metastasis) and angiogenesis [33].

Ependymoma has a high malignancy rate and is incurable in approximately 45% of patients. It is defined by the Warburg phenotype, which is characterized by increased expression of PKM2, hexokinase-2 (HK2), and pyruvate dehydrogenase kinase (PDK), as well as a high level of lactate formation [63].

Medulloblastoma (MB), the most common juvenile brain tumor, exhibits abnormal sonic hedgehog (SHH) pathway activation. Hedgehog stimulation promotes PKM2 transcription. Pharmacological inhibition of MB growth using dichloroacetate (DCA), a PK inhibitor, is effective *in vivo* and *in vitro* [64].

7. PKM2 and Glioblastoma

The most prevalent brain cancer is glioma. Gliomas include astrocytomas, oligodendrogliomas, and ependymomas. The most frequent and most dangerous malignant glioma is glioblastoma multiforme (GBM), which is classified as a grade IV glioma by the WHO. GBM has high glycolytic PKM2 and isoform flipping between PKM1 and PKM2. This is unique to GBM and not to other malignancies. PKM2 expression is upregulated in human glioma; however, this does not correspond with PK activity [33].

PKM2 controls gene transcription directly in addition to its metabolic activity. Additionally, PKM2 acts as a protein kinase, phosphorylating histone H3, which promotes carcinogenesis. When the EGFR is activated, PKM2 binds directly to histone H3 and phosphorylates it at Thr-11, resulting in the dissociation of HDAC from CCND1 and the c-Myc promoter region. PKM2 also plays a role in glioma development via its interaction with octamer-binding transcription factor 4 (Oct4), which governs cell pluripotency and apoptosis. PKM2 interacts with the RNA-binding protein HuR in the nucleus of glioblastoma cells to control HuR subcellular localization, p27 levels, and cell cycle progression, resulting in glioma development [65].

FBP works as a PKM2 modulator because it binds to the allosteric site of PKM2 dimer and induces its tetramerization; however, in the absence of FBP, PKM2 dimer was discovered to be translocated to the nucleus, which may benefit cancer cell proliferation, and metastasis. PKM2 has phosphorylated Tyr binding capabilities, lowering the quantity of HuR in the cytoplasm, which affects the translation of tumor suppressor protein p27. Thus, PKM2 is also implicated indirectly in the downregulation of the tumor suppressor protein p27, which results in the malignancy of GBM [65].

8. PKM2 in GI Cancers: Functions and Therapeutic Targets

PKM2 enhances gastric cancer cell proliferation by regulating Bcl-xL transcription. The stability of the nuclear factor-kappa B (NF-kB) component p65 was partly impacted by PKM2 knockdown, suggesting that post-translational control of p65 is one of the mechanisms employed by PKM2 to promote tumor development [66]. PKM2 is a promising therapeutic target in GI malignancies because of its essential involvement in these tumors. The prognostic and therapeutic importance of PKM2 in various GI malignancies has been discussed below because of the metabolic variety of cancer cells.

8.1 Colorectal Cancer (CRC)

CRC is the third most frequent cancer among men and the second most common among women. It has been suggested that elevated levels of fecal PKM2 might serve as a new diagnostic biomarker for the early detection of CRC with a sensitivity of 64%. Small molecule pharmacological impairment of nuclear PKM2 interaction with STAT3 modified the susceptibility of CRC cells to gefitinib, suggesting that this might be a way to overcome EGFR-tyrosine kinase inhibitor (TKI) resistance in CRC patients [67].

8.2 Pancreatic Cancer

Pancreatic cancer is the fourth most common and major cause of death from cancer worldwide. Only around 10–15% of individuals with pancreatic cancer have a probability of survival after surgical resection. As a result, finding therapeutic targets for pancreatic cancer is a critical area of study. Chronic exposure to gemcitabine alters PKM alternative splicing in pancreatic ductal adenocarcinoma cells, giving drug resistance. Combining gemcitabine with a PKM2 inhibitor to enhance chemotherapy response in pancreatic cancer is a very realistic strategy [68].

8.3 Gastric Cancer

Globally, the fifth most prevalent tumor diagnosed is gastric cancer. Progression characteristics in gastric cancer were strongly linked to the overexpression of PKM2. However, the increased PKM2 levels were not related to tumor cell differentiation. Metformin, an anti-diabetic medication, has been shown to slow the development of cancer. A potent anticancer impact of metformin is achieved by inhibiting the HIF1/PKM2 signaling pathway in gastric cancer cells SGC7901 and BGC823 [69].

8.4 Esophageal Cancer

Esophageal cancer is the sixth most prevalent type of cancer and the eighth most common type of cancer globally. Surgery is the primary treatment option for esophageal cancer, either alone or in conjunction with neoadjuvant chemoradiotherapy, adjuvant radiation, and/or adjuvant chemotherapy. A meta-analysis showed that increased PKM2 expression was related to a poor outcome in esophageal squamous carcinoma, emphasizing the prognostic relevance of this gene [70]. PKM2 was recently identified as a marker for altered and rapidly proliferating cells in Barrett's esophagus during the metaplasia-dysplasia-adenocarcinoma cycle. Tanshinone IIA has attracted interest in the past two decades because of its possible therapeutic benefits in cancer. Tanshinone IIA was studied for its anticancer effect in human esophageal cancer Ec109 cells by inhibiting PKM2 expression [71].

9. PKM2 as a Novel Cancer Therapeutic Target

The use of pharmacological therapies that specifically target PKM2 has been shown to be effective in suppressing cancer development. Two techniques are often used to control PKM2-influenced oncogenic signaling in cancer cells. PKM2 can be inhibited in its dimer form to prevent its subsequent translocation into subcellular components, such as the nucleus and mitochondria, or activated in its tetramer form to carry out its integral roles of converting PEP into pyruvate, carrying no non-metabolic activities. The various antagonists and activators of PKM2 that are used to suppress different forms of cancer are shown in Figure 6 [13, 72].

9.1 Inhibition of PMK2 Activity

Inhibiting PKM2 with a potential inhibitor has been effective in restraining a variety of processes in cancer that are known to be controlled by PKM2. Because of inhibiting PKM2, mitochondrial respiration was restored in skin cancer cells [73]. Additionally, it has been observed that inhibiting PKM2 reverses incidences of multidrug resistance in tumor cells. Numerous reports indicate that inhibiting PKM2 with a potential inhibitor reversed resistance to anticancer agents, such as paclitaxel, cisplatin, and gefitinib in various cancer types [74]. Furthermore, inhibiting PKM2 increases cell death in a number of ways. PKM2 suppression resulted in apoptosis in cholangiocarcinoma through enhancing intracellular ROS levels [75]. However, a distinct approach was used to decrease PKM2 in osteosarcoma, where inhibition of PKM2 encouraged necroptosis mediated by receptor-interacting serine/threonine

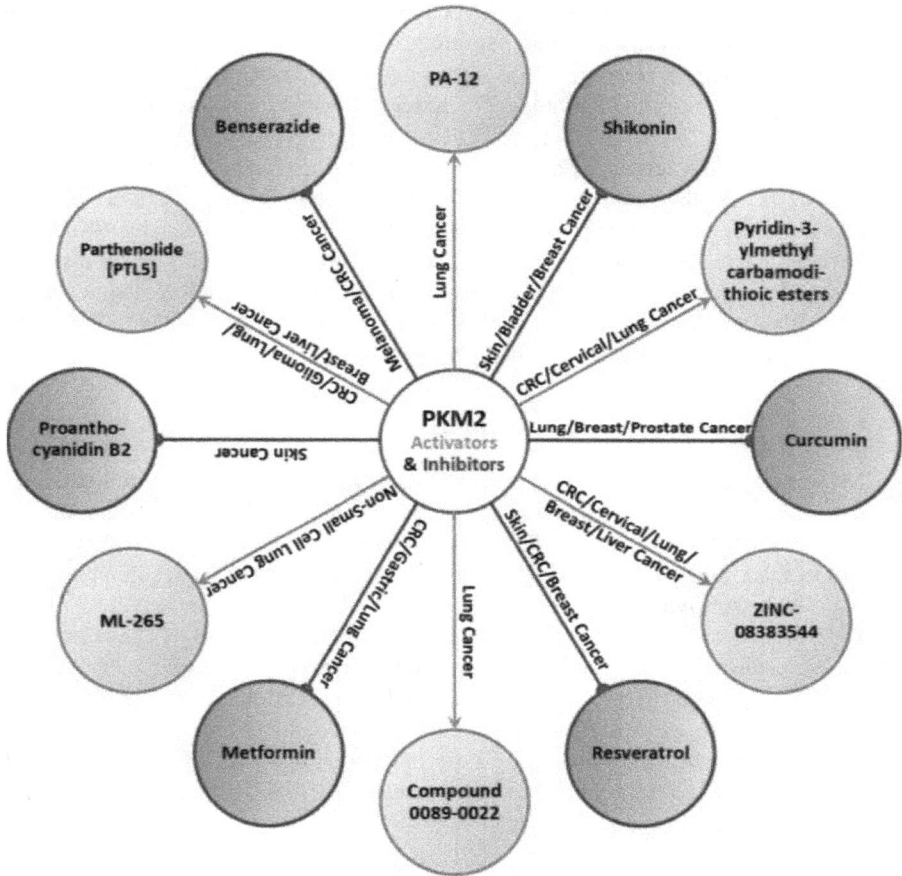

Figure 6: Different PKM2 inhibitors (blunted arrows) and activators (sharpened arrows) known to inhibit the growth of different types of cancer.

kinase 1 and 3 (RIP1 and RIP3) [76]. PKM2 inhibitors have also been found effective in regulating cell adherence to the extracellular matrix, metastasis, and migration in lung cancer [13].

9.2 Activators of PKM2

Various cancer types express PKM2 in different ways, according to their severity and the location of the tumor. From that perspective, it does not look acceptable to declare that a broad PKM2 inhibition medication is a promising way of combating cancer. Activating the PKM2 tetramer from a dimeric form is an alternative strategy for controlling the aberration in cancer cells induced by dimer PKM2. PKM2 activation has been shown in several studies to be an effective strategy for suppressing tumor development. Lung cancer progression has been shown to be inhibited by small-molecule PKM2 activators [77]. However, conflicting studies raised doubts on this front. PKM2 tetramer conversion does not seem to be a viable strategy for cancer growth control since activation of PKM2 lowers carbon flow and induces serine auxotrophy inside cancer cells, compensating for nutritional stress and maintaining their prolonged proliferation [31]. Targeting PKM isoforms in cancer may undoubtedly be a beneficial strategy if a better understanding of their function can be achieved [78].

Several PKM2 activators and inhibitors are being studied in preclinical and clinical trials, and the findings suggest that these inhibitors and activators might be effective anticancer medications. The intracellular process induced by PKM2 is, however, significantly more complicated than it sounds. As a result, further research is needed to make PKM2 a viable cancer treatment target, as well as the necessity to qualitatively and quantitatively evaluate PKM2 levels in patients with cancer in order to make PKM2 a successful therapeutic target.

Conclusion

PKM2 is a key enzyme in the metabolic rearrangement processes that are hallmarks of cancer. In addition to regulating metabolic activities, PKM2 regulates a number of non-metabolic activities that are related to oncogenic events. In this review, we have sought to shed light on these PKM2 moonlighting capabilities that are thought to be critical for the survival of cancer.

PKM2 is derived from the PKM gene by alternative splicing. It is evident that the dimeric form of PKM2 is linked to the development of cancer; at least, it has been found in a majority of cancers studied, as discussed above. However, still more study is needed to delineate its role in cancer so that PKM2 can be targeted for cancer treatment. In the future, it is likely that modulators of PKM2 activity and or in combination with inhibitors of other cancer development pathways may find application in cancer treatment.

Glossary

ADP	:	Adenosine diphosphate
ATP	:	Adenosine triphosphate
CRC	:	Colorectal cancer
DAPK	:	Death-associated protein kinase
EMT	:	Epithelial-mesenchymal transition
ENCODE	:	Encyclopedia of DNA Elements
FBP	:	Fructose 1,6-bis phosphate
GBM	:	Glioblastoma multiforme
hnRNPA	:	Heterogeneous ribonucleoprotein
lncRNA	:	Long non-coding RNA
MEG	:	Maternally expressed gene
NCBI	:	National Center for Biotechnology Information
Oct4	:	Octamer-binding transcription factor 4
PDK	:	Pyruvate dehydrogenase kinase
PEP	:	Phosphoenolpyruvate
PK	:	Pyruvate kinase

PKL	:	Liver isoform of pyruvate kinase
PKM	:	Muscle isoform of pyruvate kinase
PKR	:	Red blood cell isoform of pyruvate kinase
PS	:	Phosphatidylserine
PTB	:	Polypyrimidine-tract binding protein
RIP	:	Receptor-interacting protein
SAICAR	:	Succinyl 5'-aminoimidazole-4-carboxamide-1-ribonucleotide-5'-phosphate
TCGA	:	The Cancer Genome Atlas
TSS	:	Transcription start sites

References

[1] Christofk, H.R., Vander Heiden, M.G., Wu, N., Asara, J.M. and Cantley, L.C. (2008). Pyruvate kinase M2 is a phosphotyrosine-binding protein. Nature 452(7184): 181–6.

[2] Wu, H., Yang, P., Hu, W., Wang, Y., Lu, Y., Zhang, L., Fan, Y., Xiao, H. and Li, Z. (2016). Overexpression of PKM2 promotes mitochondrial fusion through attenuated p53 stability. Oncotarget 7(47): 78069–78082.

[3] Wong, N., De Melo, J. and Tang, D. (2013). PKM2, a central point of regulation in cancer metabolism. Int. J. Cell Biol. 2013: 242513.

[4] Schormann, N., Hayden, K.L., Lee, P., Banerjee, S. and Chattopadhyay, D. (2019). An overview of structure, function, and regulation of pyruvate kinases. Protein Sci. 28(10): 1771–1784.

[5] Dong, G., Mao, Q., Xia, W., Xu, Y., Wang, J., Xu, L. and Jiang, F. (2016). PKM2 and cancer: The function of PKM2 beyond glycolysis. Oncol. Lett. 11(3): 1980–1986.

[6] Keller, K.E., Doctor, Z.M., Dwyer, Z.W. and Lee, Y.S. (2014). SAICAR induces protein kinase activity of PKM2 that is necessary for sustained proliferative signaling of cancer cells. Mol. Cell. 53(5): 700–9.

[7] Imamura, K. and Tanaka, T. (1972). Multimolecular forms of pyruvate kinase from rat and other mammalian tissues. I. Electrophoretic studies. J. Biochem. 71(6): 1043–51.

[8] Muralidhar, V. (2016). Influence of Pyruvate Kinase Isoform Expression on Primary Cell Proliferation and Metabolism. Harvard University: Harvard Medical School.

[9] van Oirschot, B.A., Francois, J.J.J.M., van Solinge, W.W., van Wesel, A.C.W., Rijksen, G., van Amstel, H.K.P. and van Wijk, R. (2014). Novel type of red blood cell pyruvate kinase hyperactivity predicts a remote regulatory locus involved in PKLR gene expression. Am. J. Hematol. 89(4): 380–4.

[10] Gassaway, B.M., Cardone, R.L., Padyana, A.K., Petersen, M.C., Judd, E.T., Hayes, S., Tong, S., Barber, K.W., Apostolidi, M., Abulizi, A. and Rinehart, J. (2019). Distinct hepatic PKA and CDK signaling pathways control activity-independent pyruvate kinase phosphorylation and hepatic glucose production. Cell Rep. 29(11): 3394–3404 e9.

[11] Zhan, C., Yan, L., Wang, L., Ma, J., Jiang, W., Zhang, Y., Shi, Y. and Wang, Q. (2015). Isoform switch of pyruvate kinase M1 indeed occurs but not to pyruvate kinase M2 in human tumorigenesis. PLoS One 10(3): e0118663.

[12] Hsu, M.C. and Hung, W.C. (2018). Pyruvate kinase M2 fuels multiple aspects of cancer cells: From cellular metabolism, transcriptional regulation to extracellular signaling. Mol. Cancer 17(1): 35.

[13] Chhipa, A.S. and Patel, S. (2021). Targeting pyruvate kinase muscle isoform 2 (PKM2) in cancer: What do we know so far? Life Sci. 280: 119694.

[14] Dombrauckas, J.D., Santarsiero, B.D. and Mesecar, A.D. (2005). Structural basis for tumor pyruvate kinase M2 allosteric regulation and catalysis. Biochemistry 44(27): 9417–29.

[15] Clower, C.V., Chatterjee, D., Wang, Z., Cantley, L.C., Vander Heiden, M.G. and Krainer, A.R. (2010). The alternative splicing repressors hnRNP A1/A2 and PTB influence pyruvate kinase isoform expression and cell metabolism. Proc. Natl. Acad. Sci. USA 107(5): 1894–9.

[16] Chaneton, B., Hillmann, P., Zheng, L., Martin, A.C., Maddocks, O.D., Chokkathukalam, A., Coyle, J.E., Jankevics, A., Holding, F.P., Vousden, K.H., Frezza, C., O'Reilly M. and Gottlieb, E. (2012). Serine is a natural ligand and allosteric activator of pyruvate kinase M2. Nature 491(7424): 458–462.

[17] Guo, C., Li, G., Hou, J., Deng, X., Ao, S., Li, Z. and Lyu, G. (2018). Tumor pyruvate kinase M2: A promising molecular target of gastrointestinal cancer. Chin. J. Cancer Res. 30(6): 669–676.

[18] Buc, H., Demaugre, F. and Leroux, J.P. (1978). The kinetic effects of oxalate on liver and erythrocyte pyruvate kinases. Biochem. Biophys. Res. Commun. 85(2): 774–9.

[19] Dayton, T.L., Jacks, T. and Vander Heiden, M.G. (2016). PKM2, cancer metabolism, and the road ahead. EMBO Rep. 17(12): 1721–1730.

[20] Imamura, K. and Tanaka, T. (1982). Pyruvate kinase isozymes from rat. Methods Enzymol. 90 Pt E: 150–65.

[21] Imamura, K., Taniuchi, K. and Tanaka, T. (1972). Multimolecular forms of pyruvate kinase. II. Purification of M 2-type pyruvate kinase from Yoshida ascites hepatoma 130 cells and comparative studies on the enzymological and immunological properties of the three types of pyruvate kinases, L, M 1, and M 2. J. Biochem. 72(4): 1001–15.

[22] Berglund, L. and Humble, E. (1979). Kinetic properties of pig pyruvate kinases type A from kidney and type M from muscle. Arch. Biochem. Biophys. 195(2): 347–61.

[23] Hubbard, D.R. and Cardenas, J.M. (1975). Kinetic properties of pyruvate kinase hybrids formed with native type L and inactivated type M subunits. J. Biol. Chem. 250(13): 4931–6.

[24] Ikeda, Y., Tanaka, T. and Noguchi, T. (1997). Conversion of non-allosteric pyruvate kinase isozyme into an allosteric enzyme by a single amino acid substitution. J. Biol. Chem. 272(33): 20495–501.

[25] Morgan, H.P., O'Reilly, F.J., Wear, M.A., O'Neill, J.R., Fothergill-Gilmore, L.A., Hupp, T. and Walkinshaw, M.D. (2013). M2 pyruvate kinase provides a mechanism for nutrient sensing and regulation of cell proliferation. Proc. Natl. Acad. Sci. USA 110(15): 5881–6.

[26] Koler, R.D. and Vanbellinghen, P. (1968). The mechanism of precursor modulation of human pyruvate kinase I by fructose diphosphate. Adv. Enzyme Regul. 6: 127–42.

[27] Taylor, C.B. and Bailey, E. (1967). Activation of liver pyruvate kinase by fructose 1,6-diphosphate. Biochem. J. 102(2): 32C–33C.

[28] Vijayvargiya, R., Schwark, W.S. and Singhal, R.L. (1969). Pyruvate kinase: Modulation by L-phenylalanine and L-alanine. Can. J. Biochem. 47(9): 895–8.

[29] Weber, G. (1969). Inhibition of human brain pyruvate kinase and hexokinase by phenylalanine and phenylpyruvate: Possible relevance to phenylketonuric brain damage. Proc. Natl. Acad. Sci. USA 63(4): 1365–9.

[30] Israelsen, W.J. and Vander Heiden, M.G. (2015). Pyruvate kinase: Function, regulation and role in cancer. Semin. Cell Dev. Biol. 43: 43–51.

[31] Kung, C., Hixon, J., Choe, S., Marks, K., Gross, S., Murphy, E., DeLaBarre, B., Cianchetta, G., Sethumadhavan, S., Wang, X., Yan, S. and Dang, L. (2012). Small molecule activation of PKM2 in cancer cells induces serine auxotrophy. Chem. Biol. 19(9): 1187–98.

[32] Williams, R., Holyoak, T., McDonald, G., Gui, C. and Fenton, A.W. (2006). Differentiating a ligand's chemical requirements for allosteric interactions from those for protein binding. Phenylalanine inhibition of pyruvate kinase. Biochemistry 45(17): 5421–9.

[33] Verma, H., Cholia, R.P., Kaur, S., Dhiman, M. and Mantha, A.K. (2021). A short review on cross-link between pyruvate kinase (PKM2) and Glioblastoma Multiforme. Metab. Brain Dis. 36(5): 751–765.

[34] Prakasam, G., Iqbal, M.A., Bamezai, R.N. and Mazurek, S. (2018). Posttranslational modifications of pyruvate kinase M2: Tweaks that benefit cancer. Front. Oncol. 8: 22.

[35] Takenaka, M., Noguchi, T., Sadahiro, S., Hirai, H., Yamada, K., Matsuda, T., Imai, E. and Tanaka, T. (1991). Isolation and characterization of the human pyruvate kinase M gene. Eur. J. Biochem. 198(1): 101–6.

[36] Agirre, E., Oldfield, A.J., Bellora, N., Segelle, A. and Luco, R.F. (2021). Splicing-associated chromatin signatures: A combinatorial and position-dependent role for histone marks in splicing definition. Nat. Commun. 12(1): 682.

[37] Hu, Q., Greene, C.S. and Heller, E.A. (2020). Specific histone modifications associate with alternative exon selection during mammalian development. Nucleic Acids Res. 48(9): 4709–4724.

[38] Inoue, H., Noguchi, T. and Tanaka, T. (1986). Complete amino acid sequence of rat L-type pyruvate kinase deduced from the cDNA sequence. Eur. J. Biochem. 154(2): 465–9.

[39] Noguchi, T., Inoue, H. and Tanaka, T. (1986). The M1- and M2-type isozymes of rat pyruvate kinase are produced from the same gene by alternative RNA splicing. J. Biol. Chem. 261(29): 13807–12.

[40] Li, X., Kim, W., Arif, M., Gao, C., Hober, A., Kotol, D., Strandberg, L., Forsström, B., Sivertsson, Å., Oksvold, P. and Turkez, H. and Mardinoglu, A. (2021). Discovery of functional alternatively spliced PKM transcripts in human cancers. Cancers (Basel) 13(2).

[41] Bluemlein, K., Grüning, N.M., Feichtinger, R.G., Lehrach, H., Kofler, B. and Ralser, M. (2011). No evidence for a shift in pyruvate kinase PKM1 to PKM2 expression during tumorigenesis. Oncotarget 2(5): 393–400.

[42] Gao, X., Wang, H., Yang, J.J., Liu, X.and Liu, Z.R. (2012). Pyruvate kinase M2 regulates gene transcription by acting as a protein kinase. Mol. Cell. 45(5): 598–609.

[43] Kosugi, M., Ahmad, R., Alam, M., Uchida, Y. and Kufe, D. (2011). MUC1-C oncoprotein regulates glycolysis and pyruvate kinase M2 activity in cancer cells. PLoS One 6(11): e28234.

[44] Lee, J., Kim, H.K., Han, Y.M. and Kim, J. (2008). Pyruvate kinase isozyme type M2 (PKM2) interacts and cooperates with Oct-4 in regulating transcription. Int. J. Biochem. Cell Biol. 40(5): 1043–54.

[45] Luo, W., Hu, H., Chang, R., Zhong, J., Knabel, M., O'Meally, R., Cole, R.N., Pandey, A. and Semenza, G.L. (2011). Pyruvate kinase M2 is a PHD3-stimulated coactivator for hypoxia-inducible factor 1. Cell 145(5): 732–44.

[46] Mor, I., Carlessi, R., Ast, T., Feinstein, E. and Kimchi, A. (2012). Death-associated protein kinase increases glycolytic rate through binding and activation of pyruvate kinase. Oncogene 31(6): 683–93.

[47] Shimada, N., Shinagawa, T. and Ishii, S. (2008). Modulation of M2-type pyruvate kinase activity by the cytoplasmic PML tumor suppressor protein. Genes Cells 13(3): 245–54.

[48] Varghese, B., Swaminathan, G., Plotnikov, A., Tzimas, C., Yang, N., Rui, H. and Fuchs, S.Y. (2010). Prolactin inhibits activity of pyruvate kinase M2 to stimulate cell proliferation. Mol. Endocrinol. 24(12): 2356–65.

[49] Yang, W., Xia, Y., Hawke, D., Li, X., Liang, J., Xing, D., Aldape, K., Hunter, T., Yung, W.A. and Lu, Z. (2012). PKM2 phosphorylates histone H3 and promotes gene transcription and tumorigenesis. Cell 150(4): 685–96.

[50] Yang, W., Xia, Y., Ji, H., Zheng, Y., Liang, J., Huang, W., Gao, X., Aldape, K. and Lu, Z. (2011). Nuclear PKM2 regulates beta-catenin transactivation upon EGFR activation. Nature 480(7375): 118–22.

[51] Ma, R., Liu, Q., Zheng, S., Liu, T., Tan, D. and Lu, X. (2019). PKM2-regulated STAT3 promotes esophageal squamous cell carcinoma progression via TGF-beta1-induced EMT. J. Cell Biochem. 2019.

[52] Qian, Z., Hu, W., Lv, Z., Liu, H., Chen, D., Wang, Y., Wu, J. and Zheng, S. (2020). PKM2 upregulation promotes malignancy and indicates poor prognosis for intrahepatic cholangiocarcinoma. Clin. Res. Hepatol. Gastroenterol. 44(2): 162–173.

[53] Bian, Z., Zhang, J., Li, M., Feng, Y., Wang, X., Zhang, J., Yao, S., Jin, G., Du, J., Han, W., Yin, Y. and Huang, Z. (2018). LncRNA-FEZF1-AS1 Promotes tumor proliferation and metastasis in colorectal cancer by regulating PKM2 signaling. Clin. Cancer Res. 24(19): 4808–4819.

[54] Taniguchi, K., Sugito, N., Kumazaki, M., Shinohara, H., Yamada, N., Nakagawa, Y., Ito, Y., Otsuki, Y., Uno, B., Uchiyama, K. and Akao, Y. (2015). MicroRNA-124 inhibits cancer cell growth through PTB1/PKM1/PKM2 feedback cascade in colorectal cancer. Cancer Lett. 363(1): 17–27.

[55] Tang, R., Yang, C., Ma, X., Wang, Y., Luo, D., Huang, C., Xu, Z., Liu, P. and Yang, L. (2016). MiR-let-7a inhibits cell proliferation, migration, and invasion by down-regulating PKM2 in gastric cancer. Oncotarget 7(5): 5972–84.

[56] Zheng, Q., Lin, Z., Xu, J., Lu, Y., Meng, Q., Wang, C., Yang, Y., Xin, X., Li, X., Pu, H., Gui, X. and Lu, D. (2018). Long noncoding RNA MEG3 suppresses liver cancer cells growth through inhibiting beta-catenin by activating PKM2 and inactivating PTEN. Cell Death Dis. 9(3): 253.

[57] Yang, W., Zheng, Y., Xia, Y., Ji, H., Chen, X., Guo, F., Lyssiotis, C.A., Aldape, K., Cantley, L.C. and Lu, Z. (2012). ERK1/2-dependent phosphorylation and nuclear translocation of PKM2 promotes the Warburg effect. Nat. Cell Biol. 14(12): 1295–304.

[58] Hardwick, J.M. and Soane, L. (2013). Multiple functions of BCL-2 family proteins. Cold Spring Harb. Perspect. Biol. 5(2).

[59] Chen, Y.L., Song, J.J., Chen, X.C., Xu, W., Zhi, Q., Liu, Y.P., Xu, H.Z., Pan, J.S., Ren, J.L. and Guleng, B. (2015). Mechanisms of pyruvate kinase M2 isoform inhibits cell motility in hepatocellular carcinoma cells. World J. Gastroenterol. 21(30): 9093–102.

[60] Lim, S.O., Li, C.W., Xia, W., Lee, H.H., Chang, S.S., Shen, J., Hsu, J.L., Raftery, D., Djukovic, D., Gu, H., Chang, W.C. and Hung, M.C. (2016). EGFR signaling enhances aerobic glycolysis in triple-negative breast cancer cells to promote tumor growth and immune escape. Cancer Res. 76(5): 1284–96.

[61] Guan, M., Tong, Y., Guan, M., Liu, X., Wang, M., Niu, R., Zhang, F., Dong, D., Shao, J. and Zhou, Y. (2018). Lapatinib inhibits breast cancer cell proliferation by influencing PKM2 expression. Technol. Cancer Res. Treat. 17: 1533034617749418.

[62] Van Veelen, C.W.M., Verbiest, H., Staal, G.E.J. and Vlug, A.M.C. (1977). Alanine inhibition of pyruvate kinase in gliomas and meningiomas. A diagnostic tool in surgery for gliomas? Lancet 2(8034): 384–5.

[63] Mack, S.C., Agnihotri, S., Bertrand, K.C., Wang, X., Shih, D.J., Witt, H., Hill, N., Zayne, K., Barszczyk, M., Ramaswamy, V., Remke, M. and Taylor, M.D. (2015). Spinal myxopapillary ependymomas demonstrate a warburg phenotype. Clin. Cancer Res. 21(16): 3750–8.

[64] Di Magno, L., Manzi, D., D'Amico, D., Coni, S., Macone, A., Infante, P., Di Marcotullio, L., De Smaele, E., Ferretti, E., Screpanti, I., Agostinelli, E. and Canettieri, G. (2014). Druggable glycolytic requirement for Hedgehog-dependent neuronal and medulloblastoma growth. Cell Cycle 13(21): 3404–13.

[65] Mukherjee, J., Ohba, S., See, W.L., Phillips, J.J., Molinaro, A.M. and Pieper, R.O. (2016). PKM2 uses control of HuR localization to regulate p27 and cell cycle progression in human glioblastoma cells. Int. J. Cancer 139(1): 99–111.

[66] Azoitei, N., Becher, A., Steinestel, K., Rouhi, A., Diepold, K., Bobrovich, S. and Seufferlein, T. (2016). PKM2 promotes tumor angiogenesis by regulating HIF-1alpha through NF-kappaB activation. Mol. Cancer 15: 3.

[67] Li, Q., Zhang, D., Chen, X., He, L., Li, T., Xu, X. and Li, M. (2015). Nuclear PKM2 contributes to gefitinib resistance via upregulation of STAT3 activation in colorectal cancer. Sci. Rep. 5: 16082.

[68] Tian, S., Li, P., Sheng, S. and Jin, X. (2018). Upregulation of pyruvate kinase M2 expression by fatty acid synthase contributes to gemcitabine resistance in pancreatic cancer. Oncol. Lett. 15(2): 2211–2217.

[69] Chen, G., Feng, W., Zhang, S., Bian, K., Yang, Y., Fang, C., Chen, M., Yang, J. and Zou, X. (2015). Metformin inhibits gastric cancer via the inhibition of HIF1alpha/PKM2 signaling. Am. J. Cancer Res. 5(4): 1423–34.

[70] Zhu, H., Luo, H., Zhu, X., Hu, X., Zheng, L. and Zhu, X. (2017). Pyruvate kinase M2 (PKM2) expression correlates with prognosis in solid cancers: A meta-analysis. Oncotarget 8(1): 1628–1640.

[71] Zhang, H.S., Zhang, F.J., Li, H., Liu, Y., Du, G.Y. and Huang, Y.H. (2016). Tanshinone A inhibits human esophageal cancer cell growth through miR-122-mediated PKM2 down-regulation. Arch. Biochem. Biophys. 598: 50–6.

[72] Zhu, S., Guo, Y., Zhang, X., Liu, H., Yin, M., Chen, X. and Peng, C. (2021). Pyruvate kinase M2 (PKM2) in cancer and cancer therapeutics. Cancer Lett. 503: 240–248.

[73] Li, W., Liu, J. and Zhao, Y. (2014). PKM2 inhibitor shikonin suppresses TPA-induced mitochondrial malfunction and proliferation of skin epidermal JB6 cells. Mol. Carcinog. 53(5): 403–12.

[74] Wang, Z., Xu, J., Liu, Y., Chen, J., Lin, H., Huang, Y., Bian, X. and Zhao, Y. (2019). Selection and validation of appropriate reference genes for real-time quantitative PCR analysis in Momordica charantia. Phytochemistry 164: 1–11.

[75] Thonsri, U., Seubwai, W., Waraasawapati, S., Wongkham, S., Boonmars, T., Cha'on, U. and Wongkham, C. (2020). Antitumor effect of shikonin, a PKM2 inhibitor, in cholangiocarcinoma cell lines. Anticancer Res. 40(9): 5115–5124.

[76] Fu, Z., Deng, B., Liao, Y., Shan, L., Yin, F., Wang, Z., Zeng, H., Zuo, D., Hua, Y. and Cai, Z. (2013). The anti-tumor effect of shikonin on osteosarcoma by inducing RIP1 and RIP3 dependent necroptosis. BMC Cancer 13: 580.

[77] Parnell, K.M., Foulks, J.M., Nix, R.N., Clifford, A., Bullough, J., Luo, B., Senina, A., Vollmer, D., Liu, J., McCarthy, V., Xu, Y. and Kanner, S.B. (2013). Pharmacologic activation of PKM2 slows lung tumor xenograft growth. Mol. Cancer Ther. 12(8): 1453–60.

[78] Zahra, K., Dey, T., Mishra, S.P. and Pandey, U. (2020). Pyruvate kinase M2 and cancer: The role of PKM2 in promoting tumorigenesis. Front. Oncol. 10: 159.

8

Alternative Splicing and Immune Surveillance

Moumita Roy

1. Introduction

Alternative splicing in eukaryotes allows for the expression of multiple RNA and protein isoforms from a single gene while making it a major contributory event for the generation of diversified transcriptome and proteomes products. In mammalian cells, the canonical RNA splicing takes place inside the spliceosome, which is a multi-protein complex called a 'metalloribozyme', and during this process, pre-mRNA becomes mature mRNA via the excision of introns and combining of exons [1, 2]. Metalloribozyme is made up of five small nuclear riboproteins (snRNPs) that contain snRNAs and a large number of accessory proteins that help to identify the targeted pre-mRNA being spliced. During the transcription event, the assembly of the metalloribozyme complex takes place, and both transcription and splicing machinery are orchestrated events [3]. Increasing scientific research evidence has shown that ~ 94% of human genes have intronic regions during the pre-mRNA processes, and a majority of the eukaryotic genes undergo alternative splicing events controlled by the tempo-spatial-dependent manner [4].

Dysregulation of alternative splicing (AS) is widely considered a new hallmark of cancer, and this is associated with the occurrence of tumors. The abnormal changes in AS could affect tumor progression, and it could also disrupt the protein interaction pathways in tumor development [5–7]. Pre-mRNAs can be spliced into mature mRNAs by retaining specific intron regions or excluding specific exons in multi-exon genes. The mammalian transcriptome and proteome are far more diverse than expected from the 'one gene→one mRNA→one protein' paradigm. Transcriptome and proteome complexity has evolved from the acquisition of new genes along with

Postdoctoral Research fellow, Mechanical Engineering, Texas Tech University.
Email: roymoumitatx@gmail.com

alternative transcription and AS mechanisms that generate a variety of transcripts from a single gene and in a limited genome which generates structural and functional protein variants, along with the promotion of protein diversity and phenotypic complexity [8].

The human genome contains annotations for 28,526 genes that express a total of 120,145 transcripts, of which 80,932 are protein-coding and 39,213 are noncoding transcripts (Ensembl 65). According to the annotation of the current annotations, approximately 65% of the genes produce multiple transcripts, with an average of six transcript variants per gene, while the remaining 35% are single transcript genes. The AS products are widely acknowledged as potentially useful biomarkers [9]. Genome-wide studies revealed 92–95% of human exons have undergone AS [10] events, while the expression of most cellular transcripts has spatial and temporal differences [11].

2. The Splicing Machinery and Regulatory Splicing Components (Core Spliceosome Machinery, Cis-Acting Elements, and other RBPs, Main AS Patterns)

The intron removal from a gene takes place by a two-step concerted chemical process by which specific phosphodiester bonds in the polynucleotide chains of RNA are excised, and new ones are formed. The steps involve (1) the formation of an unusual 2'–5' phosphodiester bond between the 5' nucleotide of the intron and (2) a key internal adenosine residue (the branch site) located 15–30 nucleotides upstream of the 3' end of the intron (Figure 1).

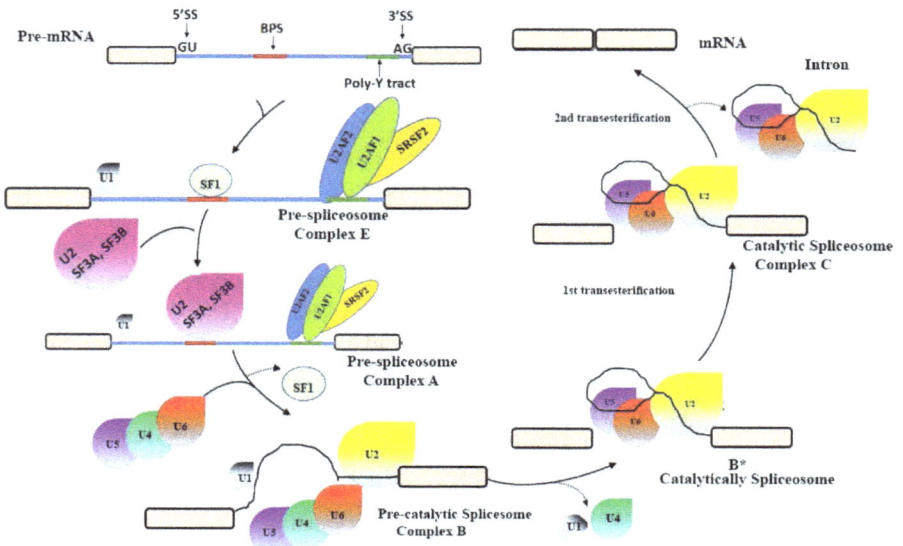

Figure 1: Splicesome assembly process: Different stages of splicesome assembly and spliced product formation are described above.

This process is mechanically identical to that of group II self-catalytic RNAs, possibly revealing the ancestral origin of pre-mRNA introns [12]. During the reaction process, the sequence of the entire group II intron has a 3D structure, while enabling their excision in the absence of cofactors, introns in pre-mRNAs harbor only short consensus sequences at the exon-intron boundaries, known as 5' and 3' splice sites. The removal of these introns relies on one of the most sophisticated macromolecular complexes of eukaryotic cells: the spliceosome [13, 14]. The essential step in the splicing process involves eukaryotic gene expression. In the case of hereditary cancer, the genes are particularly susceptible to inactivating mutations in splice sites [15]. At least one event of the AS events is common in most multi-exon genes [16], as explored by the advent of high-throughput sequencing technology, while in several situations generating two or more distinct mRNAs from the same gene with the number of alternatively spliced transcripts potentially staggering for some genes.

The different potential modes of AS events of the pre-mRNA can be divided into six different categories: (1) exon skipping, (2) intron retention, (3) alternative 5'/3' donor/acceptor sites, (4) mutually exclusive exons, (5) alternative promoters, and (6) AS and polyadenylation [17] (Figure 2).

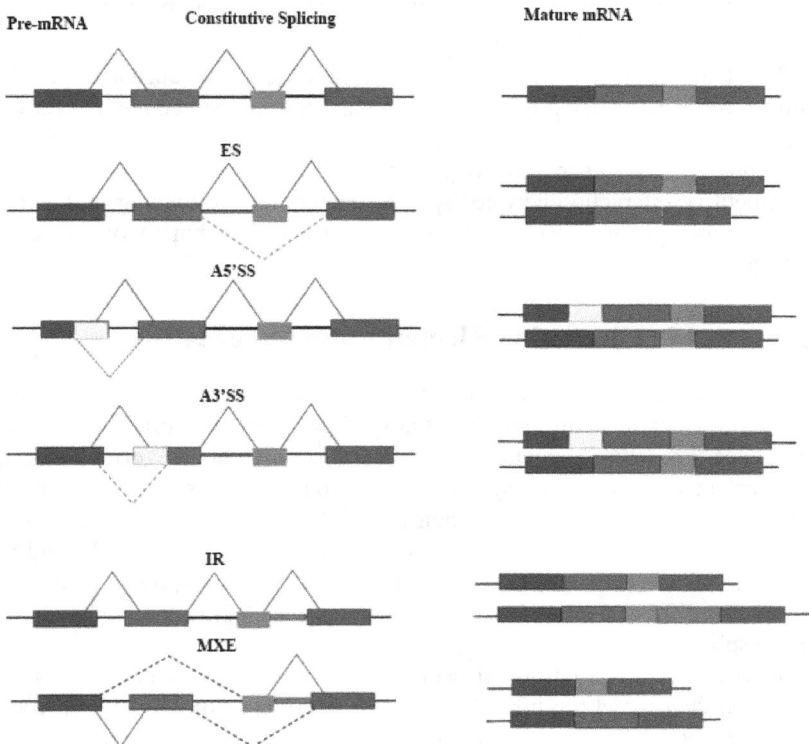

Figure 2: Categories of AS events of the pre-mRNA: (a) exon skipping, (b) intron retention, (c) alternative 5'/3' donor/acceptor sites, (d) mutually exclusive exons, (e) alternative promoters, and (f) AS and polyadenylation

3. Core Spliceosome Machinery

The highly regulated process of the AS is mainly executed by the spliceosome machinery. A simplified spliceosome assembly pathway and the core splicing factors required for exon/intron definition were summarized in Figure 1. It can be easily affected by the activity of splicing regulators—e.g., serine/arginine-rich (SR) proteins or heterogeneous nuclear ribonucleoproteins (hnRNPs), which are the most important mediators of SS recognition [18]. The essential spliceosome components are five small nuclear RNPs (snRNPs; U1, U2, U4, U5, and U6 snRNPs); each contains its own snRNA complexed to a group of 300 associated proteins [19, 20]. The AS is a highly concerted multi-step process

(I) snRNPS bind to pre-mRNA (Figure 1), U1 snRNP binds to 5'SS, and U2 (including SF3A and SF3B) binds to 3'SS and polypyrimidine sequences.

(II) U5, U4/U6 were recruited, and a catalytically active complex was formed as a result of rearrangement between snRNPs after complete intron excision and exon linkage [21, 22].

(III) The two essential auxiliary components of cis-acting RNA elements—SR proteins [23] and heterogeneous nuclear ribonucleoprotein (hnRNP) family [18] act as enhancers or repressors of the splice-site usage by the recognition of specific cis-acting RNA elements.

SR proteins primarily play a positive role in splicing regulation, and it specifically binds to exonic splicing enhancer (ESE) and intronic splicing enhancer (ISE). Antithetically, the binding of hnRNPs to an exonic splicing silencer (ESS) and intronic splicing silencer (ISS) primarily repress exon inclusion. Chances of competitive interaction between splicing factors for the binding in the context of the cis-regulatory RNA elements and/or the recruitment of spliceosome components [24].

4. Cis-Acting Elements and Other RBPs

The splicing of pre-mRNA is regulated by both cis-acting elements and trans-acting factors. According to a different locations of positions and different effects on splicing sites, cis-acting elements are divided into different categories—e.g., exon splicing enhancers, exon splicing silencers, intron splicing enhancers, and intron splicing silencers, which determine their affinity with homologous splicing factors (SFs). Several Cis-acting elements, along with RBPs like 5'SS, 3'SS, BPS, and splice enhancers (ESE and ISE) or silencer elements (ESS and ISS) in precursor mRNA are important for different varieties of splicing events, including the constitutive and regulated splicing [25].

The changes in the ability of sequence-specific RBPs to bind to cis-acting sequences in their target pre-mRNA lead to the occurrence of AS [26]. SR proteins and hnRNPs both have the ability to either promote or repress splicing when binding to different positions of pre-mRNAs [23, 25, 27], and different categories of mutations in these cis-acting splice enhancer or silencer elements affect AS [28].

Other than the above core components (snRNPs, SR proteins, and hnRNPs), there are hundreds (maybe around > 300) of RBPs genome-widely identified to bind to mRNA, which was previously not considered as RBPs [29, 30]. There are different alternative ways of competition or combination between different SFs and other RBPs [e.g., epithelial splicing regulatory protein 1 (ESRP1), RNA-binding motif proteins 4, 5, 6, 10], and the regulatory complexity of these groups of SFs contribute in diversifying the alternatively spliced RNA products.

These cis-acting elements and RPBS (as mentioned earlier), along with the splicing enhancer (ESE and ISE) and silencer elements (ESS and ISS) in precursor mRNA, can act both constitutive and regulated splicing events [25]. Additionally, mutations in these cis-acting splicing enhancers or silencer elements could largely affect AS. The ability of sequence-specific RBPs to bind to cis-acting sequences in their target pre-mRNA can cause various end products as a result of AS [26].

Trans-acting factors, including Ser/Arg-rich members and heterologous ribonucleoprotein family members, act by combining with exon splicing enhancers and silencers to activate further or inhibit specific splicing sites [31]. The sequence SFs affect the splicing site selection of the splicing regulatory complex (spliceosome) by binding to pre-mRNAs on exon splicing enhancers or silencers [32].

5. Role of AS Event in Cancer

Distinct mRNA and protein isoforms from a single gene are generated as a result of the AS event, especially in cancer. In cancer, splicing perturbations are common events and are associated with mutations in and/or altered expression of the different components of the splicing machinery. Recurrent somatic mutations in components of the human splicing machinery have occurred in human different cancers and solid tumors. Defects in AS are frequent in human tumors, while RNA splicing regulators have emerged as a new class of oncoproteins or tumor suppressors [33]. Aberrant splicing can take place because of a couple of mechanistic models; alterations in core spliceosomal components, which will lead to global splicing deregulation. That will result in a large number of aberrant products. Similarly, genomic mutations in a critical splicing motif of a single gene can alter the splicing pattern of the specific transcript. On the other hand, the alterations in an accessory splicing factor often lead to the deregulation of splicing for the limited set of transcripts where the limited sets of transcripts are required for accurate splicing.

AS dysfunction leads to alteration in the biological behaviors of tumor cells, including cell proliferation [34], apoptosis [35], tumor angiogenesis [36], and immune escape of cancer cells or tumors [37] (Figure 3). Research evidence shows that the unbalanced expression or misexpressed isomers of splicing variants is another feature of cancer [38]. Cancer-specific splicing variants can be used as diagnostic, prognostic, and therapeutic targets. Recent scientific studies indicate that alternative isoforms are regulated by a combination of genetic and epigenetic regulatory mechanisms, which are orchestrated by their balanced expression. The alternative transcription and splicing mechanisms have identified in several pro-oncogenes and oncogenes, e.g., LEF1, TP63, TP73, HNF4A, RASSF1, BCL2L1, BDNF, MYC, P53, BARD1, AR (androgen receptor), and FGFR.

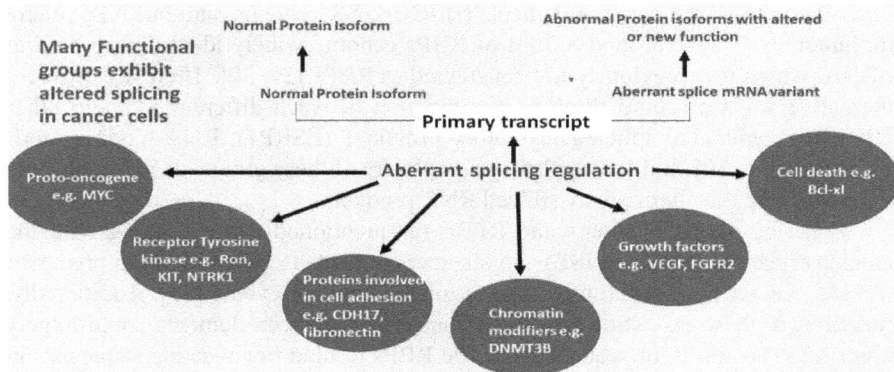

Figure 3: AS and its role in the regulation of different cellular events in cancer.

These oncogene's prevalence in almost all multi-exon genes has been recently explored with the help of the increasing application of recently developed techniques, like high-throughput experimental methods, such as exon arrays and NGS methods. It is well-known that the aberrant use of one isoform over another in some of these genes is directly linked to cancerous cell growth, proliferation, and apoptosis events [39, 40]. Regardless of the mechanism, aberrant splicing of proto-oncogenes can produce constitutively active or even gain-of-function variants that confer new survival or proliferative abilities. The regulation between antagonistic splice variants of the same gene can also be disrupted to affect proliferative pathways, interactions with proto-oncogenes and tumor suppressors, and the epithelial-to-mesenchymal transition, a path that promotes invasion and metastasis. Although, the switch between antagonistic gene isoforms in cancer illustrates critical roles for AS.

First reported AS events were used as a prognostic biomarker for non-small cell lung cancer in 2017 [41], and later studies were performed in thyroid cancer [42], colorectal cancer [43], pancreatic cancer [44], and other tumors as a prognostic biomarker for non-small cell lung cancer. In reality, the recurrent somatic mutations in components of the human splicing machinery have occurred in human solid tumors, including the brain [45], skin [46], bladder [47], breast [48], cervix [49], colon [50], kidney [51], lung [52], oral/HN [53], liver [54], stomach [55, 56], ovary [57], prostate [58], and thyroid [59] tumors, as well as hematological malignancies, including acute myeloid leukemia (AML) [60], chronic myelogenous leukemia [61], *de novo* AML [62], myelodysplastic syndrome (MDS) [63], myelodysplastic syndrome without ringed sideroblasts (MDS w/o RS) [64], myeloproliferative neoplasm MPN [66], etc.

6. Role of AS Events in Different Cellular Events

6.1 Metabolic Pathway Alteration by AS of the Pyruvate Kinase M Gene (PKM)

Alteration of 'Metabolic pathways' are a frequent event in cancer, and the Warburg effect has a wonderful contribution to the energy production in cancer cells. During the Warburg effect, a major shift happens from oxidative phosphorylation to aerobic glycolysis, leading to synthetic pathways at the expense of ATP production in cancer cells [67]. It is partly driven by AS of the pyruvate kinase M gene (PKM). Mutually exclusive exons 9 and 10 gave rise to PKM1 (exon 9), the adult isoform, or PKM2 (exon 10), the embryonic or tumor isoform [68, 69]. PKM2, a widely expressed gene variant in cancer [70], and reversal of the Warburg effect along with increases in the oxidative phosphorylation can cause the replacement of PKM2 with PKM1 reverses the Warburg effect and increases oxidative phosphorylation [71].

6.2 AS Patterns in BRCA1 and Role with Tumor Initiation

During cancer, the natural tendency of the cancer cells is to escape from cell death, which ultimately leads to the critical event for tumorigenesis. The apoptotic modulation cause alteration of the spliced genes. Different BCL2 genes of the AS variants are (a) BCL2L1, which possesses an alternative 5' splice site after exon 2 that produces long and short isoforms that are translated into the BCL-XL and BCL-XS proteins. BCL-XL has anti-apoptotic effects, while BCL-XS promotes apoptosis activities [72]. BCL-XL isoforms predominant expression was observed in cancer [73, 74]. The FAS receptor (TNR6), which acts as a cell surface receptor, can initiate cell death when bound to the ligand TNFS6 [75]. TNR6 is only active after it is subject to AS; in particular, a splice variant lacking the transmembrane domain results from exon skipping at exon 6 [76]. This shorter product, which is soluble, presumably acts by competitive binding for FAS ligands and inhibits FAS-mediated cell death [77]. High concentration of the soluble FAS is detectable in the serum of cancer patients compared with healthy individuals [78–80].

6.3 Transcription and Role of AS

Transcriptional events which involve alternative promoters and/or transcriptional termination sites often lead to multiple pre-mRNAs that can further undergo AS to generate a plethora of transcript variants corresponding to a single gene. The resulting transcript variants, majority of the time, translate to alternative proteins known as protein isoforms with various structural and functional properties or remain as noncoding transcripts. A gene can yield multiple transcript variants that differ either in their regulatory UTRs or/and protein-coding regions, thereby expanding the complexity of mammalian transcript tomes and proteomes. Interestingly, in the genes with alternative promoters, the occurrence of alternative transcriptional termination and splicing is higher, and the choice of alternative promoter and transcription termination can influence the AS pattern of the pre-mRNAs [81–84]. The differential recruitment of splicing, termination, and polyadenylation factors coupled with alternative promoter usage could be mediated through a combination of genetic and epigenetic factors, leading to variable expression of isoforms in different cellular contexts.

Some important AS events are described, e.g., the transcript variants of BDNF, which differ in their first exon (5' UTR), translate the same pre-proBDNF protein [85, 86]. SRA1 produces both NcRNA and protein isoforms that function as coactivators for transcriptional factors [87]. The AS events in the integrin subunit α6 (ITGA6) pre-mRNA lead to two distinct splice isoforms: ITGA6A and ITGA6B. The pro-proliferative ITGA6A variant was enhanced in colon cancer cells because of a process contributed to by Myc-mediated promoter activation, as well as the splicing factor epithelial splicing regulatory protein 2-mediated AS, thus promoting cell proliferation [88]. Similarly, C-Myc can upregulate polypyrimidine tract-binding protein (PTB), hnRNPA1, and hnRNPA2 to change the splicing of pyruvate kinase (PKM) and make it develop into PKM2, which could promote tumor cell proliferation [89].

7. AS Role in Mutation Events

The genetic alterations events, e.g., deletions, insertions, and epigenetic changes, such as hypo- or hyper-methylation in CpG islands, often lead to misregulation of specific isoforms leading to cancer. As an instance, the expression of MYC is controlled by two independently regulated alternative promoters, which are closely spaced. The upstream promoter P1 generates only ~ 10–25% of the MYC mRNA, while the downstream promoter P2 is the predominant promoter in normal tissues, making up 75–90% of the MYC transcripts [90]. In Burkitt's lymphoma cells, characteristic features of amplification and chromosomal translocation of the MYC locus on chromosome 8 to one of the immunoglobulin loci on chromosomes 14, 2, or 22 take place. During the chromosomal translocation event, the insertion of immunoglobulin enhancer elements renders a shift in promoter used from P2 to P1, resulting in 50–90% of MYC mRNA isoform derived by the P1 promoter [90].

8. AS Role (in Chromatin Remodeling, Histone Modification, and DNA Methylation Event, Celluler Kinases, Transcription Factors, and Different Cancer Events)

8.1 AS Role in Histone Remodeling

Nucleosomes, consisting of a single histone and 147-bp double-stranded DNA, act as the basic unit of chromatin. Histone variants are transcribed from separate genes. It has been shown to play key roles in the regulation of chromatin features and alternative RNA processing (probably affecting the co-transcriptional RNA processing). Chromatin structure is dominated by nucleosome density and positioning as well as by histone modifications and DNA methylation events [91, 92]. In mammals, five somatic H1 (histone) variants (H1.1 to H1.5) are present [93]. A recent report indicated that H1.5 deposition is observed at the splicing sites of the short exons in human lung fibroblasts (IMR90 cells), and Pol II on H1.5-marked exons exhibits greater stalling than any unmarked exons [93]. Deletion of H1.5 affects the inclusion of short exons with relatively long introns and reduces Pol II occupancy on it [92].

Chromatin-remodeling factors also affect chromosome segregation and transcription [94]. Brahma (BRM), the core adenosine triphosphatase (ATPase) subunit of the switch/sucrose nonfermenting (SWI/SNF) chromatin-remodeling complex, was first shown to facilitate the inclusion of alternative exons by interacting with the Pol II to induce its stalling [95, 96]. Analysis of transcript and protein data for 71 human genes involved in chromatin-remodeling or histone modification revealed that 70% of genes produced isoforms and 45% of the genes with altered protein domains (e.g., SUV39H2, EZH2, PRMT4, MLL 1–3, JMJD2B, and SMARCA2) [97].

9. AS Role in DNA Methylation Event

During the process of *de novo* DNA methylation event, the DNMT3B, which acts as an enzyme catalyzing the process, expresses about 40 distinct isoforms through the use of alternative promoters and splicing with many of the isoforms being catalytically inactive. Interestingly in recent scientific studies, in a DNMT3B knockout background, it also enhanced colony formation and the ectopic expression of DNMT3B3d5 in a colon cancer cell line repressed endogenous DNMT3B3 expression and reduced repetitive element methylation. Similarly, ectopic expression of another isoform, DNMT3B1, resulted in a higher incidence and increased average size of colonic micro-adenomas [98].

In human cells, the DNA-binding protein, named CCCTC-binding factor (CTCF), can promote the inclusion of weak upstream exons by mediating local RNA polymerase II pausing [99]. The process of DNA methylation inhibits CTCF binding to CD45 exon 5, which enables Pol II to transcribe more rapidly, giving rise to an exon 5 exclusion and thereby inhibit the methylation process [100]. The CTCF acts like a bifunctional regulator which influences both AS and alternative polyadenylation [101]. Removal of DNA methylation enables CTCF binding and recruitment of the cohesin complex, forming chromatin loops to promote proximal polyadenylation site usage. This mechanism suggests that DNA methylation has an important participation in RNA processing regulation. While limited current information regarding the event of how DNA-binding proteins disturb the elongation of Pol II directs that Pol II elongation in CTCF-mediated AS regulation, like the cohesin complex.

10. AS Role in Celluler Kinases

Importantly, some kinase-like Jak2, c-Met, and Aurora-B produce cancer-specific isoforms that are oncogenic. Analysis of cancer patient data revealed that Jak2 transcripts in myeloproliferative neoplasm patients identified a Jak2 transcript variant, where exon 14 is deleted, resulting in a truncated protein lacking the kinase domain in 15% of patients [102]. In the case of patients lacking the well-known V617F-activating mutation of JAK2, this transcript variant was predominant. The kinase-deficient JAK2 isoform's contribution to MPN is unknown. While the aurora kinase B isoform AURKB-2 was specifically expressed in 70% of metastatic hepatocellular carcinoma (HCC). It neither acts in the normal liver nor in non-invasive HCC [103–107].

11. AS Role in Transcription Factors

AS events can cause alteration of the biological transcription factors—e.g., KLF6 is a transcription factor that generates multiple variants with opposing biological functions. Full-length KLF6 is a tumor suppressor that displays loss of expression and is frequently inactivated through loss of heterozygosity (LOH) and mutations, while the splice variant 1, KLF6-SV1, functions as an oncogene and its overexpression in ovarian, prostate, and lung cancers are associated with poor prognosis [108].

12. AS Role in Different Cancer Events

Numerous splice variants, and their respective splicing regulators, have positively contributed to a selective advantage to tumor cells. As an example, the splicing regulators RBM5, 6, and 10 favor tumor cell proliferation and colony formation while regulating the AS of the membrane-bound protein NUMB [109]. Regulation of VEGF splicing is detrimental to the stimulation of angiogenesis [110]. The splicing factor SRSF1 after post-translational activation (also known as ASF/SF2) confers resistance to apoptosis by inducing the inclusion of the anti-apoptotic splice variant in a network of functionally related genes like Bcl-X and Mcl1 [111]. The acquisition of migratory and invasive phenotypes necessary for distal metastasis is caused by a list of well-known alternatively spliced variants related to cell adhesion (CTNND1 and CD44) and cytoskeleton organization (ENAH and FLNB) [112–114].

12.1 Angiogenesis

Angiogenesis, the critical feature of tumor progression, happens because of the formation of new blood vessels. Several proteins act as vascularization activators, including basic fibroblast growth factor, tumor necrosis factor-α, and vascular endothelial growth factor (VEGF). VEGF-A, the angiogenesis controller, is composed of 8 exons, and exons 6, 7, and 8 alternately select 3' and 5'SSs to produce isoforms. VEGF-A could lead to both angiogenic and anti-angiogenic events and might thus be utilized for anti-angiogenic therapeutics [115]. Inhibition of serine/arginine protein-specific splicing factor kinase 1 (SRPK1) by downregulating the tumor suppressor factor Wilms' tumor suppressor 1 and indirectly suppressing SRSF1 could turn the splicing of VEGF into VEGF120 with an anti-angiogenic effect and inhibiting the growth of tumor endothelial cells [116, 117].

12.2 Signal Transduction and Tumor Microenvironment

During the tumor development and metastasis event, the signals from tumor microenvironments often play important roles, and alternative splice variants of various genes act together in accumulation patterns in tumors and surrounding stromal tissues. As an example, the CD44 gene, which encodes a cell hyaluronate receptor, undergoes extensive AS in a concerted manner while functioning at its normal function and also in response to tumor-specific splicing regulation. In colorectal cancer, the full-length CD44 mRNA protein product is detected at high levels in both cancer cells and surrounding stromal tissues, whereas at least one splice variant (v6) is present in 77% of tumors but only in 17% of stromal cells [118]. Additionally, Tenascin-C (TN), an extracellular matrix glycoprotein, exists as several splice variant isoforms. Two of its isoforms are especially associated with an invasive phenotype, and these isoforms are also seen as associated with a subset of ductal carcinomas *in situ* (DCIS), which can be used as predictors of aggressive tumor behavior [119]. Fibronectin, the extracellular matrix protein, exists in multiple isoforms as generated by AS and plays roles in cell adhesion and migration. Mammary mesenchymal cells produce growth factors that can stimulate a mammary epithelial cell line to produce specific fibronectin mRNA splice variants, especially associated with cancer cell proliferation, migration, and tissue remodeling [120]. MMP-3, a stromal protein, which acts as a promoter of malignant transformation in cultured mammary epithelial cells, has been found to promote the expression of an alternative splice form of the small GTPase rac1and an increase in cellular reactive oxygen species, resulting in DNA damage and genomic instability [121].

13. Signal Transduction During Tumorigenesis and Role of AS

Cellular signaling pathways could affect the expression of SFs at either the transcriptional or the post-translational modification levels, which easily modulate the subcellular localization to the nucleus or cytoplasm in a given cell. During the process of spliceosome assembly and catalysis, the activity of SR proteins is usually regulated by phosphorylation [122]. The regulation of SR-protein phosphorylation by SRPKs in the cytoplasm and CDC-like kinases in the nucleus are guided by two families of kinases [123]. These biochemical events usually transmit signals from receptors with upstream ligand stimulating downstream effectors inside the cell. Aberrant transduction of such signaling pathways is common during tumorigenesis. The RAS/RAF/ERK pathway, which is characterized by the activation of the small GTPase RAS with a cascade stimulation of three mitogen-activated protein kinases (MAPK; namely RAF, MEK, and ERK) and acts as a hallmark event in many epithelial cell-derived tumors. The oncogenic KRAS was correlated with an ETS transcription factor-mediated aberrant-splicing regulatory network as we obtained from the transcriptomic RNA-seq data from colon adenocarcinoma or lung squamous cell carcinoma. The network involved different AS products that induce the expression of the AS factor PTBP1, which was associated with a shift in the AS of transcripts encoding the small GTPase RAC1, adapter protein NUMB, and PKM [124]. The MEK/ERK signaling pathway was also reported to mediate the phosphorylation of splicing activator DAZAP1, which is required for its cytoplasm-to-nucleus translocation, as well as splicing regulatory activity [125]. Additionally, the activation of the MAPK/ERK downstream of RAS can lead to the phosphorylation of SFs, such as signal transduction and activation of RNA metabolism 68 (SAM68). Phospho-SAM68 promotes an intron retention event in the 3' UTR transcript of the SRSF1 while binding with it, thus leading to the subsequent switch in the splicing profile, such as the RON gene transcripts [126, 127].

The phosphatidylinositol 3-kinase (PI3K)/AKT pathway is another key regulatory pathway in regulating cell survival and apoptosis. Activated AKT has been shown to phosphorylate SRSF1 in lung cancer cells, thereby generating an anti-apoptotic caspase-9b isoform through the exclusion of an exon 3, 4, 5, 6 cassette [128]. EGF signaling while acting as a splicing regulator by the phosphorylation of AKT/SRPK/SR protein148 and/or SPSB1-hnRNPA1 ubiquitination [129]. The PI3K/AKT pathway is also known to activate the key regulator of cell metabolism and growth factor mammalian target of rapamycin complex 1 (mTORC1). The activation of the SR proteins to participate in the splicing of lipogenesis-related transcripts to fuel tumor cell metabolism and the signaling of the mTORC1-S6K1 axis promotes the phosphorylation and cytoplasm-to-nucleus translocation of kinase SRPK2 [130].

The regulating development and stemness of the Wingless (Wnt) signaling pathway are well-known for and for their close association with many cancer types, especially colorectal cancer (CRC) [131]. Glycogen synthase 3β is part of the canonical Wnt pathway and is directly involved in the phosphorylation of SFs, such as SRSF2152 or PTB-associated splicing factor (PSF) [132, 133]. Activated Wnt/β-catenin signaling act as an enhancer of the transcript level of SRSF3 [134]. SRSF3 also acts as a negative regulator of the alternative exon inclusion variant of RAS-related C3 botulinum toxin substrate 1b [134].

14. Noncoding RNA Role in AS Event

Noncoding RNAs (ncRNAs) are major components of the human transcriptome [135], and these groups of RNA include miRNAs (microRNAs), lncRNAs (long noncoding RNAs), circRNAs (circular RNAs), snRNAs (small nuclear RNAs), pRNA, and tRNA and have been it have been proven to act as regulatory molecules that mediate cancer processes through alternate splicing (AS) events. LncRNAs—an RNA subgroup that is usually longer than 200 nucleotides—act as an important regulatory effect on cellular metabolism and are involved in the regulation of AS [136, 137]. For instance, lncRNA HOXB-AS3 could interact with the ErbB3-binding protein 1 (EBP1) and regulate ribosomal RNA transcription and de novo protein synthesis in NPM1-mutated AML [138].

The back-splicing process leads to endogenous circRNAs expression, which is the covalent joining of a downstream splice donor site with an upstream splice acceptor site [139]. The circRNA biogenesis is performed by the canonical spliceosome machinery and regulated by the same cis-regulatory elements and trans-acting factors as that control linear mRNA splicing; circRNAs can be regarded as an additional form of AS [140]. The abnormal expression of ncRNAs is associated with invasion, metastasis, chemoresistance, and radioresistance of different variety of cancers. It influences different gene expression levels via chromatin modification, transcription, and post-transcriptional processing. NcRNAs (noncoding RNAs) include miRNAs (microRNAs), lncRNAs (long noncoding RNAs), circRNAs (circular RNAs), and snRNAs (small nuclear RNAs) and have been proven to act as regulatory molecules that mediate cancer processes through AS. NcRNAs can directly or indirectly influence a plethora of molecular targets to regulate cis-acting elements, trans-acting factors, or pre-mRNA transcription at multiple levels, affecting the AS process and generating alternatively spliced isoforms. Consequently, ncRNA-mediated AS outcomes affects multiple cellular signaling pathways that promote or suppress cancer progression. NcRNAs regulate the individual genes and gene expression programs by changing the fundamental transcriptional mechanism or via epigenetic regulation at multiple levels, such as transcription, translation, and protein function.

Conclusions

AS, a cellular hallmark event orchestrated in modulating several cancer events, including immune surveillance and current therapeutic modalities in modulating these events, can be very helpful. In this current chapter, we mostly kept our focus on modalities of AS events and did not add much to the therapeutic modalities areas.

Acknowledgment

I would like to acknowledge Dr. Kaushik Ghose, Research Scientist, IGCAST, Plant and soil science, Texas Tech University, for helping me in creating figures and inspiring me to write this chapter.

References

[1] Berget, S.M., Moore, C. and Sharp, P.A. (1977). Spliced segments at the 5′ terminus of adenovirus 2 late mRNA. Proc. Natl. Acad. Sci. USA 74: 3171–3175.

[2] Chow, L.T., Roberts, J.M., Lewis, J.B. and Broker, T.R. (1977). A map of cytoplasmic RNA transcripts from lytic adenovirus type 2, determined by electron microscopy of RNA: DNA hybrids. Cell 11: 819–836.

[3] Herzel, L., Ottoz, D.S.M., Alpert, T. and Neugebauer, K.M. (2017). Splicing and transcription touch base: Co-transcriptional spliceosome assembly and function. Nature Reviews Molecular Cell Biology 18: 637–650. (https://doi.org/10.1038/nrm.2017.63).

[4] Baralle, F.E. and Giudice, J. (2017). Alternative splicing as a regulator of development and tissue identity. Nat. Rev. Mol. Cell Biol. 7: 437–451.

[5] Agrawal, A.A., Yu, L., Smith, P.G. and Buonamici, S. (2018). Targeting splicing abnormalities in cancer. Curr. Opin. Genet. Dev. 48: 67–74.

[6] Climente-Gonzalez, H., Porta-Pardo, E., Godzik, A. and Eyras, E. (2017). The functional impact of alternative splicing in cancer. Cell Rep. 20: 2215–2226.

[7] Leoni, G., Le Pera, L., Ferrè, F., Raimondo, D. and Tramontano, A. (2011). Coding potential of the products of alternative splicing in human. Genome Biol. 12: R9.

[8] Kelemen, O., Convertini, P., Zhang, Z., Wen, Y., Shen, M., Falaleeva, M. and Stamm, S. (2013). Function of alternative splicing. Gene 514: 1–30. doi: 10.1002/ 9783527678679.dg00350.

[9] Feng, H., Qin, Z. and Zhang, X. (2013). Opportunities and methods for studying alternative splicing in cancer with RNA-Seq. Cancer Lette. 340: 179–191. doi: 10.1016/j.canlet.2012.11.010.

[10] Wang, E.T., Sandberg, R., Luo, S., Khrebtukova, I., Zhang, L., Mayr, C., Kingsmore, S.F., Schroth, G.P. and Burge, C.B. (2008). Alternative isoform regulation in human tissue transcriptomes. Nature 456: 470. doi: 10.1038/nature07509.

[11] Haack, D.B., Yan, X., Zhang, C., Hingey, J., Lyumkis, D., Baker, T.S. and Toor, N. (2019). Cryo-EM structures of a group II intron reverse splicing into DNA. Cell 178: 612–623.

[12] Wahl, M.C., Will, C.L. and Lührmann, R. (2009). The spliceosome: Design principles of a dynamic RNP machine. Cell 136: 701–718.

[13] Hoskins, A.A. and Moore, M.J. (2012). The spliceosome: A flexible, reversible macromolecular machine. Trends Biochem. Sci. 5: 179–188.

[14] Papasaikas, P. and Valcárcel, J. (2016). The spliceosome: The ultimate RNA chaperone and sculptor. Trends Biochem. Sci. 41: 33–45.

[15] Rhine, C.L., Cygan, K.J., Soemedi, R., Maguire, S., Murray M.F., Monaghan, S. F., Fairbrother, W.G. (2018). Hereditary cancer genes are highly susceptible to splicing mutations. PLoS Genet. 14: e1007231.

[16] Pan, Q., Shai, O., Lee, L.J., Frey, B.J. and Blencowe, B.J. (2008). Deep surveying of alternative splicing complexity in the human transcriptome by high-throughput sequencing. Nat. Genet. 40: 1413–1415. doi: 10.1038/ng.259.

[17] Blencowe, B.J. (2006). Alternative splicing: new insights from global analyses. Cell 126: 37–47. doi: 10.1016/j.cell.2006.06.023. [PubMed] [Google Scholar].

[18] Busch, A. and Hertel, K.J. (2012). Evolution of SR protein and hnRNP splicing regulatory factors. Wiley Interdiscip. Rev. RNA 3: 1–12.

[19] Hoskins, A.A. et al. (2011). Ordered and dynamic assembly of single spliceosomes. Science 331: 1289–1295.

[20] Jurica, M.S. and Moore, M.J. (2003). Pre-mRNA splicing: Awash in a sea of proteins. Mol. Cell 12: 5–14.

[21] Dredge, B.K., Polydorides, A.D. and Darnell, R.B. (2001). The splice of life: Alternative splicing and neurological disease. Nat. Rev. Neurosci. 2: 43–50.

[22] Matlin, A.J., Clark, F. and Smith, C.W. (2005). Understanding alternative splicing: Towards a cellular code. Nat. Rev. Mol. Cell Biol. 6: 386–398.

[23] Howard, J.M. and Sanford, J.R. (2015). The RNAissance family: SR proteins as multifaceted regulators of gene expression. Wiley Interdiscip. Rev. RNA 6: 93–110.

[24] Martinez-Montiel, N. et al. (2018). Alternative Splicing as a target for cancer treatment. Int. J. Mol. Sci. 19: 545–572.

[25] Fu, X.D. and Ares, M. Jr. (2014). Context-dependent control of alternative splicing by RNA binding proteins. Nat. Rev. Genet. 15: 689–701.

[26] Goldammer, G. et al. (2018). Characterization of cis-acting elements that control oscillating alternative splicing. RNA Biol. 15: 1081–1092.

[27] Geuens, T., Bouhy, D. and Timmerman, V. (2016). The hnRNP family: Insights into their role in health and disease. Hum. Genet. 135: 851–867.

[28] Yoshimi, A. and Abdel-Wahab, O. (2017). Molecular pathways: Understanding and targeting mutant spliceosomal proteins. Clin. Cancer Res. 23: 336–341.

[29] Barbosa-Morais, N.L., Carmo-Fonseca, M. and Aparicio, S. (2006). Systematic genome-wide annotation of spliceosomal proteins reveals differential gene family expansion. Genome Res. 16: 66–77.

[30] Castello, A., Fischer, B., Eichelbaum, K., Horos, R., Beckmann, B.M., Strein, C., Davey, N.E., Humphreys, D.T., Preiss, T., Steinmetz, L.M., Krijgsveld, J. and Hentze, M.W. (2012). Insights into RNA biology from an atlas of mammalian mRNA binding proteins. Cell 149: 1393–1406.

[31] Dvinge, H., Kim, E. and Abdel-Wahab, O. et al. (2016). RNA splicing factors as oncoproteins and tumour suppressors. Nat. Rev. Cancer 16: 413–430. https://doi.org/10.1038/nrc.2016.51.

[32] Kornblihtt, A.R., Schor, I.E., Alló, M., Dujardin, G., Petrillo, E. and Muñoz, M.J. (2013). Alternative splicing: A pivotal step between eukaryotic transcription and translation. Nat. Rev. Mol. Cell Biol. 14: 153. doi: 10.1038/nrm3525.

[33] Babic, I., Anderson, E.S., Tanaka, K., Guo, D., Masui, K., Li, B., Zhu, S., Gu, Y., Villa, G.R., Akhavan, D., Nathanson, D., Gini, B., Mareninov, S., Li, R., Camacho, C.E., Kurdistani, S.K., Eskin, A., Nelson, S.F., Yong, W.H., Cavenee, W.K., Cloughesy, T.F., Christofk, H.R., Black, D.L. and Mischel, P.S. (2013). EGFR mutation-induced alternative splicing of max contributes to growth of glycolytic tumors in brain cancer. Cell Metab. 17: 1000–1008.

[34] Endo, T. (2019). Dominant-negative antagonists of the Ras–ERK pathway: DA-Raf and its related proteins generated by alternative splicing of Raf. Exp. Cell Res. 387: 111775. doi: 10.1016/j.yexcr.2019.111775.

[35] Pal, S., Medatwal, N., Kumar, S., Kar, A., Komalla, V., Yavvari, P.S., Mishra, D., Rizvi, Z.A., Nandan, S., Malakar, D., Pillai, M., Awasthi, A., Das, P., Sharma, R.D., Srivastava, A., Sengupta, S., Dasgupta, U. and Bajaj, A. (2019). A localized chimeric hydrogel therapy combats tumor

progression through alteration of sphingolipid metabolism. ACS Central Sci. 5: 1648–1662. doi: 10.1021/acscentsci.9b00551.

[36] Pentheroudakis, G., Mavroeidis, L., Papadopoulou, K., Koliou, G.A., Bamia, C., Chatzopoulos, K., Samanthas, E., Mauri, D., Efstratiou, I., Pectasides, D., Mekatsoris, T., Bafaloukos, D., Papakostus, P., Papatsibas, G., Bombolaki, I., Chrisafi, S., Koerea, H.P., Petraki, K., Kafiri, G., Fountzilas, G. and Kotula, V. (2019). Angiogenic and antiangiogenic VEGFA splice variants in colorectal cancer: Prospective retrospective cohort study in patients treated with irinotecan-based chemotherapy and bevacizumab. Clin. Colorectal Cancer 18: e370–e384.

[37] Yao, J., Caballero, O.L., Huang, Y., Lin, C., Rimoldi, D., Behren, A. et al. (2016). Altered expression and splicing of ESRP1 in malignant melanoma correlates with epithelial–mesenchymal status and tumor-associated immune cytolytic activity. Cancer Immunol. Res. 4: 552–561. doi: 10.1158/2326-6066.cir-15-0255.

[38] Ladomery, M. (2013). Aberrant alternative splicing is another hallmark of cancer. Int. J. Cell Biol. 2013: 463786.

[39] Davuluri, R.V., Suzuki, Y., Sugano, S., Plass, C. and Huang, T.H. (2008). The functional consequences of alternative promoter use in mammalian genomes. Trends Genet. 24: 167–177.

[40] Rajan, P., Elliott, D.J., Robson, C.N. and Leung, H.Y. (2009). Alternative splicing and biological heterogeneity in prostate cancer. Nat. Rev. Urol. 6: 454–460.

[41] Li, Y., Sun, N., Lu, Z., Sun, S., Huang, J., Chen, Z. and He, J. (2017). Prognostic alternative mRNA splicing signature in non-small cell lung cancer. Cancer Lett. 393: 40–51. doi: 10.1016/j.canlet.2017.02.016.

[42] Lin, P., He, R.Q., Huang, Z.G., Zhang, R., Wu, H.Y., Shi, L., Li, X.-J., Li, Q, Chen, G.,Yang, H. and Yun, H. (2019). Role of global aberrant alternative splicing events in papillary thyroid cancer prognosis. Aging 11: 2082. doi: 10.18632/aging.101902.

[43] Xiong, Y., Deng, Y., Wang, K., Zhou, H., Zheng, X., Si, L. and Fu, Z. (2018). Profiles of alternative splicing in colorectal cancer and their clinical significance: A study based on large-scale sequencing data. EBioMedicine 36: 183–19. doi: 10.1016/j.ebiom.2018.09.021.

[44] Yu, M., Hong, W., Ruan, S., Guan, R., Tu, L., Huang, B., Hou, B., Jian, Z., Ma, L. and Jin, H. (2019). Genome-wide profiling of prognostic alternative splicing pattern in pancreatic cancer. Front. Oncol. 9: 773. doi: 10.3389/fonc.2019.00773.

[45] Babic, I., Anderson, E.S., Tanaka, K., Guo, D., Masui, K., Li, B., Zhu, S., Gu, Y., Villa, G.R., Akhavan, D., Nathanson, D., Gini, B., Mareninov, S., Li, R., Camacho, C.E., Kurdistani, S.K., Eskin, A., Nelson, S.F., Yong, W.H., Cavenee, W.K., Cloughesy, T.F., Christofk, H.R., Black, D.L., Mischel, P.S. (2013). EGFR mutation-induced alternative splicing of Max contributes to growth of glycolytic tumors in brain cancer. Cell Metab. 17: 1000–1008.

[46] Jensen, M.A., Wilkinson, J.E. and Krainer, A.R. (2014). Splicing factor SRSF6 promotes hyperplasia of sensitized skin. Nat. Struct. Mol. Biol. 21: 189–197.

[47] Xie, R., Chen, X., Chen, Z., Huang, M., Dong, W., Gu, P., Zhang, J., Zhou, Q., Dong, W., Han, J., Wang, X., Li, H., Huang, J. and Lin, T. (2019). Polypyrimidine tract binding protein 1 promotes lymphatic metastasis and proliferation of bladder cancer via alternative splicing of MEIS2 and PKM. Cancer Lett. 449: 31–44.

[48] Anczuków, O., Rosenberg, A.Z., Akerman, M., Das, S., Zhan, L., Karni, R., Muthuswamy, S.K. and Krainer, A.R. (2012). The splicing factor SRSF1 regulates apoptosis and proliferation to promote mammary epithelial cell transformation. Nat. Struct. Mol. Biol. 19: 220–228.

[49] Liu, F., Dai, M., Xu, Q., Zhu, X., Zhou, Y., Jiang, S., Wang, Y., Ai, Z., Ma, L., Zhang, Y., Hu, L., Yang, Q., Li, J., Zhao, S., Zhang, Z. and Teng, Y. (2018). SRSF10-mediated IL1RAP alternative splicing regulates cervical cancer oncogenesis via mIL1RAP-NF-kappaB-CD47 axis. Oncogene 37: 2394–2409.

[50] Zhou, X., Li X., Chen, Y., Wu, W., Xie, Z., Xi, Q., Han, J., Wu, G., Fang, J. and Feng, Y. (2014). BCLAF1 and its splicing regulator SRSF10 regulate the tumorigenic potential of colon cancer cells. Nat. Commun. 5: 4581.

[51] Sokół, E., Kędzierska, H., Czubaty, A., Rybicka, B., Rodzik, K., Tański, Z., Bogusławska, J. and Piekiełko-Witkowska, A. (2018). MicroRNA-mediated regulation of splicing factors SRSF1, SRSF2 and hnRNP A1 in context of their alternatively spliced 3'UTRs. Exp. Cell Res. 363: 208–217.

[52] Sheng, J., Zhao, Q., Zhao, J., Zhang, W., Sun, Y., Qin, P., Lv, Y., Bai, L., Yang, Q., Chen, L., Qi, Y., Zhang, G., Zhang, L., Gu, C., Deng, X., Liu, H., Meng, S., Gu, H., Liu, Q., Coulson, J.M., Li, X., Sun, B. and Wang, Y. (2018). SRSF1 modulates PTPMT1 alternative splicing to regulate lung cancer cell radioresistance. EBioMedicine 38: 113–126.

[53] Peiqi, L., Zhaozhong, G., Yaotian, Y., Jun, J., Jihua, G. and Rong, J. (2016). Expression of SRSF3 is correlated with carcinogenesis and progression of oral squamous cell carcinoma. Int. J. Med. Sci. 13: 533–539.

[54] Duriez, M., Mandouri, Y., Lekbaby, B., Wang, H., Schnuriger, A., Redelsperger, F., Guerrera, C.I., Lefevre, M., Fauveau, V., Ahodantin, J., Quetier, I., Chhuon, C., Gourari, S., Boissonnas, A., Gill, U., Kennedy, P., Debzi, N., Sitterlin, D., Maini, M.K., Kremsdorf, D. and Soussan, P. (2017). Alternative splicing of hepatitis B virus: A novel virus/host interaction altering liver immunity. J. Hepatol. 67: 687–699.

[55] Iborra, S., Hirschfeld, M., Jaeger, M., Zur Hausen, A., Braicu, I., Sehouli, J., Gitsch, G. and Stickeler, E. (2013). Alterations in expression pattern of splicing factors in epithelial ovarian cancer and its clinical impact. Int. J. Gynecol. Cancer 23: 990–996.

[56] Liang, X., Chen, W., Shi, H., Gu, X., Li, Y., Qi, Y., Xu, K., Zhao, A. and Liu, J. (2018). PTBP3 contributes to the metastasis of gastric cancer by mediating CAV1 alternative splicing. Cell Death Dis. 9: 569.

[57] Ailiken, G., Kitamura, K., Hoshino, T., Satoh, M., Tanaka, N., Minamoto, T., Rahmutulla, B., Kobayashi, S., Kano, M., Tanaka, T., Kaneda, A., Nomura, F., Matsubara, H. and Matsushita, K. (2020). Post-transcriptional regulation of BRG1 by FIRDeltaexon2 in gastric cancer. Oncogenesis 9: 26.

[58] Fan, L. et al. (2018). Histone demethylase JMJD1A promotes alternative splicing of AR variant 7 (AR-V7) in prostate cancer cells. Proc. Natl Acad. Sci. USA 115: E4584–E4593.

[59] Liu, J., Huang, B., Xiao, Y., Xiong, H.M., Li, J., Feng, D.Q., Chen, X.M., Zhang, H.B. and Wang, X.Z. (2012). Aberrant expression of splicing factors in newly diagnosed acute myeloid leukemia. Onkologie 35: 335–340.

[60] Itzykson, R., Kosmider, O., Renneville, A., Gelsi-Boyer, V., Meggendorfer, M., Morabito, M., Berthon, C., Adès, L., Fenaux, P., Beyne-Rauzy, O., Vey, N., Braun, T., Haferlach, T., Dreyfus, F., Cross, N.C., Preudhomme, C., Bernard, O.A., Fontenay, M., Vainchenker, W., Schnittger, S., Birnbaum, D., Droin, N. and Solary, E..(2013). Prognostic score including gene mutations in chronic myelomonocytic leukemia. J. Clin. Oncol. 31: 2428–2436.

[61] Wang, H., Zhang, N., Wu, X., Zheng, X., Ling, Y. and Gong, Y. (2019). Prognostic value of U2AF1 mutant in patients with *de novo* myelodysplastic syndromes: A meta-analysis. Ann. Hematol. 98: 2629–2639.

[62] Kanagal-Shamanna, R., Luthra, R., Yin, C.C., Patel, K.P., Takahashi, K., Lu, X., Lee, J., Zhao, C., Stingo, F., Zuo, Z., Routbort, M.J., Singh, R.R., Fox, P., Ravandi, F., Garcia-Manero, G., Medeiros, L.J. and Bueso-Ramos, C.E. (2016). Myeloid neoplasms with isolated isochromosome 17q demonstrate a high frequency of mutations in SETBP1, SRSF2, ASXL1 and NRAS. Oncotarget 7: 14251–14258.

[63] Mupo, A., Seiler, M., Sathiaseelan, V., Pance, A., Yang, Y., Agrawal, A.A., Iorio, F., Bautista, R., Pacharne, S., Tzelepis, K., Manes, N., Wright, P., Papaemmanuil, E., Kent, D.G., Campbell, P.C., Buonamici, S., Bolli, N. and Vassiliou, G.S. (2017). Hemopoietic-specific Sf3b1-K700E knock-in mice display the splicing defect seen in human MDS but develop anemia without ring side-roblasts. Leukemia 31: 720–727.

[64] Schischlik, F., Jäger, R., Rosebrock, F., Hug, E., Schuster, M., Holly, R., Fuchs, E., Milosevic Feenstra, J.D., Bogner, E., Gisslinger, B., Schalling, M., Rumi, E., Pietra, D., Fischer, G., Faé, I., Vulliard, L., Menche, J., Haferlach, T., Meggendorfer, M., Stengel, A., Bock, C., Cazzola, M., Gisslinger, H. and Kralovics, R. (2019). Mutational landscape of the transcriptome offers putative targets for immunotherapy of myeloproliferative neoplasms. Blood 134: 199–210.

[65] Lin, P., He, R.Q., Huang, Z.G., Zhang, R., Wu, H.Y., Shi, L., Li, X.J., Li, Q., Chen, G., Yang, H. and He, Y. (2019). Role of global aberrant alternative splicing events in papillary thyroid cancer prognosis. Aging (Albany NY) 11: 2082–2097.

[66] Christofk, H.R., vander Heiden, M.G., Harris, M.H., Ramanathan, A., Gerszten, R.E., Wei, R., Fleming, M.D, Schreiber, S.L. and Cantley, L.C. (2008). The M2 splice isoform of pyruvate kinase is important for cancer metabolism and tumour growth. Nature 452: 230–233.

[67] Warburg, O. (1956). On the origin of cancer cells. Science 123(3191): 309–14. doi: 10.1126/science.123.3191.309. PMID: 13298683.

[68] Noguchi, T., Inoue, H. and Tanaka, T. (1986). The M1- and M2-type isozymes of rat pyruvate kinase are produced from the same gene by alternative RNA splicing. J. Biol. Chem. 261: 13807–13812.

[69] Mazurek, S., Boschek, C., Hugo, F. and Eigenbrodt, E. (2005). Pyruvate kinase type M2 and its role in tumor growth and spreading. Semin. Cancer Biol. 15: 300–308.

[70] Boise, L.H., González-García, M., Postema, C.E., Ding, L., Lindsten, T., Turka, L.A., Mao, X., Nuñez, G. and Thompson, C.B. (1993). Bcl-x, a bcl-2-related gene that functions as a dominant regulator of apoptotic cell death. Cell 74: 597–608.

[71] Xerri, L., Parc, P., Brousset, P., Schlaifer, D., Hassoun, J., Reed, J.C., Krajewski, S. and Birnbaum, D. (1996). Predominant expression of the long isoform of Bcl-x (Bcl-xL) in human lymphomas. Br. J. Haematol. 92: 900–906.

[72] Takehara, T., Liu, X., Fujimoto, J., Friedman, S.L. and Takahashi, H. (2001). Expression and role of Bcl-xL in human hepatocellular carcinomas. Hepatology 34: 55–61.

[73] Bouillet, P. and O'Reilly, L.A. (2009). CD95, BIM and T cell homeostasis. Nat. Rev. Immunol. 9: 514–519.

[74] Cascino, I., Fiucci, G., Papoff, G. and Ruberti, G. (1995). Three functional soluble forms of the human apoptosis-inducing Fas molecule are produced by alternative splicing. J. Immunol. 154: 2706–2713.

[75] Cheng, J., Zhou, T., Liu, C., Shapiro, J.P., Brauer, M.J., Kiefer, M.C., Barr, P.J. and Mountz, J.D. (1994). Protection from Fas-mediated apoptosis by a soluble form of the Fas molecule. Science 263: 1759–1762.

[76] Liu, J.H., Wei, S., Lamy, T., Li, Y., Epling-Burnette, P.K., Djeu, J.Y. and Loughran, T.P. Jr. (2002). Blockade of Fas-dependent apoptosis by soluble Fas in LGL leukemia. Blood 100: 1449–1453.

[77] Sheen Chen, S.M., Chen, H.S., Eng, H.L. and Chen, W.J. (2003). Circulating soluble Fas in patients with breast cancer. World J. Surg. 27: 10–13.

[78] Kondera Anasz, Z., Mielczarek-Palacz, A. and Sikora, J. (2005). Soluble Fas receptor and soluble Fas ligand in the serum of women with uterine tumors. Apoptosis 10: 1143–1149.

[79] Cramer, P., Pesce, C.G., Baralle, F.E. and Kornblihtt, A.R. (1997). Functional association between promoter structure and transcript alternative splicing. Proc. Natl. Acad. Sci. USA 94: 11456–11460.

[80] Timmusk, T., Palm, K., Metsis, M., Reintam, T., Paalme, V., Saarma, M. and Persson, H. (1993). Multiple promoters direct tissue-specific expression of the rat BDNF gene. Neuron 10: 475–489.

[81] Kornblihtt, A.R. (2005). Promoter usage and alternative splicing. Curr. Opin. Cell Biol. 17: 262–268.

[82] Davuluri, R.V., Suzuki, Y., Sugano, S., Plass, C. and Huang, T.H. (2008). The functional consequences of alternative promoter use in mammalian genomes. Trends Genet. 24: 167–177.

[83] Pal, S., Gupta, R., Kim, H., Wickramasinghe, P., Baubet, V., Showe, L.C., Dahmane, N. and Davuluri, R.V. (2011). Alternative transcription exceeds alternative splicing in generating the transcriptome diversity of cerebellar development. Genome Res. 21: 1260–1272.

[84] Albulescu, L.O., Sabet, N., Gudipati, M., Stepankiw, N., Bergman, Z.J., Huffaker, T.C. and Pleis, J.A. (2012). A quantitative, high-throughput reverse genetic screen reveals novel connections between Pre-mRNA splicing and 5′ and 3′ end transcript determinants. PLoS Genet. 8: e1002530.

[85] Ulveling, D., Francastel, C. and Hube, F. (2011). When one is better than two: RNA with dual functions. Biochimie 93: 633–644.

[86] Pruunsild, P., Kazantseva, A., Aid, T., Palm, K. and Timmusk, T. (2007). Dissecting the human BDNF locus: Bidirectional transcription, complex splicing, and multiple promoters. Genomics 90: 397–406.

[87] Winter, J., Kunath, M., Roepcke, S., Krause, S., Schneider, R. and Schweiger, S. (2007). Alternative polyadenylation signals and promoters act in concert to control tissue-specific expression of the Opitz syndrome gene MID1. BMC Mol. Biol. 8: 105.

[88] David, C.J., Chen, M., Assanah, M., Canoll, P. and Manley, J.L. (2010). HnRNP proteins controlled by c-Myc deregulate pyruvate kinase mRNA splicing in cancer. Nature 463(7279): 364–368. doi:10.1038/nature08697.

[89] Marcu, K.B., Bossone, S.A. and Patel, A.J. (1992). myc function and regulation. Annu. Rev. Biochem. 61: 809–60. doi: 10.1146/annurev.bi.61.070192.004113. PMID: 1497324.

[90] Duan, C.G., Wang, X., Zhang, L., Xiong, X., Zhang, Z., Tang, K., Pan, L., Hsu, C.C., Xu, H. and Tao, W.A. (2017). A protein complex regulates RNA processing of intronic heterochromatin containing genes in Arabidopsis. Proc. Natl. Acad. Sci. USA 114: E7377–E7384. doi: 10.1073/pnas.1710683114.

[91] Happel, N. and Doenecke, D. (2009). Histone H1 and its isoforms: Contribution to chromatin structure and function. Gene 431: 1–12. doi: 10.1016/j.gene.2008.11.003.

[92] Glaich, O., Leader, Y., Lev Maor, G. and Ast, G. (2019). Histone H1.5 binds over splice sites in chromatin and regulates alternative splicing. Nucleic Acids Res. 47: 6145–6159. 10.1093/nar/gkz338.

[93] Clapier, C.R. and Cairns, B.R. (2009). The biology of chromatin remodeling complexes. Annu. Rev. Biochem. 78: 273–304. doi: 10.1146/annurev.biochem.77.062706.153223. PMID: 19355820.

[94] Batsché, E., Yaniv, M. and Muchardt, C. (2006). The human SWI/SNF subunit Brm is a regulator of alternative splicing. Nat. Struct. Mol. Biol. 13(1): 22–9. doi: 10.1038/nsmb1030. Epub 2005 Dec 11. PMID: 16341228.

[95] Jancewicz, I., Siedlecki, J.A., Sarnowski, T.J. and Sarnowska, E. (2019). BRM: The core ATPase subunit of SWI/SNF chromatin-remodelling complex—A tumour suppressor or tumour-promoting factor? Epigenetics & Chromatin 12: 68. https://doi.org/10.1186/s13072-019-0315-4.

[96] Linhart, H.G., Lin, H., Yamada, Y., Moran, E., Steine, E.J., Gokhale, S., Lo, G., Cantu, E., Ehrich, M., He, T., Meissner, A. and Jaenisch, R. (2007). Dnmt3b promotes tumorigenesis *in vivo* by gene-specific *de novo* methylation and transcriptional silencing. Genes Dev. 21: 3110–3122.

[97] Shukla, S., Kavak, E., Gregory, M., Imashimizu, M., Shutinoski, B. and Kashlev, M. (2011). CTCF-promoted RNA polymerase II. Nature 479: 74–79. doi: 10.1038/nature10442.

[98] Ong, C.T. and Corces, V.G. (2014). CTCF: An architectural protein bridging genome topology and function. Nat. Rev. Genet. 15: 234–246. doi: 10.1038/nrg3663.

[99] Nanavaty, V., Abrash, E.W., Hong, C., Park, S., Fink, E.E., Li, Z., Sweet, T.J., Bhasin, J.M., Singuri, S., Lee, B.H., Hwang, T.H. and Ting, A.H. (2020). DNA methylation regulates alternative polyadenylation via CTCF and the cohesin complex. Mol. Cell 78: 752–764.e6. doi: 10.1016/j.molcel.2020.03.024.

[100] Ma, W., Kantarjian, H., Zhang, X., Wang, X., Zhang, Z., Yeh, C.H., O'Brien, S., Giles, F., Bruey, J.M. and Albitar, M. (2010). JAK2 exon 14 deletion in patients with chronic myeloproliferative neoplasms. PLoS One 5: e12165.

[101] Sistayanarain, A., Tsuneyama, K., Zheng, H., Takahashi, H., Nomoto, K., Cheng, C., Murai, Y., Tanaka, A. and Takano, Y. (2006). Expression of Aurora-B kinase and phosphorylated histone H3 in hepatocellular carcinoma. Anticancer Res. 26: 3585–3593.

[102] Narla, G., Heath, K.E., Reeves, H.L., Li, D., Giono, L.E., Kimmelman, A.C., Glucksman, M.J., Narla, J., Eng, F.J., Chan, A.M., Ferrari, A.C., Martignetti, J.A. and Friedman, S.L. (2001). KLF6, a candidate tumor suppressor gene mutated in prostate cancer. Science 294: 2563–2566. doi: 10.1126/science.1066326.

[103] Narla, G., Difeo, A., Reeves, H.L., Schaid, D.J., Hirshfeld, J., Hod, E., Katz, A., Isaacs, W.B., Hebbring, S., Komiya, A., McDonnell, S.K., Wiley, K.E., Jacobsen, S.J., Isaacs, S.D., Walsh, P.C., Zheng, S.L., Chang, B.L., Friedrichsen, D.M., Stanford, J.L., Ostrander, E.A., Chinnaiyan, A.M., Rubin, M.A., Xu, J., Thibodeau, S.N., Friedman, S.L. and Martignetti, J.A. (2005a). A germline DNA polymorphism enhances alternative splicing of the KLF6 tumor suppressor gene and is associated with increased prostate cancer risk. Cancer Res. 65: 1213–1222. doi: 10.1158/0008-5472.CAN-04-4249.

[104] Narla, G., DiFeo, A., Yao, S., Banno, A., Hod, E., Revees, H.L., Qiao, R.F, Camacho-Venegras, O., Levine, A., Kirschenbaum, A., Chan, A.M., Friedman, S.L. and Martignetti, J.A. (2005b). Targeted inhibition of the KLF6 splice variant, KLF6 SV1, suppresses prostate cancer cell growth and spread. Cancer Res. 65: 5761–5768. doi: 10.1158/0008-5472.CAN-05-0217.

[105] Narla, G., DiFeo, A., Fernandez, Y., Dhanasekaran, S., Huang, F., Sangodkar, J., Hod, E., Leake, D., Friedman, S.L., Hall, S.J., Chinnaiyan, A.M., Gerald, W.L., Rubin, M.A. and Martignetti, J.A. (2008). KLF6-SV1 overexpression accelerates human and mouse prostate cancer progression and metastasis. J. Clin. Invest. 118: 2711–2721. doi: 10.1172/JCI34780.

[106] Bechara, E.G., Sebestyén, E., Bernardis, I., Eyras, E. and Valcárcel, J. (2013). RBM5, 6, and 10 differentially regulate NUMB alternative splicing to control cancer cell proliferation. Mol. Cell 52(5): 720–33. https://doi.org/10.1016/j.molcel.2 013.11.010.

[107] Amin, E.M, Oltean, S., Hua, J., Gammons, M.V., Hamdollah-Zadeh, M., Welsh, G.I., Cheung, M.K., Ni, L., Kase, S., Rennel, E.S., Symonds, K.E., Nowak, D.G., Royer-Pokora, B., Saleem, M.A., Hagiwara, M., Schumacher, V.A., Harper, S.J., Hinton, D.R., Bates, D.O. and Ladomery, M.R. (2011). WT1 mutants reveal SRPK1 to be a downstream angiogenesis target by altering VEGF splicing. Cancer Cell. 20(6): 768–80. https://doi.org/10.1016/j.ccr.2011.10.016.

[108] Moore, M.J., Wang, Q., Kennedy, C.J. and Silver, P.A. (2010). An alternative splicing network links cell-cycle control to apoptosis. Cell 142(4): 625–36. https://doi. org/10.1016/j.cell.2010.07.019.

[109] Sebestyén, E., Zawisza, M. and Eyras, E. (2015). Detection of recurrent alternative splicing switches in tumor samples reveals novel signatures of cancer. Nucleic Acids Res. 43(3): 1345–56. https://doi.org/10.1093/nar/gku1392.

[110] Xu, Y., Gao, X.D., Lee, J.H., Huang, H., Tan, H., Ahn, J., Reinke, L.M., Peter, M.E., Feng, Y., Gius, D., Siziopikou, K.P., Peng, J., Xiao, X. and Cheng, C. (2014). Cell type-restricted activity of hnRNPM promotes breast cancer metastasis via regulating alternative splicing. Genes Dev. 28(11): 1191–203. https://doi.org/10.1101/gad.241968.114.

[111] Brown, R.L., Reinke, L.M., Damerow, M.S., Perez, D., Chodosh, L.A., Yang, J. and Cheng, C. (2011). CD44 splice isoform switching in human and mouse epithelium is essential for epithelial-mesenchymal transition and breast cancer progression. J. Clin. Investig. 121(3): 1064–74. https:// doi.org/10.1172/JCI44540.

[112] Li, J., Choi, P.S., Chaffer, C.L., Labella, K., Hwang, J.H., Giacomelli, A.O., Kim, J.W., Ilic, N., Doench, J.G., Ly, S.H., Dai, C., Hagel, K., Hong, A.L., Gjoerup, O., Goel, S., Ge, J.Y., Root, D.E., Zhao, J.J., Brooks, A.N., Weinberg, R.A. and Hahn, W.C. (2018). An alternative splicing switch in FLNB promotes the mesenchymal cell state in human breast cancer. eLife 7: 1–28.

[113] Ranieri, D., Rosato, B., Nanni, M., Magenta, A., Belleudi, F. and Torrisi, M.R. (2016). Expression of the FGFR2 mesenchymal splicing variant in epithelial cells drives epithelial-mesenchymal transition. Oncotarget 7(5): 5440–60. https:// doi.org/10.18632/oncotarget.6706.

[114] Luo, J., Qu, J., Wu, D.K., Lu, Z.L., Sun, Y.S. and Qu, Q. (2017). Long non-coding RNAs: A rising biotarget in colorectal cancer. Oncotarget 8(13): 22187–22202.

[115] Harper, S.J. and Bates, D.O. (2008). VEGF-A splicing: the key to anti-angiogenic therapeutics? Nat. Rev. Cancer 8: 880–887.

[116] Wagner, K.D., El Maï, M., Ladomery, M., Belali, T., Leccia, N., Michiels, J.F. and Wagner, N. (2019). Altered VEGF splicing isoform balance in tumor endothelium involves activation of splicing factors Srpk1 and Srsf1 by the Wilms' tumor suppressor Wt1. Cells 8: 41–58.

[117] Furuta, K., Zahurak, M., Goodman, S.N. et al. (1998). CD44 expression in the stromal matrix of colorectal cancer: association with prognosis. Clin. Cancer Res. 4: 21–29.

[118] Adams, M., Jones, J.L., Walker, R.A., Pringle, J.H. and Bell, S.C. (2002). Changes in tenascin-C isoform expression in invasive and preinvasive breast disease. Cancer Research 62(11): 3289–3297.

[119] Blaustein, M., Pelisch, F., Coso, O.A., Bissell, M.J., Kornblihtt, A.R. and Srebrow, A. (2004). Mammary epithelial-mesenchymal interaction regulates fibronectin alternative splicing via phosphatidylinositol 3-kinase. J. Biol. Chem. 14; 279(20): 21029–37. doi: 10.1074/jbc. M314260200.

[120] Radisky, D.C., Levy, D.D., Littlepage, L.E., Liu, H., Nelson, C.M., Fata, J.E., Leake, D., Godden, E.L., Albertson, D.G., Nieto, M.A., Werb, Z. and Bissell, M.J. (2005). Rac1b and reactive oxygen species mediate MMP-3-induced EMT and genomic instability. Nature 36(7047): 123–7. doi: 10.1038/nature03688.

[121] Fu, X.D. and Ares, M. Jr. (2014). Context-dependent control of alternative splicing by RNA-binding proteins. Nat. Rev. Genet. 15: 689–701.

[122] Martín Moyano, P., Němec, V. and Paruch, K. (2020). Cdc-Like Kinases (CLKs): Biology, chemical probes, and therapeutic potential. International Journal of Molecular Sciences 21(20): 7549. https://doi.org/10.3390/ijms21207549.

[123] Zhou, Z. and Fu, X.D. (2013). Regulation of splicing by SR proteins and SR protein-specific kinases. Chromosoma 122: 191–207.

[124] Hollander, D., Donyo, M., Atias, N., Mekahel, K., Melamed, Z., Yannai, S., Lev-Maor, G., Shilo, A., Schwartz, S., Barshack, I., Sharan, R. and Ast, G. (2016). A network-based analysis of colon cancer splicing changes reveals a tumorigenesis-favoring regulatory pathway emanating from ELK1. Genome Res. 26: 541–553.

[125] Choudhury, R., Ghose Roy, S., Tsai, Y.S., Tripathy, A., Graves, L.M. and Wang, Z. (2014). The splicing activator DAZAP1 integrates splicing control into MEK/Erk-regulated cell proliferation and migration. Nat. Commun. 5: 3078.

[126] Weg-Remers, S., Ponta, H., Herrlich, P. and Konig, H. (2001). Regulation of alternative premRNA splicing by the ERK MAP-kinase pathway. EMBO J. 20: 4194–4203.

[127] Shultz, J.C., Goehe, R.W., Wijesinghe, D.S., Murudkar, C., Hawkins, A.J., Shay, J.W., Minna, J.D. and Chalfant, C.E. (2010). Alternative splicing of caspase 9 is modulated by the phosphoinositide 3-kinase/Akt pathway via phosphorylation of SRp30a. Cancer Res. 70: 9185–9196.

[128] Wang, F., Fu, X., Chen, P., Wu, P., Fan, Z., Li, N., Zhu, H., Jia, T.T., Ji, H., Wang, Z., Wong, C.C.L., Hu, R. and Hui, J. (2017). SPSB1-mediated HnRNP A1 ubiquitylation regulates alternative splicing and cell migration in EGF signaling. Cell Res. 27: 540–558.

[129] Lee, G., Zheng, Y., Cho, S., Jang, C., England, C., Dempsey, J.M., Yu, Y., Liu, X., He, L., Cavaliere, P.M., Chavez, A., Zhang, E., Isik, M., Couvillon, A., Dephoure, N.E., Blackwell, T.K., Yu, J.J., Rabinowitz, J.D., Cantley, L.C. and Blenis, J. (2017). Post-transcriptional regulation of *de novo* lipogenesis by mTORC1- S6K1-SRPK2 signaling. Cell 171: 1545–1558 e1518.

[130] Zhan, T., Rindtorff, N. and Boutros, M. (2017). Wnt signaling in cancer. Oncogene 36: 1461–1473.

[131] Hernández, F., Pérez, M., Lucas, J.J., Mata, A.M., Bhat, R. and Avila, J. (2004). Glycogen synthase kinase-3 plays a crucial role in tau exon 10 splicing and intranuclear distribution of SC35. Implications for Alzheimer's disease. J. Biol. Chem. 279: 3801–3806.

[132] Heyd, F. and Lynch, K.W. (2010). Phosphorylation-dependent regulation of PSF by GSK3 controls CD45 alternative splicing. Mol. Cell 40: 126–137.

[133] Goncalves, V., Matos, P. and Jordan, P. (2008). The beta-catenin/TCF4 pathway modifies alternative splicing through modulation of SRp20 expression. RNA 14: 2538–2549.

[134] Goncalves, V., Matos, P., Jordan, P. and Antagonistic, S.R. (2009). Proteins regulate alternative splicing of tumor-related Rac1b downstream of the PI3-kinase and Wnt pathways. Hum. Mol. Genet. 18: 3696–3707.

[135] Romero-Barrios, N., Legascue, M.F., Benhamed, M., Ariel, F. and Crespi, M. (2018). Splicing regulation by long noncoding RNAs. Nucleic Acids Res. 46: 2169–2184.

[136] Gawronski, A.R., Uhl, M., Zhang, Y., Lin, Y.Y., Niknafs, Y.S., Ramnarine, V.R., Malik, R, Feng, F., Chinnaiyan, A.M., Collins, C.C., Sahinalp, S.C. and Backofen, R. (2018). MechRNA: Prediction of lncRNA mechanisms from RNARNA and RNA-protein interactions. Bioinformatics 34: 3101–3110.

[137] Chen, L.L. (2016). The biogenesis and emerging roles of circular RNAs. Nat. Rev. Mol. Cell Biol. 17: 205–211.

[138] Papaioannou, D., Petri, A., Dovey, O.M., Terreri, S., Wang, E., Collins, F.A., Woodward, L.A., All Walker, A.E., Nicolet, D. Pepe, F., Kumchala, P., Bill, M., Walker, C.J., Karunasiri, M., Mrózek, K., Gardner, M.L., Camilotto, V., Zitzer, N., Cooper, J.L., Cai, X., Rong-Mullins, X., Kohlschmidt, J., Archer, K.J., Michael, A., Freitas, M.A. and Garzon, R. (2019). The long non-coding RNA HOXB-AS3 regulates ribosomal RNA transcription in NPM1-mutated acute myeloid leukemia. Nat. Commun. 10: 5351.

[139] Zhang, X.O., Dong, R., Zhang, Y., Zhang, J.L., Luo, Z., Zhang, J., Chen, L.L. and Yang, L. (2016). Diverse alternative back-splicing and alternative splicing landscape of circular RNAs. Genome Res. 26: 1277–1287.

[140] Feng, J., Chen, K., Dong, X., Xu, X., Jin, Y., Zhang, X., Chen, W., Yujing Han, Y., Shao, L., Gao, Y. and He, C. (2019). Genome-wide identification of cancer-specific alternative splicing in circRNA. Mol. Cancer 18: 35.

9

Therapeutic Targeting of Alternative Splicing in Cancers

Sabra Parveen,[1] Bilal Rah,[2] Waseem Akram,[3] Shahid Ali,[4] Abdul Basit Baba,[2] Shahnawaz A. Wani,[2] Amrita Bhat,[5] Itty Sethi[5],* and Gh. Rasool Bhat[2],*

1. Introduction

The term alternative splicing (AS) goes back to a time in 1978 when it was first identified by Walter. This domain of research attracted prominent attention as there were new insights into the production of multiple proteins from a single gene [1]. The fascinating fact about the human body is that the most complex mechanisms are yet to be discovered [2]. Before pitching the concept of AS, we need to understand what is splicing? Genes contain information that is transcribed into mRNA and then translated into proteins. Widely studied concepts of transcription and translation collectively are responsible for attaining this state. Once the pre-mRNA is formed in the process of transcription, it is encoded with both intronic as well as exonic regions. The earlier concept that hit around was only exons participate in the gene regulation process, while the intronic region was considered junk DNA. But with extensive studies and advancements in genetic/molecular methodology, it was found that introns play a critically important role in gene regulation. Hence, these sequences are equally essential and more significant.

[1] CSIR Indian Institute of Integrative Medicine, Jammu, Jammu & Kashmir, India.
[2] Advanced Centre for Human Genetics, Sher-I- Kashmir Institute of Medical Sciences, Soura, Kashmir, India.
[3] Department of General Medicine, Sher-I- Kashmir Institute of Medical Sciences, Soura, Kashmir, India.
[4] Department of Hospital Administration, Sher-I- Kashmir Institute of Medical Sciences, Soura, Kashmir, India.
[5] Instuite of Human Genetics, University of Jammu, Jammu and Kashmir, India.
* Corresponding authors: seithbhat11@gmail.com; bhavya2288@gmail.com

The concept of splicing arises with the fact that genetic coding sequences called exons, separated by noncoding sequences called introns, are meant to be connected to decipher entire information in them to make a gene product. This is brought about by the excision of intronic sequences and the joining of exonic sequences, a process known as splicing. This highly regulated process is carried out by an important complex called the spliceosome, made up of five small nuclear ribonucleoproteins (snRNP) particles, namely, U1, U2, U4, U5, and U6. The complex performs the splicing process and assembles around a sequence-specific location, termed a splice site. These splice sites are located across exons and introns. Each splice point is comprised of a consensus sequence that wraps around each exon-intron pair, and this junction is recognized by the spliceosome [3, 4]. Apart from this, other sequences play a critical role in regulation either by enhancing or suppressing the splicing, known as enhancers or silencers, respectively. These sequences are assumed to be a result of the existence of multiple cis-regulatory components that serve the purpose of initiating or suppressing a splicing reaction. Splicing regulatory elements (SREs) work by employing trans-acting splicing factors that activate or suppress splice site recognition or assemblage of spliceosomes via diverse mechanisms [5].

AS is one of the most fundamental methods for producing a vast range of mRNA and isoforms of proteins from a single gene. Unlike promoter activity, which essentially controls the number of transcripts produced, AS alters the structure of transcripts and the proteins they encode. AS factors control pre-mRNAs, and the discovery of potential physiological targets suggests that a single splicing factor controls pre-mRNAs with similar biological roles. AS organizes physiologically significant changes in protein isoform production and is a critical mechanism for multicellular animals to produce their complex and diverse proteome [6].

The role of AS is so prominent that any dysregulated pattern in this process can be responsible for generating a diverse protein range with the capability of disturbing normal physiology and developing neoplastic growth. Similarly, splicing has been associated with the process of oncogenesis, and aberrant alterations in AS have the capability of changing an entire function of proteins, hence contributing to the disease progression [7]. Various human solid tumors, such as the bladder, brain, breast, cervix, colon, kidney, liver, lung, oral, ovary, prostate, skin, stomach, and thyroid tumors, as well as hematological malignancies, such as acute myeloid leukemia, myelodysplastic syndrome, chronic leukemia, etc., have been found to occur because of recurring somatic mutations in a preview of the human splicing process [8–16]. Thus, alteration in AS is usually associated with increased cell proliferation and metastasis, which accounts for 90% of all cancer deaths in humans [17]. According to several recent studies, mutations impact the splicing of critical cancer-related genes, as well as copy-number changes disturbing spliceosome assembly and hence trigger the earlier process of tumorigenesis. There is mounting evidence that certain cancer subtypes rely heavily on wild-type splicing functions for the survival of the cell. Splicing catalysis, splicing regulatory proteins, or changed splicing events have all become favorable targets in cancer therapeutics [18]. This review will focus briefly on the mechanism of splicing and AS and its implications for cancer and therapeutic potential.

2. Splicing and its Regulatory Mechanism

Splicing is a cellular mechanism whereby a gene that is encoded by exons as well as their intervening sequence, called introns, are reassembled in a way that introns are removed, and an intact product (transcript) is yielded by the joining of different exons. Splicing of mRNAs (pre-mRNA) is a crucial step in the production of eukaryotic genes. The correct removal of introns and identification of the actual splice sites are required for the synthesis of appropriate encoded proteins along with their isoforms. Introns are not present only in eukaryotes, but they are widespread among other species as well; they generally exhibit variability in size, number, and distribution [19]. The process of splicing is carried out by a complex called the spliceosome. This is a huge complex that consists of 5 snRNPs (small nuclear ribonucleoproteins) commonly designated as U1, U2, U4, U5, and U6 [20]. Apart from these snRNPs, several other proteins are also associated with this spliceosome complex, and they play an important role in the proper occurrence of splicing [21]. These snRNPs are associated with their specific small RNAs (snRNA) as well. This remark introduces the main elements for the process of splicing, including splice site, snRNAs, snRNPs, branch point, and spliceosome.

2.1 Splice Site

These comprise the DNA sequences present on the boundaries of introns and exons. The conserved sequences are recognized by the spliceosome for initiating the splicing machinery. They are usually present on the 5' end and 3' end of the exons.

2.2 Branch Point Sequence

These are critically important sequences referring to the segments that are present near the 3' end of introns on the nuclear mRNA. They contain adenosine residues that show base pairing with the guanosine residues at the time of intron excision.

2.3 Spliceosome

A complex unit formed by the aggregation and assembly of various splicing factors and proteins that reside sequentially at their specific splice site location and carry out splicing in a regulated and stepwise manner.

2.4 Splicing

Splicing chemistry consists of two sequential SN2-type transesterification processes. In pre-mRNA splicing, many cis-acting sequences are involved, some of which control the splicing response, and others govern AS. The simplest among these sequences are those that use direct splicing. The 5' splice site has the sequence AG/GURAGU, and the 3' splice site has to have a poly-pyrimidine tract followed by an AG dinucleotide at the actual 3' splice site. The branch point sequence, which is located upstream of the 3' splice site, encloses the nucleophile for the first stage of splicing [22]. The phosphodiester bond between the 5' exon and the intron is split during the first catalytic step by a nucleophilic approach from the branch point adenosine's 2' hydroxyl. The very first phase of catalysis leads to the production of a lariat intermediate, (a 2', 5' branched RNA) and a free 5' exon. Conformational changes inside the active region of the spliceosome subsequently allow the free 3' hydroxyl of the 5' exon to target the intron-3' exon junction, resulting in exon ligation and elimination of the lariat intron [23]. Multiple introns (on the order of eight per gene) are found in human genes, with an average length of 11,000 nucleotides but a significant diversity in length. The average length of exons in humans is 200 nucleotides long.

Human gene splicing is a complicated system and multi-step process. It is a co-transcriptional process, where introns appear once the RNA polymerase is eliminated. The main regulator of splicing machinery is the spliceosome. Spliceosome forms a complex in the sequential pattern to chop off and ligate the exon-intron regions of the mRNA. This complex system is highly dynamic and involves the utility of specific spliceosome proteins and other regulatory factors. Five ribonucleoproteins (RNP) complexes, each containing a U-rich short nuclear RNA (snRNA), sharing core components, and numerous unique proteins, constitute the spliceosome. RNA polymerase II transcribes U1, U2, U4, and U5 snRNAs, which have a tri-methyl-guanosine cap; however, RNA polymerase III transcribes U6 snRNA, as it has a monomethyl guanosine cap. Seven homologous Sm proteins form a network around the U-rich sequence known as the Sm site, which is situated near the 3' terminus of U1, U2, U4, and U5 snRNAs, whereas a U-rich sequence near the 3' end of U6 snRNA threads through a ring of seven paralogous LSm proteins (LSm2–8). Each one of these snRNAs bonds to a distinct collection of extra proteins, resulting in the formation of a small nuclear ribonucleoprotein (snRNP) particle. Splicing is further aided by several non-snRNP-associated proteins and eight ATP-dependent helicases. The snRNAs have critical functions in catalysis and substrate recognition inside the spliceosome [24–27].

Table 1: Designation of small nuclear riboproteins and their specific RNAs, along with the mRNA features.

S. No.	snRNP	snRNA	Transcribed By	Capping
1.	U1	U1	RNA pol III	Tri-methyl guanosine cap
2.	U2	U2	RNA pol III	Tri-methyl guanosine cap
3.	U4	U4	RNA pol III	Tri-methyl guanosine cap
4.	U5	U5	RNA pol III	Tri-methyl guanosine cap
5.	U6	U6	RNA pol II	Monomethyl guanosine cap

As listed in Table 1, the snRNA component of the complexes, known as 'snRNPs' or 'snurps', is designated as U1, U2, U4, U5, and U6. Additionally, spliceosomes contain a complex involved with the Prp19 protein (NTC for 19 complexes) and a multitude of some of the other proteins. These factors work together so they detect homologous cis-RNA sequences; then organize the catalytic spliceosome and make splicing composition easier [28]. The U1 and U2 snRNPs are associated with the pre-spliceosome formation and they recognize and mark an intron in the early stages of spliceosome construction. This step is regulated extensively in higher eukaryotes, both by cis-acting sequence elements and trans-acting splicing factors, and it is thought to be the important element influencing splice site selection throughout AS since it determines an intron to be excised [29, 30]. The snRNPs and their association with spliceosome is listed in Table 2 pre-spliceosome or 'A' complex formation by the U1 and U2 snRNPs, read the 5' SS and BP sequences, respectively. To create the fully assembled spliceosome, the pre-spliceosome connects well with pre-configured U4/U6 and U5 tri-snRNP [31]. Within the tri-snRNP, U6 snRNA is significantly base-paired to U4 snRNA, which eventually folds to create the active site of the spliceosome. Prp28, a DEAD-box helicase, liberates the 5' SS from U1 snRNP and transports it to the ACAGAGA box inside U6 snRNA [32].

Table 2: snRNPs and their catalytic association for splicing machinery.

snRNPs	Association with Spliceosome
U1 and U2	➤ Associated with pre-spliceosome complex. ➤ Recognize and mark intron in earlier stages.
U4/U5/6	➤ U6 and U4 form the active site of spliceosome. ➤ Interact with Prp28 and DEAD-BOX helicase.

The RNA helicase Brr2 eventually splits U4 snRNA from U6 snRNA, allowing the U6 snRNA sequence proximal to the 5' SS-bound ACAGAGA box to scrunch around and join with a portion of U2 snRNA to form the active site, which contains two catalytic metal ions [33]. The metal ion M1, however, is occupied by the 5' SS. The branching process creates the sliced 5' exon as well as the lariat intron intermediate once the BP adenosine is bound into the active site. The 5' exon stays there in the active site; however, the BP adenosine must therefore leave the active site to make room for the approaching 3' SS site in order to allow the exon ligation event to occur [34]. The 5' and 3' exons would then be ligated, and then the resultant mRNA (ligated exons) gets liberated from the active site. The spliceosome orchestrates these substrates' complex motions into and out of the active site [35].

These different events are designated by the complexes they form, such as complex E, complex A, complex B, and complex A. At the branch site, complex E is transformed into complex A by an ATP-dependent transfer of SF1 for U2. Irrespective, the development of complex A is usually preceded by the recruitment of the U4/U6. U5 tri-snRNP binds to the pre-mRNA substrate to produce the catalytically inactive complex B spliceosome, including all five U snRNPs. Complex B creation is followed by Brr2-dependent activation and the production of BACT. Since this activation requires several rearrangements that result in U6 replacing U1 at the 5' splice site, creation of the U6 ISL, and production of duplexes between the U2 and U6 snRNAs. The DExD/H box protein Prp2 promotes the first phase of catalysis, and spliceosomes capable of carrying out the chemistry of splicing are known as C complexes. This is followed by the several reconfigurations within the active site facilitated by the DExD/H box protein Prp16 that relocate the components for the second step of splicing [36]. Considering the need for precise alignment, the spliceosome active site had to be flexible because it is modified during the first and second phases of splicing to allow for the proximity of the 5' exon and 3' splice site for joining of the exon [37].

Table 3: Different events and complex formations orchestrate the action of the spliceosome.

S. No.	Splicing Complex	
1.	Complex E	➤ Formation of branch site.
2.	Complex A	➤ ATP-dependent. ➤ Transformed from Complex E. ➤ Proceeded by recruitment of U6/U4.
3.	Complex B	➤ Initially catalytically inactive. ➤ Formed by binding of U5 to pre-mRNA substrate.
4.	Complex C	➤ Replacement of U1 by U6 at the 5' splice site.

Although many introns in eukaryotic genes are removed constitutively off the pre-mRNA transcript, the excision of other introns can also be subject to significant layers of regulation. Multi-exonic transcripts can be spliced together to create several isoforms of the same gene by changing the splice site selections during the process of splicing, which is referred to as AS. AS is predicted to occur in 95% of gene transcripts in mammalian cells. The production of several mRNA isoforms from a single pre-mRNA is a strong way for enhancing the genome's protein-encoding capacity and increasing the genome's size.

AS is a significant source of proteome variation and diversity [38]. The light on this concept was first shed by Gilbert in 1978 [39]. It is also essential for cellular differentiation and development of the organisms. The regulating program of AS is a complicated phenomenon involving many interacting components, such as cis-acting elements and trans-acting factors, and is led by the functional coupling of transcription and splicing. Higher eukaryotic species have been found to have a greater proportion of alternatively spliced genes, suggesting that the process plays an important role in evolution. AS regulates a wide range of biological activities throughout an organism's existence, from conception to death [40]. Therefore, it plays a role practically in every aspect of protein function, such as protein-ligand interaction, nucleic acid as well as membrane binding, localization, and enzymatic activities. Taken as a whole, AS is a critical component of gene expression [41]. So far, statistics have shown major forms of AS. AS happens in five distinct ways to add variety to the proteome; exon skipping is the most prevalent kind of AS in which certain exons are skipped resulting in the shifting of the open reading frame (ORF). The exon cassettes might be introduced at various locations on the mRNA. Another kind of AS is mutually exclusive exons in which one exon is preserved while the other is deleted or vice versa, but not both, resulting in various transcript combinations and, ultimately, different proteins. Part of the exon is cleaved at an intra-exonic 5' donor splice site and 3' acceptor site, respectively, and connected to the following exon in alternative 5' and alternative 3' splice sites. In vertebrates, intron retention is uncommon; it includes attaching noncoding portions of the gene (introns) to the mRNA, thereby producing shortened and less functional protein. In the vertebrates and invertebrates, the cassette-type alternate exon (exon skipping) is the most common pattern (30%), but in lower metazoans, it is intron retention as shown in Figure 1. Intron retention in human transcripts is typically seen in the untranslated regions (UTRs) and has been linked to poorer splice sites, shorter intron lengths, and the control of cis-regulatory elements [42, 43].

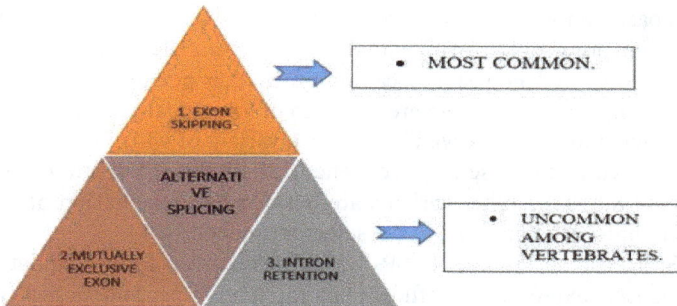

Figure 1: Shows different aspects of alternate splicing and its occurrence.

An alternative selection of 5' or 3' sites of splicing across exon sequences (25 percent) may result in minor variations in the coding sequence, and mutually exclusive alternative exons provide an extra degree of intricacy. FGFR2 is one example of a transcript that undergoes AS, which results in protein variation. Differential splicing machinery in various cell types, as well as distinct cis-acting regions in the FGFR2 pre-mRNA, result in changed tissue-specific options that produce either FGFR2III or FGFR2IIIc mature transcripts [44].

Alternative polyadenylation of mRNA impacts the coding potential or 3' UTR length by changing the binding availability of microRNA or RNA. It should be noted that each sort of AS can work stochastically, and distinct splice site identification and processing methods do not necessarily occur at the same frequency throughout all biological kingdoms. The interplay of cis-acting elements and trans-acting factors completely defines the exons that end up in the mature mRNA throughout the process of AS [45]. Exonic splicing enhancers (ESEs) and intronic splicing enhancers (ISE) are confined by positive trans-acting factors, such as SR proteins (serine/arginine-rich family of nuclear phosphoproteins), while exonic splicing silencers (ESSs) and intronic splicing silencers (ISE) are obligated by negative acting factors, such as heterogeneous nuclear ribonucleoproteins (hnRNPs). Interaction between these components promotes or inhibits spliceosome assembly of the weak splice sites, accordingly. In general, cis-acting components work additively [46]. The boosting elements are more significant in controlling constitutive splicing, whereas the silencers are more important in controlling AS. Due to the processes of physical competition, long-range RNA pairing, a structural splice code, and co-transcription splicing, a stable stem-loop structure as small as 7 base pairs in an RNA transcript have been demonstrated to eliminate enhancer function. In addition, the specificity of cis-acting enhancer elements for introns or exons demonstrated that an ESE function as an ISE depending on whether it is located in an exon or intron. Also, HnRNPs have been largely conserved from nematodes to mammals, and they play several important roles in pre-mRNA maturation. Their role is to attach to the ESS and prevent SR proteins from binding. Exonic sequester from the remainder of the pre-mRNA transcript is caused by looping out pre-mRNA.

HnRNPs A/B are a class of RNA-binding proteins with diverse functions in AS modulation based on varying affinities for their associated nucleic acids. Secondary structures within pre-mRNA can potentially impact splice site selection. The extensive AS identified in D. melanogaster Dscam pre-mRNA is perhaps the most prominent example. Dscam's exon 6 clusters are made up of 48 mutually exclusive exons. The docking site and another conserved sequence, a variant of which is located upstream of each exon 6 variant (the selector sequence), allow the incorporation of only one exon 6 variant 89; some other variants are omitted by the adhesion of hrp36, a D. melanogaster hnRNPA homolog, to the selector sequence [47].

Core Spliceosome Proteins (CSPs) are also involved in the regulation of AS. Microarray-based expression profiles from mice, chimps, and tissues within the human body indicated that snRNPs are expressed differently in various tissues. This is congruent with the findings of an RNA interference (RNAi) screen in D. melanogaster, which revealed that increasing CSP levels cause changes in AS. These CSPs contain U1, U2, and U4/U6 snRNP components, as well as the U2AF heterodimer. RNAi knockdown of U2AF35 isoforms, U2AF35a and/or U2AF35b, and an SF3b component, SAP155 (also known as SF3B1), in human cells offered further evidence [48–51].

3. Alternate Splicing in Cancer

Splicing is commonly altered in cancer. Mutations impacting the splicing of important cancer-associated genes, as well as copy-number changes affecting spliceosomal proteins themselves, are prevalent in cancer. Simultaneously, there is mounting evidence that some molecular subcategories of cancer rely heavily on wild-type splicing functions for cell survival. As a result of these, there is a rising interest in using splicing catalysis, splicing regulatory proteins, and/or particular critical changed splicing events in cancer therapeutics [52]. Since AS allows cells for diversification of their proteome, current research has identified several mechanisms in which splicing may be pathologically changed to promote the development and/or maintenance of cancer. It included mutations influencing splicing regulatory sequences of critical cancer-associated genes, and also mutations and gene expression changes that will affect core or accessory components of the spliceosome complex. AS could be influenced at several stages, resulting in alterations that could be associated with an oncogenic state, such as changed activity, overexpression, or perhaps mutations in regulatory splicing components. This shift might occur as a result of changes in post-translational modifications, such as phosphorylation, methylation, and sumoylation of various splicing components, having implications not just for splicing control but also for other aspects of cell biology. Furthermore, somatic mutations in genes encoding elements of the splicing machinery may lead to tumor growth. As per data deposited in the International Cancer Genome Consortium (ICGC) databases, it appears that nearly 300 splicing-related genes are altered in all types of cancer, with the most frequently mutated genes, including several *hnRNP* (*NOVA1, hnRNP M, hnRNP C, hnRNP A2/B1, hnRNP F,* and *RALY*) and SR proteins (SRSF4, RBM39, Tra2, and Tra2) together with R-protein kinases (SRPK1 and SRPK2) and RBM proteins (RBM4 and RBM5) [18, 53, 54]. Furthermore, comprehensive transcriptomic analyses of various cancer types have revealed the widespread alterations in alternative and constitutive splicing relative to normal tissue counterparts. These observations raise the prospect that splicing modification might give therapeutic benefits in cancer. Splicing needs many protein-to-protein and protein-to-RNA interactions and is guided by a variety of trans-acting proteins, which are subsequently regulated by post-translational modifications and protein/RNA interactions to operate normally [55–57, 16]. In line with this, various transcriptome investigations across cancer types have found extensive changes in alternative and constitutive splicing compared to normal tissue equivalents [58, 59]. Over 2,000 splicing mutations have been identified, affecting 303 genes and associated with 370 diseases. As a result, it has now become critical to investigate how this activity is controlled and how it might become dysregulated in the disease. While cancer is the most widely associated disorder with AS dysregulation in several genes, there are numerous in-depth studies of harmful/aberrant splice variants in diseases ranging from neuromuscular disorders to diabetes to cardiomyopathies [60, 61, 54]. Even before the public release of high-depth mRNA sequencing (RNA-seq) analyses from a wide range of cancer types, investigation of cancer transcriptomes utilizing expressed sequence tag libraries claimed cancer cells have "noisier" splicing than healthy tissues counterparts. This statement suggests that cancer cells produce a higher frequency of transcripts with premature termination codons (PTCs), which is associated with a higher incidence of missplicing compared to normal tissues. This data was analyzed by separating genes into oncogenes and tumor suppressor genes (TSGs), which found a far higher incidence of PTCs in transcripts encoding TSGs than oncogenes, indicating that this process is not spontaneous [62].

The Cancer Genome Atlas (TCGA) RNA-seq data analysis across 16 unique cancer types found how cancer cells and their normal tissue counterparts had similar RNA processing abnormalities. Surprisingly, substantially all forms of cancer have anomalies in intron retention (IR) considerably more frequently than in other types of splicing events, such as cassette exon splicing or 5' or 3' splice site identification. Increased IR was found to affect the constitutive and alternatively spliced introns [58]. Somatic single nucleotide variations (SNVs) influencing splicing are more frequently related to IR than cassette exon splicing or other types of AS.

Furthermore, SNVs producing IR were enhanced in TSGs over oncogenes, possibly because the preponderance of IR occurrences was predicted to lead to the formation of a PTC, whereas only 50% of cassette exon splicing events were projected to result in the generation of a PTC. That is also constant with several precisely defined and clinically significant mutations that activate proto-oncogenes by altering their splicing, such as *MET* exon 14 splicing mutations in lung cancer and mutations stimulating a cryptic splice site in chronic lymphocytic leukemia to result in an aberrantly active form of *NOTCH1* [63–65].

The epidermal growth factor receptor (EGFR) has a splice variant that lacks exon 4 and is widely expressed in certain malignancies; this exon loss makes the protein constitutively active. Also, K-Ras contains two alternative exons, 4A and 4B, and their integration or exclusion shows a substantial divergent correlation with various types or locations of colon cancer. Further, the tumor suppressor p53 contains two splice isoforms, p53beta and p53gamma, which are the product of two alternative exons; these isoforms influence the activity of the main isoform and how it regulates apoptosis in different settings. Finally, another important example would be a thoroughly studied tumor suppressor protein, retinoblastoma protein, for which more than 15% of the mutations identified in different malignancies are associated with splicing [18, 66–68]. Table 4 lists some common genes that are dysregulated in the splicing mechanism in cancer.

Table 4: Common genes and their splicing dysregulations leading to cancer.

S. No.	Genes	Alternative Splicing Associated with Cancer
1.	EGFR	When it lacks exon 4
2.	K-Ras	Has two alternative exons, 4A and 4B Found in colon cancer
3.	P53	P53 beta and P53 gamma isoforms are associated with cancer
4.	Rb pt.	15% cancer malignancies are associated with splicing
5.	hnRNP	Involves NOVA1, hnRNP M, hnRNP C, hnRNP A2/B1, hnRNP F, and RALY
6.	SR protein	Involves SRSF4, RBM39, Tra2, and Tra2) together with R-protein kinases (SRPK1 and SRPK2) and RBM proteins (RBM4 and RBM5)

4. Therapeutic Intervention

Alternative splicing is an open reason to enhance cancer development; there is a huge scope of therapeutic interventions related to AS. Various research findings suggest that dysregulation of AS events might serve as indicators and therapeutic targets for a variety of malignancies. To reprogram splicing events relevant to cancer diseases, a variety of therapeutic techniques are used. The discovery of the regulatory mechanisms involved in cancer-associated splicing events will aid in the development of suitable and particular therapeutic therapies for various cancer types. Modulating splicing by core spliceosome inhibition, as well as splicing regulatory proteins promote oncogenesis via overexpression and loss-of-function mutations, have emphasized the potential for therapeutic targeting of these proteins as new cancer therapeutics [69]. Specific molecular inhibitors that act at various stages of the process of splicing, were first identified as chemical probes for studying splicing regulation *in vivo* and *in vitro* studies. Nevertheless, several of these have shown effective in the treatment of a variety of human ailments, including cancer [70].

There are several studies suggesting various factors act as either inhibitors or anti-cancer agents; a few among them are mentioned here further. As briefly mentioned in Table 5. The natural substance FR901464 is regarded as the prototype chemical for splicing antagonists with anti-cancer efficacy. It has been reported that FR901464 suppresses pre-mRNA splicing in HeLa cells with an IC50 of 0.05M by acting on splicing factor 3b (SF3b). FR901464 does have a significant anti-proliferative impact on numerous human cancer cell lines, including MCF7 breast cancer, A549 lung adenocarcinoma, HCT116 colon cancer, SW480 colon cancer, and P388 murine leukemia, with IC50 values of 1.8, 1.3, 0.61, 1.0, and 3.3 nM, respectively [17, 71]. Furthermore, FR901464 inhibited tumor growth in various xenograft models at doses of 0.056–1 mg/kg against human solid tumors implanted in mice, and this inhibitor of splicing promoted G1 and G2/M phase arrest in the cell cycle by inhibiting p27 splicing and suppressed transcription of some endogenous genes such as C-Myc [72, 73].

Spliceostatin-A (SSA), a Fand01464 derivative that possesses anti-tumor and anti-cancer properties by influencing the splicing sequences of cell cycle regulators such as cyclin a2 and aurora A kinase, causing the concentration of cells in the G2/M stages of the cell cycle. With an IC50 of 0.01 M in HeLa cells, SSA suppresses splicing. SSA A has been studied to suppress both *in vivo* and *in vitro* splicing and increases-mRNA deposition via a non-productive induction of U2 snRNP of subunit SF3b. As a result, SSA limits spliceosome assembly by delaying complex A to B conversion. This suppression needs functioning cis-elements, such as the 50 splice site and branch point adenosine in the pre-mRNA with decoy sequences upstream of their productive binding site at the branch point sequence, as well as the trans-acting factors U1 and U2 snRNPs in the presence of ATP, allowing the insertion of U2 snRNP into the spliceosome. In a fission yeast strain lacking the multidrug resistance protein Pmd1, SSA has also been studied to suppress splicing and nuclear retention of pre-mRNA by the complex SF3b, lending support to this mechanism [74, 75]. Studies have proven that meayamycin acts as a potential analog for FR901474 and the GI50 values reported in breast cancer MCF-7 cells were 10 pM for the analog meayamycin and 1.1 nM for FR901464, indicating that the analog was 100 times more potent than the chemical represented initially throughout this cellular setting. The duo of meayamycin B and the Bcl-XL inhibitor ABT-737 was used to study *MCL1* splicing in A549 and H1299 cells and found that it effectively modulated AS. This combination therapy evoked apoptosis among both cell types at the same time. In the HEK-293 cell line, meayamycin also blocked pre-mRNA splicing. Meayamycin B's target includes SF3b1, and it functions as a splicing inhibitor by interfering with the transition from complex H to complex A [74, 76, 77]. Though meayamycin had no direct effect on alternative RNA splicing, throughout the system, this chemical may interact with several other splicing regulatory mechanisms. SAP155, for example, is one of the numerous potential targets for FR901464. It binds to ceramide-responsive RNA cis-element 1 and controls Bcl-x AS. As a result, meayamycin may influence Bcl-x AS [78].

A synthetic or artificially curated analog of FR901464 called Sudemycin E has been reported to show the effect of AS on alternative pre-messenger RNA globally and this somehow inhibited the tumor growth in model mice. This was done by targeting widely the cancerous cells and most commonly pre-messenger RNAs that showed effect by this molecule were genes: *RPp30, DUSP11, SRRM1, PAPOLG, MLH3,* and *IBTK*. In this, the splicing is inhibited by its connection with the U2 component SF3b1, which fails to sustain a tri-methylated state in actively transcribed genes. Furthermore, Sudemycin E causes widespread alterations in gene expression as well as cell cycle arrest in the G2 phase. In conjunction with ibrutinib, this medication produces selective cytotoxicity in primary chronic lymphocytic leukemia (CLL) cells. In a separate investigation, Sudemycin E had an anti-proliferative impact on HeLa, HEK293, and Rh18 cells after 48 hours of treatment with IC50 values of 0.16, 12.85, and 1.12 m, respectively [79–81].

Furthermore, for *in vitro* studies, a feented components seen as potential anti-cancer drugs and splicing inhibitors are Thailanstatin A (TST-A), Thailanstatin B (TST-B), Thailanstatin C (TST-C), and Thailanstatin D (TST-D), upon being isolated from the *Burkholderia thailandensis* MSMB43 fermentation broth. In *in vitro*, the splicing machinery associated with SFEb sub-complex spliceosome particle of U2 snRNP was inhibited by TSTs and thus prevented base-pairing interaction between the sequences located at the branch point at the 5' end. Reportedly, TSTs displayed activation of anti-proliferation in the cancerous cell lines [82, 83]. One study explored the 4Bhwe for its anti-cancer property. This molecule is a derivative of *Physalis peruviana*. This molecule acts as an inhibitor and its inhibitory effect has been seen on a wide range of genes that are associated with apoptosis, including *HIPK3, SMAC/ DIABLO, SURVIVIN, AIMP2, BCL2L11, BIRC5, CASP3, CEACAM1, CPE, FGFR2, FN1, FPGS, HIF1A, KLF6, MCL1, MDM2, MKNK2, TERT,* and *VEGFA* [84].

Table 5: AS associated with various derivatives and factors that act in the therapeutics of cancer.

Drug	Effect
FR901464	❖ Splicing antagonist. ❖ Profound anti-cancer efficacy. ❖ Acts on splicing factor SF3b.
SSA	❖ Regulator of cyclin a2 and aurora a kinase. ❖ Suppresses spliceosome assembly. ❖ Delays complex A conversion to B.
SUDEMYCINE E	❖ Inhibitor of tumor growth. ❖ Targets pre-mRNA. ❖ Arrests G2-phase.
Thailanstatin (A, B, C, D)	❖ TSA-C inhibits the association of the splicing machinery of SFEB with u2SNRNP. ❖ Prevents base pair matching. ❖ TSA-A, activator of antiproliferative agents.
Oligo-Nucleotides	❖ Utilized in shifting events of splicing toward an anti-proliferative isoform.
Site-splicing Oligo	❖ Associated with pro-apoptotic and chemo-sensitizing effects. ❖ Sensitize a549 cells to cis-platinum-induced apoptosis.

Another widely studied mechanism for targeting AS associated with therapeutics for cancer is "Antisense Oligonucleotide Technology". DNA encodes RNA, which is ultimately translated into proteins, according to the basic dogma of molecular biology. In the past few years, there has been increased interest in the use of drugs that may bind messenger RNAs (mRNAs), as suppression of protein production can assist regulate the course of inflammatory and neoplastic illnesses. For describing ASOs, they are usually 15 to 25 long bases, corresponding to those sequences that are complementary to the specific transcripts of RNAs. The antisense oligonucleotides (ASOs) to block mRNA translation and the oligonucleotides that work via the RNA interference (RNAi) pathway are the two primary therapeutic methods in present fields of research [85, 86]. The purpose of the antisense technique is to downregulate a molecular target, which is generally accomplished by inducing RNase H endonuclease activity, which binds the RNA-DNA hetero-duplex and significantly reduces target gene translation.

Other ASO-driven methods include 5' cap inhibition, alternative modification of splicing (splice-switching), and steric obstruction of activity of ribosomes [87, 88]. An ASO might be directed to areas at or near a splice site to affect an alternative splicing event, concealing normal or aberrant splicing events that lead to exon exclusion or their inclusion. ASOs are considered to be a very versatile tool that may be utilized to change RNA expression for therapeutic purposes due to their excellent specificity. Antisense oligonucleotides can also be utilized in cancer. This can be done by shifting a splicing event forward toward an anti-proliferative isoform in a highly precise and guided manner [89, 90]. In support of this theory, an example gene BCL-x can be put forward. In many cancers, including prostate cancer, the expression of the BCL-x isoform, i.e., Bcl-xl is overexpressed. As BCL-x has two isoforms (Bcl- Xl and Xs), where xL confers anti-apoptotic, while Bcl-xs confers pro-apoptotic signal. The presence of both these isoforms is required generally in the normal cellular activities. But an overexpression or under-expression of a particular variant leads to signs and initiation of cancer [91].

Among the methods, this synthesis of a splice site-switching oligo (SSO), which is RNA that selectively hybridizes with the splice site and cis-element, interfering with the interaction of responsive elements of trans-acting factors. By shifting BCL-x pre-mRNA splicing from the Bcl-XL to the Bcl-xs transcript, this SSO method was shown to have pro-apoptotic and chemo-sensitizing effects on a variety of cell types, particularly glioma cells. In a research experiment by Taylor et al., the targeting of specific sequences that were present on the adjacent splice site of Bcl-XL with a 2'-*O*-methyl oligonucleotide has been reported to induce the positive shifting of anti-apoptotic signals to pro-apoptotic signals in the pancreatic cancer cells. There was a dose-dependent and sequence-specific increase in Bcl-xS expression, with the best shifting happening at 24 hours and 100–200 nM oligonucleotide. The newly generated Bcl-xS transcripts were then translated into a protein with success. The ensuing pro-apoptotic signal did not encourage cell death, but it did sensitize A549 cells to cis-platinum-induced apoptosis. Mercatante et al. on the other hand found that an antisense 2'-O-methyl oligonucleotide targeting the xL splice site alone caused death in PC3 prostate cancer cells. The maximum splicing shifting occurred with 80 nM oligonucleotide roughly 16–24 hours after cationic lipid transfection [92].

Similarly, another study paved the way leading to recognizing and understanding the role of this technology in the treatment of cancer. A well-studied receptor IL-5 has an isoform, IL-5R, and is produced by the AS method and inclusion or exclusion of exon 9, which is responsible for encoding transmembrane domains of the particular protein [93, 94]. The other type might be a camouflage receptor with no physiological function other than to block IL-5 signaling. As a result, blocking the insertion of exon 9 would reduce not only the number of functional IL-5Rs but also the accessible concentration of the agonist, IL-5. This method also has shown therapeutic benefits for inflammatory illnesses such as asthma. In a study, antisense generated 2'-*O*-MOE oligonucleotides upon being targeted to a certain region across exon 9, which were delivered to the cells of B-cell lymphoma by electroporation method. This experiment showed an elevation in the number of soluble IR-5R and subsequent downfall of insoluble IL-5R, determining that the pathway associated with the IL-5R was successfully interrupted and showing the effect on the AS of the exon [95].

Another unexpected result has been reported in the C-myc, which is a potent oncogene upon overexpression or mutation and is associated with many cancer types. By utilizing the technique of ASOs, its downregulation was noted. A 28-mer oligomer, upon being directed toward the codon AUG is shown to cause aberrant splicing and additionally, this oligonucleotide was reported to block the exon 2 intron-exon junction. This resulted in the omission of the spliced message, further activating a cryptic 3' splice site located upstream of the exon. As a result, the portion of exon 2 containing the start codon was left out of the spliced message, resulting in translation initiation from a downstream AUG and the production of a truncated C-myc protein. This protein has been detected with an antibody called C-myc, and they did not appear to be treated with random or sense control oligomers [96]. Apart from this, interference mediated by SSOs by binding onto hnRNPA1 (heterogeneous nuclear ribonucleoprotein A1) or SRSF1 (serine and arginine-rich splicing factor 1) responsive elements has been identified by the use of nucleotide resolution crosslinking immunoprecipitation (iCLIP) and the ensuing influence on the reprogramming splicing profile indicate a possible use to selectively target cancer-related splicing events [97, 98].

The other applications related to this technology are Exon-specific splicing enhancement by small chimeric effectors or shortly called ESSENCE, and targeted oligonucleotide silencers of splicing, commonly called as S molecules. The approach of ESSENCE uses the oligonucleotide sequences that are bound at the 3' end of a peptide RS-domain. This is designed to induce the exclusion of exons mediated by SR proteins. The utility of this method has been reported to be effective in the correction of the AS error of exon 18 of BRCA1 and BCL-x [99–101].

Though still under surveillance, there are many potential results mediated by the ASO strategies. In recent studies *in vitro* cells, an ASO (AZD9150) was shown to have a restrictive effect on signal transducer and activator of transcription 3 (STAT3) expression by directly targeting its transcripts, which have anti-cancer effects on lung cancer and lymphomas. Another ASO (AZD4785) that targets the K-RAS gene has been shown to reduce the proliferative activity of K-RAS-driven cancer types. Exclusion of MDM4 exon 6 by ASO causes a reduction in MDM4 abundance via the AS-NMD pathway, which improves drug sensitivity and death in melanoma cells. These findings suggest that instead of altering the splicing profile for cancer therapies, it may be possible to directly eliminate cancer-specific transcripts. Nonetheless, developing a particular delivery method for malignant cells remains a significant hurdle in the development of oligonucleotide-based therapeutics [102–104].

5. Conclusion and Future Perspectives

The functional impact of the splicing alteration process, as well as mutations or modulations in the activity of splicing factors in cancer. Monitoring splicing alterations can provide abundant and effective biomarkers for use in the diagnosis, prognosis, therapeutics, and monitoring of cancer patients, in addition to variation in the important steps of RNA processing, such as 3′ end formation, RNA editing, and RNA modifications, which all contribute to cancer etiology. To realize this promise, highly sensitive, specific, and cost-effective assays for the detection of alternatively spliced isoforms are the need of the hour as a remedy to the challenges of detecting transcript variants in single cells and implementing cost-effective next-generation sequencing technologies in clinical practice.

In order to give an integrated perspective of the expression profiling landscape of malignancies and characterize the major regulatory networks and hubs that may show particularly critical therapeutic vulnerabilities, systems biology approaches are essentially required.

Whether the reversal of a pathogenic state can be achieved by targeting a small number of essential transcriptome modifications or whether therapeutic targets would need to flip broader regulatory programs is critical to our knowledge to understand. The molecular enigma of whether cancer-relevant mutations in splicing factors function through unifying mechanisms and common targets, which could help in the identification of broad therapeutic modalities, is related to this concept. The poorly conserved mechanisms governing AS between mice and humans, the conception of humanized animal models and human organoids will be critical for modeling cancer start and progression, as well as testing innovative, splicing-based therapeutic methods.

Another significant topic of concern for future research is the identification of small-molecule RNA structural modulators proficient in inducing splice site selection switches. More research on biochemical changes that enhances antisense oligonucleotide stability, specificity, and delivery, as well as deeper insights into the mechanisms of their cellular uptake.

Future tailored therapies and personalized medicine could potentially take advantage and use such agents' specificity and adaptability by using 'cocktails' of antisense oligonucleotides that target the specific profile of splicing changes found in each tumor or patient.

References

[1] Fei Qi, Yong Li, Xue Yang, Yan-Ping Wu, Lian-Jun Lin and Xin-Min Liu. (2020). Significance of alternative splicing in cancer cells. Chin. Med. J. (Engl.) 133(2): 221–228.

[2] Simon Braun, Mihaela Enculescu, Samarth T. Setty, Mariela Cortés-López, Bernardo P. de Almeida, F.X. Reymond Sutandy, Laura Schulz, Anke Busch, Markus Seiler, Stefanie Ebersberger, Nuno L. Barbosa-Morais and Stefan. (2018). Decoding a cancer-relevant splicing decision in the RON proto-oncogene using high-throughput mutagenesis. Nat. Commun. 9(1): 3315.

[3] Matera, A.G. and Wang, Z. (2014). A day in the life of the spliceosome. Nat. Rev. Mol. Cell Biol. 15(2): 108–21.

[4] Xiaofeng Zhang, Chuangye Yan, Xiechao Zhan, Lijia Li, Jianlin Lei and Yigong Shi. (2018). Structure of the human activated spliceosome in three conformational states. Cell Res. 28(3): 307–322.

[5] Matlin, A.J., Clark, F. and Smith, C.W. (2005). Understanding alternative splicing: Towards a cellular code. Nat. Rev. Mol. Cell Biol. 6(5): 386–98.

[6] Stefan Stamm, Shani Ben-Ari, Ilona Rafalska, Yesheng Tang, Zhaiyi Zhang, Debra Toiber, Thanaraj, T.A. and Hermona Soreq. (2005). Function of alternative splicing. Gene 344: 1–20.

[7] Yuanjiao Zhang, Jinjun Qian, Chunyan Gu and Ye Yang. (2021). Alternative splicing and cancer: A systematic review. Signal Transduction and Targeted Therapy 6(1): 78.

[8] Ivan Babic, Erik S. Anderson, Kazuhiro Tanaka, Deliang Guo, Kenta Masui, Bing Li, Shaojun Zhu, Yuchao Gu, Genaro R. Villa, David Akhavan, David Nathanson, Beatrice Gini, Sergey Mareninov, Rui Li, Carolina Espindola Camacho, Siavash K. Kurdistani, Ascia Eskin, Stanley F. Nelson, William H. Yong, Webster K. Cavenee, Timothy F. Cloughesy, Heather R. Christofk, Douglas L. Black and Paul S. Mischel. (2013). EGFR mutation-induced alternative splicing of max contributes to growth of glycolytic tumors in brain cancer. Cell Metabolism 17(6): 1000–1008.

[9] Olga Anczuków, Avi Z. Rosenberg, Martin Akerman, Shipra Das, Lixing Zhan, Rotem Karni, Senthil K. Muthuswamy and Adrian R. Krainer. (2012). The splicing factor SRSF1 regulates apoptosis and proliferation to promote mammary epithelial cell transformation. Nature Structural & Molecular Biology 19(2): 220–228.

[10] Fei Liu, Miao Dai, Qinyang Xu, Xiaolu Zhu, Yang Zhou, Shuheng Jiang, Yahui Wang, Zhihong Ai, Li Ma, Yanli Zhang, Lipeng Hu, Qin Yang, Jun Li, Shujie Zhao, Zhigang Zhang and Yincheng Teng. (2018). SRSF10-mediated IL1RAP alternative splicing regulates cervical cancer oncogenesis via mIL1RAP-NF-κB-CD47 axis. Oncogene 37(18): 2394–2409.

[11] Xuexia Zhou, Xuebing Li, Yuanming Cheng, Wenwu Wu, Zhiqin Xie, Qiulei Xi, Jun Han, Guohao Wu, Jing Fang and Ying Feng. (2014). BCLAF1 and its splicing regulator SRSF10 regulate the tumorigenic potential of colon cancer cells. Nat. Commun. 5: 4581.

[12] Elżbieta Sokół, Hanna Kędzierska, Alicja Czubaty, Beata Rybicka, Katarzyna Rodzik, Zbigniew Tański, Joanna Bogusławska and Agnieszka Piekiełko-Witkowska. (2018). microRNA-mediated regulation of splicing factors SRSF1, SRSF2 and hnRNP A1 in context of their alternatively spliced 3'UTRs. Exp. Cell Res. 363(2): 208–217.

[13] Junxiu Sheng, Qingzhi Zhao, Jinyao Zhao, Wenjing Zhang, Yu Sun, Pan Qin, Yuesheng Lv, Lu Bai, Quan Yang, Lei Chen, Yangfan Qi, Ge Zhang, Lin Zhang, Chundong Gu, Xiaoqin Deng, Han Liu, Songshu Meng, Hong Gu, Quentin Liu, Judy M. Coulson, Xiaoling Li, Bing Sun and Yang Wang. (2018). SRSF1 modulates PTPMT1 alternative splicing to regulate lung cancer cell radioresistance. EBioMedicine 38: 113–126.

[14] Liu Peiqi, Guo Zhaozhong, Yin Yaotian, Jia Jun, Guo Jihua and Jia Rong. (2016). Expression of SRSF3 is correlated with carcinogenesis and progression of oral squamous cell carcinoma. Int. J. Med. Sci. 13(7): 533–9.

[15] Severine Iborra, Marc Hirschfeld, Markus Jaeger, Axel Zur Hausen, Iona Braicu, Jalid Sehouli, Gerald Gitsch and Elmar Stickeler. (2013). Alterations in expression pattern of splicing factors in epithelial ovarian cancer and its clinical impact. Int. J. Gynecol. Cancer 23(6): 990–6.

[16] Jensen, M.A., Wilkinson, J.E. and Krainer, A.R. (2014). Splicing factor SRSF6 promotes hyperplasia of sensitized skin. Nat. Struct. Mol. Biol. 21(2): 189–97.

[17] Nancy Martinez-Montiel, Nora Hilda Rosas-Murrieta, Maricruz Anaya Ruiz, Eduardo Monjaraz-Guzman and Rebeca Martinez-Contreras. (2018). Alternative splicing as a target for cancer treatment. Int. J. Mol. Sci. 19(2).

[18] Lee, S.C. and Abdel-Wahab, O. (2016). Therapeutic targeting of splicing in cancer. Nat. Med. 22(9): 976–86.

[19] Gehring, N.H. and Roignant, J.Y. (2021). Anything but ordinary - emerging splicing mechanisms in eukaryotic gene regulation. Trends Genet. 37(4): 355–372.

[20] Wilkinson, M.E., Charenton, C. and Nagai, K. (2020). RNA splicing by the spliceosome. Annu. Rev. Biochem. 89: 359–388.

[21] Wahl, M.C., Will, C.L. and Lührmann, R. (2009). The spliceosome: Design principles of a dynamic RNP machine. Cell 136(4): 701–18.

[22] McManus, C.J. and Graveley, B.R. (2011). RNA structure and the mechanisms of alternative splicing. Curr. Opin. Genet. Dev. 21(4): 373–9.

[23] Moore, M.J. and Proudfoot, N.J. (2009). Pre-mRNA processing reaches back totranscription and ahead to translation. Cell 136(4): 688–700.

[24] Girish C. Shukla, Andrea J. Cole, Rosemary C. Dietrich and Richard A. Padgett. (2002). Domains of human U4atac snRNA required for U12-dependent splicing *in vivo*. Nucleic Acids Research 30(21): 4650–4657.

[25] Leung, A.K., Nagai, K. and Li, J. (2011). Structure of the spliceosomal U4 snRNP core domain and its implication for snRNP biogenesis. Nature 473(7348): 536–9.

[26] Séraphin, B. (1995). Sm and Sm-like proteins belong to a large family: Identification of proteins of the U6 as well as the U1, U2, U4 and U5 snRNPs. Embo. J. 14(9): 2089–98.

[27] Lerner, M.R. and Steitz, J.A. (1979). Antibodies to small nuclear RNAs complexed with proteins are produced by patients with systemic lupus erythematosus. Proc. Natl. Acad. Sci. USA 76(11): 5495–9.

[28] Jurica, M.S. and Roybal, G.A. (2013). RNA splicing, in encyclopedia of biological chemistry. pp. 185–190. *In*: Lennarz, W.J. and Lane, M.D. (eds.). (Second Edition) Academic Press: Waltham.

[29] Wilkinson, M.E., Charenton, C. and Nagai, K. (2020). RNA splicing by the spliceosome. Annu. Rev. Biochem. 89(1): 359–388.

[30] Clemens Plaschka, Pei-Chun Lin, Clément Charenton and Kiyoshi Naga. (2018). Prespliceosome structure provides insights into spliceosome assembly and regulation. Nature 559(7714): 419–422.

[31] Steitz, T.A. and Steitz, J.A. (1993). A general two-metal-ion mechanism for catalytic RNA. Proc. Natl. Acad. Sci. USA 90(14): 6498–502.

[32] Bringmann, P. and Lührmann, R. (1986). Purification of the individual snRNPs U1, U2, U5 and U4/U6 from HeLa cells and characterization of their protein constituents. Embo. J. 5(13): 3509–16.

[33] Sebastian M. Fica, Nicole Tuttle, Thaddeus Novak, Nan-Sheng Li, Jun Lu, Prakash Koodathingal, Qing Dai, Jonathan P. Staley and Joseph A. Piccirilli. (2013). RNA catalyses nuclear pre-mRNA splicing. Nature 503(7475): 229–34.

[34] Ruixue Wan, Chuangye Yan, Rui Bai, Gaoxingyu Huang and Yigong Shi. (2016). Structure of a yeast catalytic step I spliceosome at 3.4 Å resolution. Science 353(6302): 895–904.

[35] Zamore, P.D., Patton, J.G. and Green, M.R. (1992). Cloning and domain structure of the mammalian splicing factor U2AF. Nature 355(6361): 609–14.

[36] Smith, D.J., Konarska, M.M. and Query, C.C. (2009). Insights into branch nucleophile positioning and activation from an orthogonal Pre-mRNA splicing system in yeast. Molecular Cell 34(3): 333–343.

[37] Nilsen, T.W. and Graveley, B.R. (2010). Expansion of the eukaryotic proteome by alternative splicing. Nature 463(7280): 457–463.

[38] Leff, S.E., Rosenfeld, M.G. and Evans, R.M. (1986). Complex transcriptional units: Diversity in gene expression by alternative RNA processing. Annu. Rev. Biochem. 55: 1091–117.

[39] Leff, S.E., Rosenfeld, M.G. and Evans, R.M. (1986). Complex transcriptional units: Diversity in gene expression by alternative RNA processing. 55(1): 1091–1117.

[40] Yan Wang, Jing Liu, Bo Huang, Yan-Mei Xu, Jing Li, Lin-Feng Huang, Jin Lin, Jing Zhang, Qing-Hua Min, Wei-Ming Yang and Xiao-Zhong Wang. (2015). Mechanism of alternative splicing and its regulation. Biomed. Rep. 3(2): 152–158.

[41] Guoliang Fu, Kirsty C. Condon, Matthew J. Epton, Peng Gong, Li Jin, George C. Condon, Neil I. Morrison, Tarig H. Dafa'alla and Luke Alphey. (2007). Female-specific insect lethality engineered using alternative splicing. Nat. Biotechnol. 25(3): 353–7.

[42] Olga Kelemen, Paolo Convertini, Zhaiyi Zhang, Yuan Wen, Manli Shen, Marina Falaleeva and Stefan Stamm. (2013). Function of alternative splicing. Gene 514(1): 1–30.

[43] Pedro Alexandre Favoretto Galante, Noboru Jo Sakabe, Natanja Kirschbaum-Slager and Sandro José de Souza. (2004). Detection and evaluation of intron retention events in the human transcriptome. RNA 10(5): 757–65.

[44] Sakabe, N.J. and de Souza, S.J. (2007). Sequence features responsible for intron retention in human. BMC Genomics 8: 59.

[45] Goldstrohm, A.C., Greenleaf, A.L. and Garcia-Blanco, M.A. (2001). Co-transcriptional splicing of pre-messenger RNAs: Considerations for the mechanism of alternative splicing. Gene 277(1-2): 31–47.

[46] Wang, Z. and Burge, C.B. (2008). Splicing regulation: From a parts list of regulatory elements to an integrated splicing code. RNA 14(5): 802–13.

[47] Jin, Y., Yang, Y. and Zhang, P. (2011). New insights into RNA secondary structure in the alternative splicing of pre-mRNAs. RNA Biol. 8(3): 450–7.

[48] Chen, M. and Manley, J.L. (2009). Mechanisms of alternative splicing regulation: Insights from molecular and genomics approaches. Nat. Rev. Mol. Cell Biol. 10(11): 741–54.

[49] Guil, S., Long, J.C. and Cáceres, J.F. (2006). hnRNP A1 relocalization to the stress granules reflects a role in the stress response. Mol. Cell Biol. 26(15): 5744–58.

[50] Willemien van der Houven van Oordt, María T. Diaz-Meco, José Lozano, Adrian R. Krainer, Jorge Moscat and Javier F. Cáceresa. (2000). The MKK(3/6)-p38-signaling cascade alters the subcellular distribution of hnRNP A1 and modulates alternative splicing regulation. J. Cell Biol. 149(2): 307–16.

[51] Hui Zhu, Melissa N. Hinman, Robert A. Hasman,, Priyesh Mehta and Hua Lou. (2008). Regulation of neuron-specific alternative splicing of neurofibromatosis type 1 pre-mRNA. Mol. Cell Biol. 28(4): 1240–51.

[52] Jung W. Park, Katherine Parisky, Alicia M. Celotto and Brenton R. Graveley. (2004). Identification of alternative splicing regulators by RNA interference in Drosophila. Proc. Natl. Acad. Sci. USA 101(45): 15974–9.

[53] Zhou, Z. and Fu, X.D. (2013). Regulation of splicing by SR proteins and SR protein-specific kinases. Chromosoma 122(3): 191–207.

[54] Oltean, S. and Bates, D.O. (2014). Hallmarks of alternative splicing in cancer. Oncogene 33(46): 5311–8.

[55] Kaneb, H.M., Dion, P.A. and Rouleau, G.A. (2012). The FUS about arginine methylation in ALS and FTLD. Embo. J. 31(22): 4249–51.

[56] Fran Supek, Belén Miñana, Juan Valcárcel, Toni Gabaldón and Ben Lehner. (2014). Synonymous mutations frequently act as driver mutations in human cancers. Cell 156(6): 1324–1335.

[57] Hyunchul Jung, Donghoon Lee, Jongkeun Lee, Donghyun Park, Yeon Jeong Kim, Woong-Yang Park, Dongwan Hong, Peter J. Park and Eunjung Lee. (2015). Intron retention is a widespread mechanism of tumor-suppressor inactivation. Nat. Genet. 47(11): 1242–8.

[58] Chen, L., Tovar-Corona, J.M. and Urrutia, A.O. (2011). Increased levels of noisy splicing in cancers, but not for oncogene-derived transcripts. Human Molecular Genetics 20(22): 4422–4429.

[59] Rotem Karni, Elisa de Stanchina, Scott W. Lowe, Rahul Sinha, David Mu and Adrian R. Krainer. (2007). The gene encoding the splicing factor SF2/ASF is a proto-oncogene. Nature Structural & Molecular Biology 14(3): 185–193.

[60] Olga Anczuków, Avi Z. Rosenberg, Martin Akerman, Shipra Das, Lixing Zhan, Rotem Karni, Senthil K. Muthuswamy and Adrian R. Krainer. (2012). The splicing factor SRSF1 regulates apoptosis and proliferation to promote mammary epithelial cell transformation. Nat. Struct. Mol. Biol. 19(2): 220–8.

[61] Juan Wang, Jie Zhang, Kaibo Li, Wei Zhao and Qinghua Cui. (2012). SpliceDisease database: Linking RNA splicing and disease. Nucleic Acids Res. 40(Database issue): D1055-9.

[62] Wei Guo, Sebastian Schafer, Marion L. Greaser, Michael H. Radke, Martin Liss, Thirupugal Govindarajan, Henrike Maatz, Herbert Schulz, Shijun Li, Amanda M. Parrish, Vita Dauksaite,

Padmanabhan Vakeel, Sabine Klaassen, Brenda Gerull, Ludwig Thierfelder, Vera Regitz-Zagrosek, Timothy A. Hacker, Kurt W. Saupe, G. William Dec, Patrick T. Ellinor, Calum A. MacRae, Bastian Spallek, Robert Fischer, Andreas Perrot, Cemil Özcelik, Kathrin Saar, Norbert Hubner and Michael Gotthardt. (2012). RBM20, a gene for hereditary cardiomyopathy, regulates titin splicing. Nat. Med. 18(5): 766–73.

[63] Dvinge, H. and Bradley, R. (2015). Widespread intron retention diversifies most cancer transcriptomes. Genome Medicine 7.

[64] Xose S. Puente, Silvia Beà, Rafael Valdés-Mas, Neus Villamor, Jesús Gutiérrez-Abril, José I. Martín-Subero, Marta Munar, Carlota Rubio-Pérez, Pedro Jares, Marta Aymerich, Tycho Baumann, Renée Beekman, Laura Belver, Anna Carrio, Giancarlo Castellano, Guillem Clot, Enrique Colado, Dolors Colomer, Dolors Costa, Julio Delgado, Anna Enjuanes, Xavier Estivill, Adolfo A. Ferrando, Josep L. Gelpí, Blanca González, Santiago González, Marcos González, Marta Gut, Jesús M. Hernández-Rivas, Mónica López-Guerra, David Martín-García, Alba Navarro, Pilar Nicolás, Modesto Orozco, Ángel R. Payer, Magda Pinyol, David G. Pisano, Diana A. Puente, Ana C. Queirós, Víctor Quesada, Carlos M. Romeo-Casabona, Cristina Royo, Romina Royo, María Rozman, Nuria Russiñol, Itziar Salaverría, Kostas Stamatopoulos, Hendrik G. Stunnenberg, David Tamborero, María J. Terol, Alfonso Valencia, Nuria López-Bigas, David Torrents, Ivo Gut, Armando López-Guillermo, Carlos López-Otín and Elías Campo. (2015). Non-coding recurrent mutations in chronic lymphocytic leukaemia. Nature 526(7574): 519–24.

[65] Peter S. Hammerman, Michael S. Lawrence, Douglas Voet, Rui Jing, Kristian Cibulskis, Andrey Sivachenko, Petar Stojanov, Aaron McKenna, Eric S. Lander, Gad Getz, Marcin Imielinski, Elena Helman, Bryan Hernandez, Nam H. Pho, Matthew Meyerson, Gordon Saksena, Andrew D. Cherniack, Stephen E. Schumacher, Barbara Tabak, Scott L. Carter, Nam H. Pho, Huy Nguyen, Andrew Crenshaw, Rameen Beroukhim, Wendy Winckler, Peter S. Hammerman, Gad Getz, Matthew Meyerson, Hailei Zhang, Sachet Shukla, Lynda Chin, Gad Getz, Michael Noble, Doug Voet, Gordon Saksena, Nils Gehlenborg, Daniel DiCara, Hailei Zhang, Spring Yingchun Liu, Michael S. Lawrence, Lihua Zou, Andrey Sivachenko, Pei Lin, Petar Stojanov, Rui Jing, Juok Cho, Marc-Danie Nazaire, Jim Robinson, Helga Thorvaldsdottir, Jill Mesirov, Lynda Chin, Matthew Meyerson, Gad Getz, Peter S. Hammerman, Bryan Hernandez, Marcin Imielinski, Michael S. Lawrence, Andrey Sivachenko, Peter S. Hammerman, Gad Getz, Andrey Sivachenko and Matthew Meyerson. (2012). Comprehensive genomic characterization of squamous cell lung cancers. Nature 489(7417): 519–25.

[66] Hai Wang, Min Zhou, Bizhi Shi, Qingli Zhang, Hua Jiang, Yinghao Sun, Jianhua Liu, Keke Zhou, Ming Yao, Jianren Gu,, Shengli Yang, Ying Mao and Zonghai Li. (2011). Identification of an exon 4-deletion variant of epidermal growth factor receptor with increased metastasis-promoting capacity. Neoplasia 13(5): 461–71.

[67] Jehad Abubaker, Prashant Bavi, Wael Al-Haqawi, Mehar Sultana, Sayer Al-Harbi, Nasser Al-Sanea, Alaa Abduljabbar, Luai H. Ashari, Samar Alhomoud, Fouad Al-Dayel, Shahab Uddin and Khawla S. Al-Kuraya (2009). Prognostic significance of alterations in KRAS isoforms KRAS-4A/4B and KRAS mutations in colorectal carcinoma. J. Pathol. 219(4): 435–45.

[68] Marcel, V., Fernandes, K., Terrier, O., Lane, D.P., Bourdon, J.-C. (2014). Modulation of p53β and p53γ expression by regulating the alternative splicing of TP53 gene modifies cellular response. Cell Death Differ. 21(9): 1377–87.

[69] Katherine Zhang, Inga Nowak, Diane Rushlow, Brenda L. Gallie and Dietmar R. Lohmann. (2008). Patterns of missplicing caused by RB1 gene mutations in patients with retinoblastoma and association with phenotypic expression. 29(4): 475–484.

[70] Eunhee Kim, Janine O. Ilagan, Yang Liang, Gerrit M. Daubner, Stanley C.-W. Lee, Aravind Ramakrishnan, Yue Li, Young Rock Chung, Jean-Baptiste Micol, Michele E. Murphy, Hana Cho, Min-Kyung Kim, Ahmad S. Zebari, Shlomzion Aumann, Christopher Y. Park, Silvia Buonamici, Peter G. Smith, H. Joachim Deeg, Camille Lobry, Iannis Aifantis, Yorgo Modis, Frederic H.-T. Allain, Stephanie Halene, Robert K. Bradley and Omar Abdel-Wahab. (2015). SRSF2 mutations contribute to myelodysplasia by mutant-specific effects on exon recognition. Cancer Cell 27(5): 617–30.

[71] Nakajima, H., Hori, Y., Terano, H., Okuhara, M., Manda, T., Matsumoto, S. and Shimomura, K. (1996). New antitumor substances, FR901463, FR901464 and FR901465. II. Activities against experimental tumors in mice and mechanism of action. J. Antibiot. (Tokyo) 49(12): 1204–11.

[72] Sebestyén, E., Zawisza, M. and Eyras, E. (2015). Detection of recurrent alternative splicing switches in tumor samples reveals novel signatures of cancer. Nucleic Acids Research 43(3): 1345–1356.

[73] Daisuke Kaida, Hajime Motoyoshi, Etsu Tashiro, Takayuki Nojima, Masatoshi Hagiwara, Ken Ishigami, Hidenori Watanabe, Takeshi Kitahara, Tatsuhiko Yoshida, Hidenori Nakajima, Tokio Tani, Sueharu Horinouchi and Minoru Yoshida. (2007). Spliceostatin A targets SF3b and inhibits both splicing and nuclear retention of pre-mRNA. Nat. Chem. Biol. 3(9): 576–83.

[74] Sami Osman, Brian J. Albert, Yanping Wang, Miaosheng Li, Nancy L. Czaicki and Kazunori Koide. (2011). Structural requirements for the antiproliferative activity of pre-mRNA splicing inhibitor FR901464. Chemistry 17(3): 895–904.

[75] Brian J. Albert, Ananthapadmanabhan Sivaramakrishnan, Tadaatsu Naka, Nancy L. Czaicki and Kazunori Koide. (2007). Total syntheses, fragmentation studies, and antitumor/antiproliferative activities of FR901464 and its low picomolar analogue. J. Am. Chem. Soc. 129(9): 2648–59.

[76] Michael J. Moore, Qingqing Wang, Caleb J. Kennedy and Pamela A. Silver. (2010). An alternative splicing network links cell-cycle control to apoptosis. Cell 142(4): 625–36.

[77] Letai, A.G. (2008). Diagnosing and exploiting cancer's addiction to blocks in apoptosis. Nat. Rev. Cancer 8(2): 121–32.

[78] Brian J. Albert, Peter A. McPherson, Kristine O'Brien, Nancy L. Czaicki, Vincent Destefino, Sami Osman, Miaosheng Li, Billy W. Day, Paula J. Grabowski, Melissa J. Moore, Andreas Vogt and Kazunori Koide. (2009). Meayamycin inhibits pre-messenger RNA splicing and exhibits picomolar activity against multidrug-resistant cells. Mol. Cancer Ther. 8(8): 2308–18.

[79] Mammen, A.L. and Tiniakou, E. (2015). Intravenous immune globulin for statin-triggered autoimmune myopathy. N. Engl. J. Med. 373(17): 1680–2. doi: 10.1056/NEJMc1506163. PMID: 26488714; PMCID: PMC4629845.

[80] Sílvia Xargay-Torrent, Mónica López-Guerra, Laia Rosich, Arnau Montraveta, Jocabed Roldán, Vanina Rodríguez, Neus Villamor, Marta Aymerich, Chandraiah Lagisetti, Thomas R. Webb, Carlos López-Otín, Elias Campo and Dolors Colomer. (2015). The splicing modulator sudemycin induces a specific antitumor response and cooperates with ibrutinib in chronic lymphocytic leukemia. 6(26).

[81] Chandraiah Lagisetti, Gustavo Palacios, Tinopiwa Goronga, Burgess Freeman, William Caufield and Thomas R. Webb. (2013). Optimization of antitumor modulators of pre-mRNA splicing. J. Med. Chem. 56(24): 10033–44.

[82] Paolo Convertini, Manli Shen, Philip M. Potter, Gustavo Palacios, Chandraiah Lagisetti, Pierre de la Grange, Craig Horbinski, Yvonne N. Fondufe-Mittendorf, Thomas R. Webb and Stefan Stamm. (2014). Sudemycin E influences alternative splicing and changes chromatin modifications. Nucleic Acids Res. 42(8): 4947–61.

[83] Xiangyang Liu, Hui Zhu, Sreya Biswas and Yi-Qiang Chenga. (2016). Improved production of cytotoxic thailanstatins A and D through metabolic engineering of Burkholderia thailandensis MSMB43 and pilot scale fermentation. Synth. Syst. Biotechnol. 1(1): 34–38.

[84] Xiangyang Liu, Sreya Biswas, Michael G. Berg, Christopher M. Antapli, Feng Xie, Qi Wang, Man-Cheng Tang, Gong-Li Tang, Lixin Zhang, Gideon Dreyfuss and Yi-Qiang Cheng. (2013). Genomics-guided discovery of thailanstatins A, B, and C As pre-mRNA splicing inhibitors and antiproliferative agents from Burkholderia thailandensis MSMB43. J. Nat. Prod. 76(4): 685–93.

[85] Clare V. Lefave, Massimo Squatrito, Sandra Vorlova, Gina L. Rocco, Cameron W. Brennan, Eric C. Holland, Ying-Xian Pan and Luca Cartegni. (2011). Splicing factor hnRNPH drives an oncogenic splicing switch in gliomas. Embo. J. 30(19): 4084–97.

[86] Chery, J. (2016). RNA therapeutics: RNAi and antisense mechanisms and clinical applications. Postdoc. J. 4(7): 35–50.

[87] Crooke, S.T. (2004). Progress in antisense technology 55(1): 61–95.

[88] Frank C. Bennett, Brenda F. Baker, Nguyen Pham, Eric Swayze and Richard S. Geary. (2017). Pharmacology of antisense drugs 57(1): 81–105.

[89] Chan, J.H., Lim, S. and Wong, W.F. (2006). Antisense oligonucleotides: From design to therapeutic application 33(5-6): 533–540.

[90] Yimin Hua, Kentaro Sahashi, Gene Hung, Frank Rigo, Marco A. Passini, C. Frank Bennett and Adrian R. Krainer. (2010). Antisense correction of SMN2 splicing in the CNS rescues necrosis in a type III SMA mouse model. Genes Dev. 24(15): 1634–44.

[91] Christine P. Donahue, Christina Muratore, Jane Y. Wu, Kenneth S. Kosik and Michael S. Wolfe. (2006). Stabilization of the tau exon 10 stem loop alters pre-mRNA splicing. J. Biol. Chem. 281(33): 23302–6.

[92] Krajewska, M., Krajewski, S., Epstein, J.I., Shabaik, A., Sauvageot, J., Song, K., Kitada, S. and Reed, J.C. (1996). Immunohistochemical analysis of bcl-2, bax, bcl-X, and mcl-1 expression in prostate cancers. Am. J. Pathol. 148(5): 1567–76.

[93] Mercatante, D.R., Bortner, C.D., Cidlowski, J.A. and Kole, R. (2001). Modification of alternative splicing of Bcl-x pre-mRNA in prostate and breast cancer cells analysis of apoptosis and cell death. J. Biol. Chem. 276(19): 16411–7.

[94] Kotsimbos, A.T. and Hamid, Q. (1997). IL-5 and IL-5 receptor in asthma. Mem. Inst. Oswaldo Cruz 92 Suppl 2: 75–91.

[95] Karras, J.G., McKay, R.A., Dean, N.M. and Monia, B.P. (2000). Deletion of individual exons and induction of soluble murine interleukin-5 receptor-alpha chain expression through antisense oligonucleotide-mediated redirection of pre-mRNA splicing. Mol. Pharmacol. 58(2): 380–7.

[96] Giles, R.V., Spiller, D.G., Clark, R.E. and Tidd, D.M. (1999). Antisense morpholino oligonucleotide analog induces missplicing of C-myc mRNA. Antisense Nucleic Acid Drug Dev. 9(2): 213–20.

[97] Gitte H. Bruun, Thomas K. Doktor, Jonas Borch-Jensen, Akio Masuda, Adrian R. Krainer, Kinji Ohno and Brage S. Andresen. (2016). Global identification of hnRNP A1 binding sites for SSO-based splicing modulation. BMC Biol. 14: 54.

[98] Olga Anczuków, Martin Akerman, Antoine Cléry, Jie Wu, Chen Shen, Nitin H. Shirole, Amanda Raimer, Shuying Sun, Mads A. Jensen, Yimin Hua, Frédéric H.-T. Allain and Adrian R. Krainer. (2015). SRSF1-regulated alternative splicing in breast cancer. Mol. Cell 60(1): 105–17.

[99] Sanford, J.R., Ellis, J. and Cáceres, J.F. (2005). Multiple roles of arginine/serine-rich splicing factors in RNA processing. Biochem. Soc. Trans. 33(Pt 3): 443–6.

[100] Cartegni, L. and Krainer, A.R. (2003). Correction of disease-associated exon skipping by synthetic exon-specific activators. Nat. Struct. Biol. 10(2): 120–5.

[101] Wilusz, J., Devanney, S. and Caputi, M. (2005). Chimeric peptide nucleic acid compounds modulate splicing of the bcl-x gene *in vitro* and *in vivo*. Nucleic Acids Research 33: 6547–54.

[102] Michael Dewaele, Tommaso Tabaglio, Karen Willekens, Marco Bezzi, Shun Xie Teo, Diana H.P. Low, Cheryl M. Koh, Florian Rambow, Mark Fiers, Aljosja Rogiers, Enrico Radaelli, Muthafar Al-Haddawi, Soo Yong Tan, Els Hermans, Frederic Amant, Hualong Yan, Manikandan Lakshmanan, Ratnacaram Chandrahas Koumar, Soon Thye Lim, Frederick A. Derheimer, Robert M. Campbell, Zahid Bonday, Vinay Tergaonkar, Mark Shackleton, Christine Blattner, Jean-Christophe Marine and Ernesto Guccione. (2016). Antisense oligonucleotide-mediated MDM4 exon 6 skipping impairs tumor growth. J. Clin. Invest. 126(1): 68–84.

[103] Shinsuke Araki, Ryo Dairiki, Yusuke Nakayama, Aiko Murai, Risa Miyashita, Misa Iwatani, Toshiyuki Nomura and Osamu Nakanishi. (2015). Inhibitors of CLK protein kinases suppress cell growth and induce apoptosis by modulating pre-mRNA splicing. PLoS One 10(1): e0116929.

[104] Sarah J. Ross, Alexey S. Revenko, Lyndsey L. Hanson, Rebecca Ellston, Anna Staniszewska, Nicky Whalley, Sanjay K. Pandey, Mitchell Revill, Claire Rooney, Linda K. Buckett, Stephanie K. Klein, Kevin Hudson, Brett P. Monia, Michael Zinda, David C. Blakey, Paul D. Lyne and Robert Macleod, A. (2017). Targeting KRAS-dependent tumors with AZD4785, a high-affinity therapeutic antisense oligonucleotide inhibitor of KRAS. Sci. Transl. Med. 9(394).

10

Alternative Splicing in Cancer Drug Resistance

Ashna Gupta,[1] Ravi Chauhan,[1] Tarang Sharma,[1]
Muzafar A. Macha,[2] Tariq Masoodi,[3] Ammira S. Al-Shabeeb Akil,[4]
Ajaz A. Bhat[4], and Mayank Singh[1],**

1. Introduction

Drug resistance is known to arise when conventional treatment strategies can no longer treat a pathological condition. This phenomenon, which was first in bacteria, has also been found to occur in non-infectious diseases like cancer. Presently, drug resistance is a major limiting issue in successfully treating several types of cancers. The failure of standard therapy is attributed to factors such as the downregulation of apoptosis, epithelial-mesenchymal transition (EMT), drug inactivation, and drug efflux. Several cancers susceptible to chemotherapy eventually become drug resistant by different mechanisms, such as mutations in DNA, alternative splicing (AS), post-translational modifications (PTMs), etc. AS is the mechanism that processes pre-mRNA through different splicing modes to produce a variety of mature mRNAs with different structures and functions [1, 2]. AS is a deviation from the preferred sequence where certain exons are skipped resulting in various forms of mature mRNA. This process is mediated by a dynamic and flexible macromolecular machine, spliceosome, which works synergistically and is antistatic.

[1] Department of Medical Oncology (Lab), Bhim Rao Ambedkar Institute Rotary Cancer Hospital (BRAIRCH), All India Institute of Medical Sciences (AIIMS), New Delhi, Delhi, India 10029.
[2] Watson-Crick Centre for Molecular Medicine, Islamic University of Science and Technology, Kashmir, Jammu and Kashmir, India.
[3] Laboratory of Cancer immunology and genetics, Sidra Medicine, Qatar.
[4] Department of Human Genetics-Precision Medicine in Diabetes, Obesity and Cancer Research Program, Sidra Medicine, Doha, Qatar.
* Corresponding authors: Mayank.osu@gmail.com; abhat@sidra.org

Thus, AS has a role in almost every aspect of protein function, including binding between proteins and ligands, nucleic acids or membranes, localization, and enzymatic properties. AS has emerged as a central element in gene expression [3]. If the AS pathway goes unchecked, it can lead to protein expression disorders and various diseases, including cancer, neurodegenerative disorders, muscular dystrophies, and cardiovascular and immunologic diseases [4]. Recent research suggests a significant role of abnormal AS events in the occurrence of different types of tumors and their progression [5]. AS involves numerous *cis*-acting elements such as silencer elements (ESS and ISS), splice enhancers (ESE and ISE), and trans-factors like RNA-binding proteins (RBPs). Here we discuss how abnormal pre-mRNA splicing drives oncogenesis and enables cancerous cells to evade cytotoxic drugs. We also describe novel approaches targeting aberrant splicing events as a potential therapy in cancer treatment.

2. Role of Anomalous Splicing in Cancer and Drug Resistance

Uncontrolled proliferation is one of the hallmarks of tumor cells. This occurs when there is an imbalance between cell death and cell proliferation. The aberrant splicing within tumor cells results in abnormal apoptosis and cell proliferation. It is because of the increased frequency generation of splicing isoforms and corresponding protein isoforms. The alterations in splicing factors detected in tumors and other missplicing events (i.e., long non-coding and circular RNAs) in tumorigenesis have also been reported, which has led to the development of therapeutic approaches targeting splicing catalysis and splicing regulatory proteins to modulate pathogenically spliced variants (including tumor-specific neo-antigens for cancer immunotherapy) [6]. Specifically, tumor cells can express abnormal proteins that promote cancer progression through strange splicing events. Additionally, uncommon splicing variants develop drug resistance by mediating the loss of antigen targets or changing the target of drug action in cancer [4]. AS regulates the expression of several splicing factors, tumor suppressor genes, and oncogenes, thereby contributing to typical hallmarks of cancer, such as lack of differentiation, sustained proliferation, and resistance to apoptosis [7]; we will discuss this further.

2.1 Role of Splice Variants in Cytotoxic Drug Resistance, Drug Metabolism, and Signaling in Different Cancers

2.1.1 BRCA Splice Variant and PARPi Resistance in Ovarian Cancer

Loss-of-function mutations in *BRCA1* or *BRCA2* genes, which encode tumor suppressor proteins, have been reported to play a role in an increased risk of ovarian and breast cancer [8]. Poly (ADP-ribose) Polymerase (PARP), a single-strand DNA repair enzyme, is considered to maintain genome integrity because of dysfunctional *BRCA1* or *BRCA2* protein. Therefore, tumor cells with a mutation in *BRCA1* and/or *BRCA2* could be targeted by using PARP inhibitors (PARPi) [9, 10]. Although improved survival is seen in patients that are given PARPi therapy, some patients eventually develop resistance to PARPi [11]. The abnormal pre-mRNA splicing has been reported to be the potential underlying mechanism for developing resistance to PARPi therapy. Reportedly, the resistance to PARPi therapy is induced by the *BRCA1-Δ11q* splice variant that can bypass germline mutations [12]. The splicing event can be silenced, and PARPi resistance can be reversed by a small molecule inhibitor that inhibits the U2 snRNP spliceosome machinery. Moreover, cancers associated with *BRCA2* mutations can acquire chemoresistance on relapse. In a modeled acquired cross-linker resistance with a FA-derived *BRCA2*-mutated acute myeloid leukemia (AML) platform, the acquired cross-linker resistance was the expression of a functional BRCA2 protein variant lacking exon 5 and exon 7 (*BRCA2ΔE5+7*), implying a role for *BRCA2* splicing for acquired chemoresistance. A BCRA2 protein isoform with an internal deletion of 55 amino acids is encoded by a novel splice variant $BRCA2^{ΔE5+7}$, that has missing exons 5 and 7. This in-frame protein isoform contributes to the resistance to the drug mitomycin C. Evaluation of this model with clinical data shows substantial overlap with FA-hematopoiesis and BRCA2-associated ovarian cancer [13].

2.2 Role of Splice Variants in Altered Drug Metabolism and Signaling in Different Cancers

2.2.1 BRAF Splicing and Vemurafenib Resistance in Multiple Myeloma

The *BRAF* gene is known to encode a serine/threonine kinase that plays a critical role in the RAS-RAF-MEK-ERK-MAPK pathway. BRAF V600E mutation leads to a single valine-to-glutamate substitution, resulting in most BRAF mutations among melanoma patients [14]. Thus, BRAF acts as a therapeutic target for melanoma and can be inhibited by vemurafenib, specifically targeting V600E mutation in melanomas [15]. A BRAF splice variant encodes a 61 kDa BRAF (V600E) isoform with a missing RAS-binding domain (RBD) and is found to be expressed in 30% of vemurafenib-resistant melanoma patients. This resistance arises because of constitutive isoform dimerization in a RAS-independent manner. Moreover, patients with resistance to vemurafenib (PLX4032, RG7204) express a 61-kDa variant form of BRAF(V600E), p61BRAF(V600E), which lacks exons 4–8, a region that encompasses the RBD. The p61BRAF(V600E) shows enhanced dimerization in cells with low levels of RAS activation compared to full-length BRAF(V600E). In cells expressing p61BRAF(V600E) endogenously or ectopically, ERK signaling is resistant to the RAF inhibitor. Moreover, a mutation that abolishes the dimerization of p61BRAF(V600E) restores its sensitivity to vemurafenib [16, 17]. Thus, targeting the pre-mRNA in Multiple Myeloma splicing machinery is a potential therapeutic approach for treating vemurafenib resistance (Figure 1A).

2.2.3 Folylpolyglutamate Synthetase Spice Variant and Antifolate Resistant in Acute Lymphoblastic Leukemia

Methotrexate (MTX), a folate antagonist, is used to treat acute lymphoblastic leukemia (ALL). This medication acts by inhibiting an enzyme that prevents

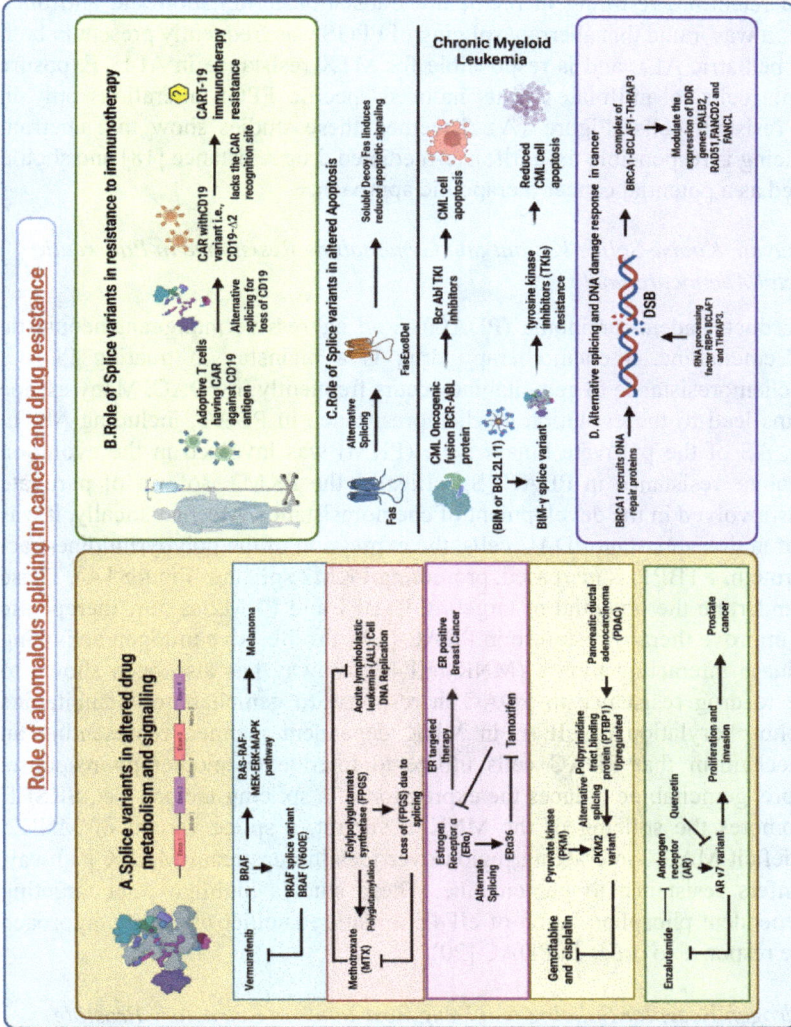

Figure 1: Role of alternate splicing in cancer and drug resistance.

(A) Splice variants as modulators of drug resistance by altering cellular cancer signaling and drug metabolism. (B) Emerging Role of Splice variants in resistance to immunotherapy in cancer. (C) Splice variants and their role in resistance to apoptosis in cancer cells. (D) Alternative splicing and its role in DNA damage response in cancer.

de novo nucleotide biosynthesis and thus DNA replication, which eventually halts the proliferation of cancer cells. The folylpolyglutamate synthetase (FPGS) enzyme allows folates and antifolates to accumulate intracellularly via polyglutamylation. Polyglutamylation of MTX is crucial for its pharmacological activity in leukemia. In ALL, loss of FPGS results in reduced levels of polyglutamylation-dependent accumulation of antifolates resulting in the development of drug resistance. Loss of FPGS is associated with irregular splicing of FPGS mRNA, including exon skipping and intron retention, resulting in premature translation termination and antifolate resistance. It was found that aberrant splicing of FPGS was frequently present in both adult and pediatric ALL and is responsible for MTX resistance in ALL. Exposure of leukemia cells to antifolate agents induced specific FPGS alterations only in antifolate-resistant cells (Figure 1A). Together, these studies show that aberrant FPGS splicing is responsible for antifolate-mediated drug resistance [18] and should be explored as a potential cancer therapeutic approach.

2.2.4 Pyruvate Kinase Splice Variant and Gemcitabine Resistance in Pancreatic Ductal Adenocarcinoma

Pancreatic ductal adenocarcinoma (PDAC) is an aggressive malignant neoplastic disease. Gemcitabine, a chemotherapy drug is a mainstay in treating PDAC; however, chemoresistance to gemcitabine occurs frequently in PDAC. Many escape mechanisms lead to the evolution of chemoresistance in PDAC, including AS. In one study, AS of the pyruvate kinase gene (PKM) was involved in the evolution of gemcitabine resistance in PDAC. Specifically, the PKM2 isoform of pyruvate kinase was involved in the development of chemoresistance. Mechanistically, it was shown that in drug-resistant PDAC cells, the expression of the polypyrimidine-tract binding protein, PTBP1, is increased, promoting PKM2 splicing (Figure 1A). These findings underline the potential of targeting PTBP1 and PKM2 as new therapeutic targets to improve therapy response in PDAC [19]. Proliferative mitogen activating protein kinase interacting kinase (MNK)/eIF4E pathway has also been shown to contribute to drug resistance in PDAC in response to gemcitabine. Gemcitabine induces phosphorylation of eIF4E in MNK dependent manner, representing an escape mechanism that PDAC cells utilize to tolerate chemotherapy response. Furthermore, gemcitabine induces the expression of a splicing factor—i.e., SRSF1, which promotes the splicing of the MNK2b variant, a splice variant of MNK2. MNK2b deficit MAPK-interacting region overrides the upstream MAPK pathway, which confers resistance to gemcitabine. These studies highlight that targeting MNK2-dependent phosphorylation of eIF4E might be another promising approach to increase response to drugs in PDAC [20].

2.2.5 Androgen Receptor Splicing and Drug Resistance in Refractory Prostate Cancer

The primary treatment for prostate cancer is blocking androgen signaling. This can be achieved by blocking androgen receptors (AR) or the deprivation of androgens.

Although patients with prostate cancer initially respond to deprivation therapy, resistance develops to the treatment as the disease progresses, leading to castration-resistance prostate cancer. One mechanism for acquiring resistance in prostate cancer is impaired AR splicing. Impaired AR splicing generates truncated splicing variants that work independently of androgen signaling, resulting in continuous expression of androgen receptor target genes [21]. The most common splice variant of AR is ARv7; this variant is produced by the inclusion of an exon called CE3, which is located within intron 3. This splice variant generates a truncated protein that lacks the ligand-binding domain, results in constitutive presence in the nucleus, and activates AR activity in the androgen-depleted cells. AR-V7-expression in prostate cancer patients is associated with resistance to antiandrogen drugs, like enzalutamide and abiraterone. Splicing factor hnRNPA1 has been shown to play a vital role in the generation of ARv7 [22]. Exploration of therapeutic options revealed that quercetin, an anti-cancer drug, targets EF-1α and HNRNPA 1 by suppressing HNRNPA 1 expression, thereby preventing the formation of AR-V7 and sensitizing the prostate cancer cell to enzalutamide (Figure 1A). Quercetin disrupts the complex of EF-1α-GTP-tRNA by binding to EF-1α. The quercetin binding to HNRNPA1 prevents its nuclear translocation, interfering with the tRNA machinery required to produce this isoform [23]. These studies shed light on how specific inhibitors can suppress AS of androgen receptor, which ultimately have important implications in treating tumor-resistant cases.

2.2.6 TP53 Splice Isoforms Predict Prognosis in Breast Cancer and its Involvement in Cisplatin Resistance in Renal Cell Carcinoma and Melanoma

The tumor suppressor TP53 is considered as guardian of the genome and is a crucial regulator in cancer prevention by maintaining genomic stability. AS of the TP53 gene gives rise to multiple splice isoforms, such as P53β, Δ40p53, and Δ133p53, whose expression has been linked to tumor progression [24–26]. In recent years, isoforms of p53 are differentially expressed in the tumor compared to normal tissue, suggesting their role in oncogenesis [27]. In normal breast tissues, there is an expression of p53α, p53β, and p53γ isoform, while in 60% of cases of breast tumors, there is a loss of expression of p53β and p53γ isoform, resulting in a poor prognosis of the disease. Δ40p53, P53β, and Δ133p53 are the small molecular weight variants of full-length *P53*. In renal cell carcinoma (RCC), isoforms such as p53β and Δ133p53 are overexpressed in tumor cells compared to normal cells [25]. In the presence of cisplatin (DNA alkylating agent), isoforms P53β and Δ40p53 are a part of the signaling cascade, which differentially regulates downstream targets in RCC. In melanoma, the p53β isoform induces protein transcription of *PUMA* and *P21* genes in a p53-dependent manner, while the Δ40p53 isoform inhibits the transcription of *PUMA* and *p21* genes, contributing to resistance to cisplatin [28].

2.2.7 Estrogen Receptor Splicing and Tamoxifen Resistance in Breast Cancer

High estrogen levels are associated with a higher risk of breast cancer. There are two isoforms of estrogen receptors, ERα and ERβ. Classical ER (ERα) has an activation domain (AF-1), a DNA-binding domain (DBD), and a hormone-binding domain (HBD). Cancers like endometrial carcinoma and breast cancer have different splice variants of ERα, which can either have a dominant positive or negative effect on cancer cells. One such splice variant of exon ER is ERαΔ3, which functions as a dominant-negative receptor in breast cancer cells by suppressing transcriptional activity induced by estrogen. Another variant of ERα is ERα46 which lacks the 173 amino acids, resulting in an aberrant protein isoform associated with the plasma membrane marker. This AS event leads to a change in the location of this protein, which mediates the rapid estrogen signaling and activation of its downstream target [29].

Another example is hERα–36. Splice forms of hERα–36 lack the AF-1 and some part of the AF-2 domain; it has a unique domain of 27 amino acids, which is encoded by exon 7 and exon 8 of hERα–66, and ERα36 overexpression causes a decrease in ERα66. Classically for the treatment of ER-positive breast cancer, tamoxifen has been a drug of choice, but patients with tumors expressing ERα36 received less benefit from tamoxifen as compared to patients expressing ERα66 demonstrating the significance of these splicing events in drug resistance [30] (Figure 1A).

2.3 Role of Splice Variants in Resistance to Immunotherapy

2.3.1 CD19 Splice Variant and CAR-T-19 Immunotherapy Resistance in B-ALL

One highly successful approach for treating B-ALL is adoptive T cells expressing chimeric antigen receptors (CAR-T) cells. Patients with B-ALL are being treated with adoptive T cells having CAR against CD19 antigen. CD19 is expressed mostly in all B cell malignancies, and CAR-T-based adoptive cell therapy has been documented to show remarkable efficacy in B-ALL with an overall 70–90% remission rate [31]. However, relapse has been observed in many patients treated with CAR-T cells, exhibiting a loss of cell surface expression—i.e., CD19 epitope. Multiple mechanisms have been proposed for the loss of surface expression, resulting in treatment failure in patients, including AS. The AS form of CD19 skipped exon 2, resulting in the expression of the N-terminally truncated CD19 variant—i.e., *CD19-Δ2*, lacks the CAR recognition site and prevents the killing of B-ALL cells by CAR-T cells [32] (Figure 1B). Different approaches have been explored to address these issues, like developing novel CARs that target alternative CD19 ectodomains or developing CARs that target different domains of the same antigen.

2.4 Role of Splice Variants in Altered Apoptosis

2.4.1 Fas Death Signaling and its Variants in Altered Apoptosis

Apoptosis, a critical phenomenon of programmed cell death, involves eliminating damaged cells from the body. The imbalance between the pro- and anti-apoptotic factors can lead to a reduced response of cancer cells to chemotherapeutic drugs, which predominantly work by inducing extrinsic or intrinsic apoptosis pathways. Interestingly AS may play a role in cell survival and cell death, as differential splicing of apoptotic regulators can give rise to protein isoforms that exhibit antagonistic activity [33]. During the extrinsic pathway of apoptosis, binding of specialized ligands such as FasL, TRAIL, and Apo3L to the Fas death receptor occurs, initiating the apoptotic cascade. Due to AS, different isoforms of death ligand, i.e., Fas, have been identified that lack the transmembrane domains. These variants inhibit cellular death by inhibiting the membrane-bound variant's activity, which cannot induce apoptotic signaling. One variant is FasExo8Del, which is a Fas variant that lacks an intracellular death domain due to exon skipping. FasExo8Del was identified in the lymphoma cell line showing resistance to FAS-mediated apoptosis and providing a survival advantage to tumor cells [34].

Another example is FasExo6Del, a soluble splice variant of the Fas ligand that lacks a transmembrane domain because of AS of the exon 6 competing with the Fas, which is membrane-bound and acts as a soluble decoy (Figure 1C). Higher levels of soluble Fas can be detected in other diseases like systemic lupus erythematosus, immunological disorders, etc. [35]. Three soluble functional forms of the human apoptosis-inducing Fas molecule are produced by AS. Therefore, the detection of soluble Fas and its associated variants makes it a potential marker to assess drug response in cancer that target this pathway.

2.4.2 BIM Splicing and TKI Resistant Chronic Myeloid Leukemia

BCL2-like 11 (*BIM* or *BCL2L11*) belongs to the BCL-2 protein family and is a pro-apoptotic protein. Its upregulation is required by tyrosine kinase inhibitors (TKIs) to induce apoptosis. TKIs like Imatinib and Dasatinib target the fusion BCR-ABL protein in chronic myelogenous leukemia (CML). One of the proposed mechanisms for the TKI resistance is linked to AS of BCR-ABL pre-mRNA. Additional mechanisms have also been discovered for TKI resistance which involves AS of BIM. A splice variant of BIM, BIM—γ, which has a deletion polymorphism because of the exon 4 to exon 3 switch, lacks the BH3 domain, which is important for the pro-apoptotic function of BIM. Individuals with the BIM-γ splice variant display poorer responses to TKIs [36]. As a therapeutic option, antisense oligonucleotides (ASOs) for splice-switching have been identified with ASO walking; 67 ASOs were isolated, which reverted the effect of aberrant *BIM* splicing, thus restoring the TKI sensitivity and sensitizing leukemic cell for TKI treatment [37] (Figure 1C). The efficacy of ASOs in correcting splicing variants is discussed later in this chapter.

2.5 AS of DNA Damage Response Genes in Cancer

DNA damage promotes oncogenesis by compromising genome integrity. Therefore, cells employ complex mechanisms to detect and activate DNA damage response (DDR) pathways. Despite the lack of direct evidence for the role of splice variants in DDR or the emergence of cancer drug resistance, DDR pathways are regulated by AS events. Evidence shows that the involvement of the DDR pathway is altered by RNA splicing [38, 39]. These splicing-based regulatory events are likely to contribute to developing resistance to therapy. BRCA1 mediated pathway is a key pathway that maintains genomic stability and prevents oncogenic transformation. It has been shown that BRCA1 recruits DNA repair proteins in response to the DNA double-strand break (DSB), which interacts with the splicing machinery through interaction with the RNA processing factors BCLAF1 and THRAP3. In response to DNA damage, BRCA1-BCLAF1-THRAP3 mediates BRAC1 phosphorylation in ATM dependent manner. BRCA1-BCLAF1-THRAP3 was found to modulate the expression of DDR genes (*BRCA2, PALB2, RAD51, FANCD2,* and *FANCL*) through AS which regulates export of mRNA of target genes (Figure 1D). Mutations in THRAP3 gene greatly alters the DSB repair capability suggesting that THRAP3 mutant tumors could be an excellent target to achieve synthetic lethality [40].

3. Strategies to Modulate the Aberrant RNA Splicing Error in Cancer

Oncogenic splicing error may be corrected by targeting impaired protein isoforms or targeting the various splicing factors involved in the machinery. In this section, we will discuss existing or development strategies to target altered reliance on spliceosome components and dysregulated splicing errors in cancer.

3.1 Oligonucleotide-Based Therapy

There has been increased interest in targeting specific splicing events with oligonucleotide-based therapies. Oligonucleotides specifically hybridize with RNA and disrupt the splicing regulatory sequences. Oligonucleotide-based therapies have been successful in correcting splicing events in non-cancerous disorders. Such therapies have been particularly effective for patients affected by spinal muscular atrophy (SMA) and Duchenne muscular dystrophy (DMD) and are currently in late-stage clinical trials. These strategies can potentially correct splicing aberrations that might lead to the emergence of cancer. A multifaced tool to manipulate splicing is ASOs, which are synthetic molecules mimicking RNA structures that specifically bind to an RNA target. Bauman and colleagues first demonstrated the anti-tumor activity of ASOs in a metastatic melanoma mouse model. They demonstrated that Bcl-X_L ASO targets BCL2 pre-mRNA of exon 2. High levels of the Bcl-X_L splice variant have been associated with the aggressiveness of cancer; however, Bcl-X_L ASO efficiently decreases the anti-apoptotic isoform—i.e., Bcl-X_L, with a simultaneous increase in the levels of the pro-apoptotic splice variant, such as Bcl-X_S [41] (Figure 2).

Another example that demonstrates the use of ASOs is in modifying the splicing *STAT3* gene involved in apoptosis. Mutation in *STAT3* pre-mRNA results in the formation of truncated protein STAT3β, which leads to aberrant signaling. However, a modified ASO targets a splicing enhancer element in the STAT3 gene, leading to tumor regression (42) (Figure 2A). Splicing generates abnormal soluble receptor tyrosine kinases (RTKs), which result in the generation of dominant-negative regulators and hence inactivate tumor signaling pathways. Interestingly, an important protein involved in the modulation of the angiogenic VEGF signaling

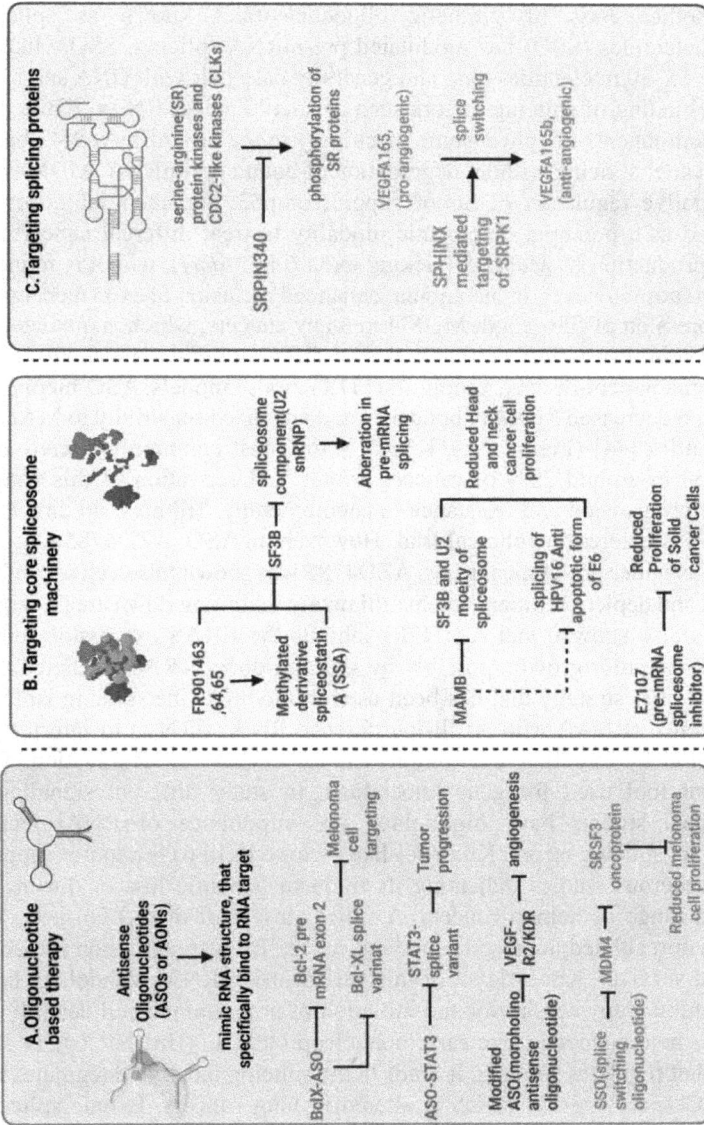

Figure 2: Strategies to modulate the aberrant RNA splicing error in cancer. (A) Different Oligonucleotide based therapy based approaches to correct splicing variants in cancer (B) Treating splicing defects by targeting core spliceosome machinery (C) Modulation of splicing by targeting splicing proteins.

is a modified ASO that induces the expression of VEGFR2/KDR. This dominant-negative secreted isoform inhibits the process of angiogenesis (Figure 2A). Splicing modulations generated by this oligonucleotide lead to the downregulation of angiogenic signaling [43].

Another class of synthetic oligonucleotides known as splice-switching oligonucleotides (SSO) has modulated pre-mRNA splicing. SSOs, like ASOs, are usually 15–30 nucleotides long and generally base pair with DNA and block the site for the binding of splicing factors such as small nuclear RNAs, RBPs, and various other components of spliceosome machinery to the pre-mRNA. SSO-based therapy aims to alter splicing without degradation of bound pre-mRNA. MDM4 and MDM2 are negative regulators of tumor suppressor p53, and their inhibitors have been explored as a potential therapeutic modality to treat different cancers. AS results in the production of *Mdm4-S* (lacking exon 6 in *Mdm4*), which is found in normal adult tissues; however, in melanoma, enhanced inclusion of exon 6 occurs, leading to the expression of full-length MDM4 in many cancers, which is mediated by SRSF3 oncoprotein. It was further shown that in different human melanoma cell lines and melanoma patient-derived xenograft (PDX) mouse models, ASO-mediated skipping of exon 6 decreased MDM4 abundance and enhanced sensitivity to MAPK-targeting therapeutics [44] (Figure 2A). KRAS is the most commonly altered gene having mutation in around 20% of cancers. Mutational activation of this gene promotes disease progression and resistance to chemotherapy. Till now, no direct inhibitor of KRAS has entered the clinical trial. However, an ASO, AZD4785 targeting KRAS mRNA is under pre-clinical trials. AZD4785 was shown to selectively inhibit KRAS mRNA and deplete cellular protein, ultimately inhibiting downstream targets. *In vivo* analysis also showed that AZD4785 inhibits the KRAS expression in tumors and suppresses tumor growth, possibly by shutting down KRAS-mediated oncogenesis [45]. Another strategy that has been used to modulate the splicing isoform is RNA interference (RNAi) with small interference RNA (siRNA) to interfere with gene expression by targeting the complementary mRNA for degradation. RNAi is an excellent tool used for gene knockdown to study different signaling cascades. Additional studies have highlighted the importance of RNAi technology in correcting splicing errors. Krueppel-like factor 6 (KLF6) is a tumor suppressor gene with numerous studies indicating its frequent genomic loss or downregulation in a broad range of human cancers. A splice variant of the KLF6 gene, i.e., KLF6-SV1, is upregulated in prostate and lung cancer. Its overexpression is associated with reduced survival. Knockdown of this variant using RNAi technology helps restore the chemotherapy sensitivity and initiation of programmed cell death [46]. Another protein, heterogeneous nuclear ribonucleoprotein L (HnRNP L), is an essential factor that regulates splicing. It binds to the splicing factor and regulates the splicing of the Caspase-9 gene, which is altered in lung cancers. Exonic splicing silencer (ESS) regulates caspase-9 pre-mRNA processing in NSCLC cells. Heterogeneous nuclear ribonucleoprotein L (hnRNP L) interacted with this ESS and downregulated hnRNP L expression, increasing the caspase-9a/9b ratio with caspase 9a being pro-apoptotic while caspase 9b being anti-apoptotic. Overexpression of hnRNP L, on the other, lowered the caspase-9a/9b ratio in NSCLC cells. Furthermore, Ser52 was identified as a necessary modification regulating the caspase-9a/9b ratio. In the mouse xenograft model, downregulation of hnRNP L in NSCLC cells induced a complete loss of tumorigenic capacity because of a change in the pattern of caspase-9 pre-mRNA processing [47].

In another interesting study, race-related tumor aggressiveness was explored based on splicing variants in prostate cancer. This study identified novel genome-wide, race-specific RNA splicing events as critical drivers of prostate cancer aggressiveness and therapeutic resistance in African American (AA) men. Enriched splice variants of *PIK3CD*, *FGFR3*, *TSC2*, and *RASGRP2* contributed to more significant oncogenic potential than corresponding European American (EA)-expressing variants. The difference in splice variants of these genes has been hypothesized as an underlying mechanism for prostate cancer health disparities between AA and EA men. In AA prostate cancer, splice variants of PIK3D, i.e., *PIK3CD-L* and *PIK3CD-S*, are expressed. PIK3CD-S has higher oncogenic potential and confers resistance to PI3Kδ inhibitor. High expression of the PIK3CD-S variant in prostate cancer is furthermore associated with poor survival. siRNA specifically targets exons 19 and 21 junctions, inhibiting the proliferative potential of the PIK3CD-S variant [48], thereby overcoming chemoresistance.

3.2 Targeting Core Spliceosome Machinery

Over the past few years, many active biomolecules obtained from bacteria that target various splicing sites, repressors, or enhancer regulators, which ultimately alter mRNA splicing patterns, have been discovered. Early modulators of splicing like FR901464, FR901463, and FR901465 were obtained from the *Pseudomonas* species. It was found that these compounds bind and inhibit the SF3B component of spliceosome machinery. FR901464, along with its methylated derivative spliceostatin A (SSA), binds to the spliceosome component SF3B1 and interferes with the binding of the spliceosome component (U2 snRNP) to the branch point, thus interrupting recognition of U2 snRNA branch point resulting in aberration in pre-mRNA splicing [49] (Figure 2B). Many other splicing modulators are present, which are isolated from different bacterial strains like GEX1, isolated from *Streptomyces* sp. [50], and Pladienolides, isolated from *Streptomyces platensis* [51]. Interestingly, all these modulators have the same mechanism of action as SSA. SF3B1 is a common target for different products obtained from bacteria suggested to share a common mode of action. Other compounds, such as E7107 (pladienolide analog), sudemycins, and meayamycin B (MAMB), have been shown to have the same anti-splicing properties as above. The complete mechanism of action for most of these compounds is still not very clear, however, above findings confirmed that these modulators interact with SF3B1 spicing factor, which is the largest subunit of spliceosome complex responsible for recognition of 3' branch site and associated with U2 that might play a pivot role in modulating splicing [52]. Different splicing modulators have different anti-tumor effects. Treatment with FR901464 in solid tumors reduced tumor size by 80%. Another analog of FR901464, meayamycin B (MAMB) inhibits splicing of anti-apoptotic isoform of E6 protein, which in turn reduces the amount of MCL1L, a variant of Myeloid leukemia 1 in head and neck cancer which is driven by HPV16 [53] (Figure 2B). A clinical trial of E7107 (pladienolide analog) which was carried out in patients having different types of tumors, including esophageal, colorectal, renal, gastric, and pharma dynamics analysis, revealed that splicing inhibition with E7107 was achieved as E7107 inhibited tumor growth, however, at higher doses it causes of toxicity [54, 55] (Figure 2B).

Mutations in the serine/threonine kinase *BRAF* are present in more than 60% of melanomas, with a mutation in BRAF(V600E) being the most prevalent. This mutation is associated with constitutive activation of downstream MAPK signaling. As discussed earlier, vemurafenib is an RAF kinase inhibitor with remarkable clinical activity in BRAF(V600E)-positive melanoma tumors. However, resistance rapidly develops to this inhibitor because of the emergence of BRAF AS isoforms leading to the elimination of the RBD. It was shown that small molecule pre-mRNA splicing modulators reduce BRAF3–9 production, limit *in vitro* cell growth of vemurafenib-resistant cells, and slow down vemurafenib-resistant tumors in xenograft models [17].

3.3 Targeting Splicing Proteins

The splicing factor has a role in driving oncogenesis, and identifying various splicing proteins has emerged as a new tool for correcting splicing errors. Targeting splicing factor kinases, such as serine-arginine (SR) protein kinases and CDC2-like kinases (CLKs), have been identified as a potential way to modulate splicing kinases that are important for exon selection. The discovery of inhibitors against these kinases has helped delineate the splicing mechanism. CLK1-4 is a member of the Cdc2-like kinase (CLK) family, which plays a key role in splicing by regulating the phosphorylation of SR proteins. SR proteins usually have an RNA recognition motif located at the N terminus, while at the C terminus, it has the RS domain that regulates protein-protein interaction. Hypo or hyperphosphorylation of SR protein generally affects splicing [49]. Consistent with the idea, studies have shown that SRPIN340, an inhibitor of SR protein phosphorylation, inhibits SR protein kinases 1 and 2 (SRPK1 and SRPK2) and also downregulates SRp75, which increases HIV expression [56] (Figure 2C). In another study, targeting SRPK1 with a new generation modulator, "SPHINX", caused splice switching from VEGFA165 pro-angiogenic to VEGFA165b anti-angiogenic in prostate cancer. SRPK1 has been shown to be an excellent target for cancer therapeutic development by SPHINX or other inhibitors. [57]. By screening thousands of compounds, TG003 was identified, which inhibits Clk1, 2, and 4 activities, thereby affecting AS; however, genome-wide effects of this drug are not yet well characterized. However, this drug affects the expression of Clk-1 isoforms, an alteration that might be an important therapeutic target [58] (Figure 2C).

MYC oncoprotein is also regulated by pre-mRNA splicing, and cancer cells driven by MYC can be affected by inhibition of spliceosome machinery. MYC is an essential component in cancer pathogenesis and is dysregulated in around 50% of cancers. So different splicing modulators can work as anti-cancer agents driven by

MYC. It has been suggested that T-025, a potent oral Clk-1 inhibitor, can be used as a novel target in MYC-driven cancer. T-025 inhibits phosphorylation of CLK, thereby suppressing cell growth and induction of death both *in vitro* and *in vivo* [59]. In a separate study, intensive screening of compounds reveals three inhibitors of CLK1-2 that inhibit the phosphorylation of various splicing factors, such as SRSF4, SRSF6, and SRSF1. Additionally, splicing alteration of the S6K (ribosomal protein kinase) gene promotes cell apoptosis and reduces cell proliferation. However, more studies are required to understand the relevance of CLK inhibitors in cancer therapeutics and explore if particular genetic subtypes of cancer are more affected by cell killing by these inhibitors [60].

Conclusion

In the past years, it has been shown that new chemotherapeutic agents, specific targeted therapy and immunotherapy, have changed the treatment outcome response in cancer patients. However, one of the biggest challenges that persist in cancer treatment is resistance to therapeutic drugs. Much evidence has suggested that dysregulated splicing mechanism is an important factor in cancer development. The splicing process occurs post-transcriptionally, so the generation of novel transcripts occurs rapidly in response to genotoxic stress. Splice variants of PI3K, TP53, BRAF, BCL21, Cyclin D1, and CASP2 drive various cancers. In addition, exon skipping or intron retention changes the complete protein domain, resulting in a loss of sensitivity to targeted therapies. Many studies show a link between aberrant splice variants and the aggressiveness of cancer. However, the exact molecular events leading to splice variants and the link between splice variants and the emergence of resistance need in-depth studies [61, 62]. To overcome splicing-mediated resistance and develop new therapeutic strategies, it is important to understand the dysregulated splicing pattern in detail with a deeper mechanistic understanding.

Several chemical compounds have been discovered that inhibit splicing catalysis either by targeting spliceosome assembly or inhibiting SR protein phosphorylation, like sudemycins, spliceostatin A, and E7107, which binds to SF3B1 and inhibit splicing at an early stage of spliceosome formation. In contrast, several other compounds like N-Palmitoyl-l-leucine and benzothiazole-4,7-dione inhibit splicing catalysis at later stages of the splicing pathway. Another popular way to modulate splicing is the use of oligonucleotides. Oligonucleotide therapy is very effective in cells that carry spliceosome gene mutations. ASOs, particularly SSOs, provide a better treatment approach that can alter splicing and control gene expression, surpassing the effects of a specific mutation. Manipulation of splicing and gene expression favors using SSOs as a drug platform for disease treatment.

Further biological studies and characterization of altered splice events are necessary to link the relationship between mutated splicing factors and cancer type, which is required to develop new therapeutic approaches. More efforts are needed to understand the effects of compounds on the splicing events that will help us understand how aberrant splicing variants are generated in the presence of cytotoxic cancer drugs. Thus, it is essential to understand further the splicing events in cancer, thereby aiding in developing smart therapies that target AS variants.

Declarations

Ethics approval and consent to participate
Not applicable.

Consent for publication
All authors consent to publication.

Availability of data
Not applicable

Funding

This study was supported by IITD Grant (AI-34) from All India Institute of Medical Sciences (AIIMS) New Delhi, and the Department of Biotechnology (DBT) Ministry of Science and Technology Government of India Grant (BT/PR40649/MED/31/441/2020) to Mayank Singh. Sidra Medicine Precision Program funding to Ajaz A. Bhat (SDR400105) and Ammira Al-Shabeeb Akil (SDR400175). Dr. Muzafar A. Macha is funded by Ramalingaswami Fellowship (Grant number: DO NO.BT/HRD/35/02/2006) from the Department of Biotechnology & Core Research Grant (CRG/2021/003805) from Science and Engineering Research Board (SERB), Government. of India, New Delhi.

Author Contributions

A.G., R.C, T.S., and M.S. wrote the manuscript and generated figures. M.S., A.A.B, and M.A.M. contributed to the concept and design and critically edited the manuscript. T.M. and A.A.S.A performed critical revision and editing of the scientific content. All authors read and approved the final manuscript.

Acknowledgments

Figures were created using BioRender.

Abbreviations

VEGF	:	Vascular endothelial growth factor,
EMT	:	Epithelial-mesenchymal transition
AS	:	Alternative splicing
ESE	:	Exonic splicing silencer
ISS	:	Intronic splicing silencer
ESS	:	Exonic splicing enhancer

ISE	:	Intronic splicing enhancer
RBP	:	RNA-binding proteins
BRAC	:	Breast Cancer gene
PARP	:	Poly (ADP) ribose polymerase
PARPi	:	PARP inhibitors
AML	:	Acute myeloid leukemia
ERK	:	Extracellular signal-regulated kinase
MTX	:	Methotrexate
FPGS	:	Folylpolyglutamate synthetase
PDAC	:	Pancreatic ductal adenocarcinoma
PKM	:	Pyruvate kinase muscle isozyme
MNK	:	MAPK-interacting kinase
SRSF1	:	Serine/arginine-rich splicing factor 1
AR	:	Androgen receptor
hnRNPs	:	Heterogeneous nuclear ribonucleoproteins
ALL	:	Acute lymphoblastic leukemia
CAR	:	Chimeric antigen receptors
TKI	:	Tyrosine kinase inhibitors
CML	:	Chronic myelogenous leukemia
ASO	:	Antisense oligonucleotides
SSO	:	Splice-switching oligonucleotides
RNAi	:	RNA interference
siRNA	:	Small interfering RNA
KLF6	:	Krüppel like factor 6
MAMB	:	Meayamycin B

References

[1] Lee, Y. and Rio, D.C. (2015). Mechanisms and regulation of alternative Pre-mRNA splicing. Annu. Rev. Biochem. 84: 291–323.

[2] Darnell, J.E. (2013, Apr). Reflections on the history of pre-mRNA processing and highlights of current knowledge: A unified picture. RNA N Y N 19(4): 443–60.

[3] Wang, Y., Liu, J., Huang, B., Xu, Y.-M., Li, J., Huang, L.-F. et al. (2015, Mar). Mechanism of alternative splicing and its regulation. Biomed. Rep. 3(2): 152–8.

[4] Deng, K., Yao, J., Huang, J., Ding, Y. and Zuo, J. (2021, Jun). Abnormal alternative splicing promotes tumor resistance in targeted therapy and immunotherapy. Transl. Oncol. 14(6): 101077.

[5] Climente-González, H., Porta-Pardo, E., Godzik, A. and Eyras, E. (2017, Aug). The functional impact of alternative splicing in cancer. Cell Rep. 20(9): 2215–26.

[6] Zhang, Y., Qian, J., Gu, C. and Yang, Y. (2021, Feb). Alternative splicing and cancer: A systematic review. Signal Transduct Target Ther. 6(1): 78.

[7] Oltean, S. and Bates, D.O. (2014, Nov). Hallmarks of alternative splicing in cancer. Oncogene 33(46): 5311–8.

[8] Friedman, L.S., Ostermeyer, E.A., Szabo, C.I., Dowd, P., Lynch, E.D., Rowell, S.E. et al. (1994, Dec). Confirmation of BRCA1 by analysis of germline mutations linked to breast and ovarian cancer in ten families. Nat Genet. 8(4): 399–404.

[9] Farmer, H., McCabe, N., Lord, C.J., Tutt, A.N.J., Johnson, D.A., Richardson, T.B. et al. (2005, Apr). Targeting the DNA repair defect in BRCA mutant cells as a therapeutic strategy. Nature 434(7035): 917–21.

[10]   Bryant, H.E., Schultz, N., Thomas, H.D., Parker, K.M., Flower, D., Lopez, E. et al. (2005, Apr). Specific killing of BRCA2-deficient tumours with inhibitors of poly(ADP-ribose) polymerase. Nature 434(7035): 913–7.

[11]   Kim, Y., Kim, A., Sharip, A., Sharip, A., Jiang, J., Yang, Q. et al. (2017, Feb). Reverse the resistance to PARP inhibitors. Int. J. Biol. Sci. 13(2): 198–208.

[12]   Wang, Y., Bernhardy, A.J., Cruz, C., Krais, J.J., Nacson, J., Nicolas, E. et al. (2016, May). The BRCA1-Δ11q Alternative splice isoform bypasses germline mutations and promotes therapeutic resistance to PARP inhibition and cisplatin. Cancer Res. 76(9): 2778–90.

[13]   Meyer, S., Stevens, A., Paredes, R., Schneider, M., Walker, M.J., Williamson, A.J.K. et al. (2017, Jun). Acquired cross-linker resistance associated with a novel spliced BRCA2 protein variant for molecular phenotyping of BRCA2 disruption. Cell Death Dis. 8(6): e2875.

[14]   Davies, H., Bignell, G.R., Cox, C., Stephens, P., Edkins, S., Clegg, S. et al. (2002, Jun). Mutations of the BRAF gene in human cancer. Nature 417(6892): 949–54.

[15]   Flaherty, K.T., Puzanov, I., Kim, K.B., Ribas, A., McArthur, G.A., Sosman, J.A. et al. (2010, Aug). Inhibition of mutated, activated BRAF in metastatic melanoma. N. Engl. J. Med. 363(9): 809–19.

[16]   Poulikakos, P.I., Persaud, Y., Janakiraman, M., Kong, X., Ng, C., Moriceau, G. et al. (2011, Nov). RAF inhibitor resistance is mediated by dimerization of aberrantly spliced BRAF(V600E). Nature 480(7377): 387–90.

[17]   Salton, M., Kasprzak, W.K., Voss, T., Shapiro, B.A., Poulikakos, P.I. and Misteli, T. (2015, May). Inhibition of vemurafenib-resistant melanoma by interference with pre-mRNA splicing. Nat. Commun. 6: 7103.

[18]   Wojtuszkiewicz, A., Raz, S., Stark, M., Assaraf, Y.G., Jansen, G., Peters, G.J. et al. (2016, Apr). Folylpolyglutamate synthetase splicing alterations in acute lymphoblastic leukemia are provoked by methotrexate and other chemotherapeutics and mediate chemoresistance. Int. J. Cancer 138(7): 1645–56.

[19]   Calabretta, S., Bielli, P., Passacantilli, I., Pilozzi, E., Fendrich, V., Capurso, G. et al. (2016, Apr). Modulation of PKM alternative splicing by PTBP1 promotes gemcitabine resistance in pancreatic cancer cells. Oncogene 35(16): 2031–9.

[20]   Adesso, L., Calabretta, S., Barbagallo, F., Capurso, G., Pilozzi, E., Geremia, R. et al. (2013, Jun). Gemcitabine triggers a pro-survival response in pancreatic cancer cells through activation of the MNK2/eIF4E pathway. Oncogene 32(23): 2848–57.

[21]   Munkley, J., Livermore, K., Rajan, P. and Elliott, D.J. (2017, Sep). RNA splicing and splicing regulator changes in prostate cancer pathology. Hum Genet. 136(9): 1143–54.

[22]   Cao, B., Qi, Y., Zhang, G., Xu, D., Zhan, Y., Alvarez, X. et al. (2014, Mar). Androgen receptor splice variants activating the full-length receptor in mediating resistance to androgen-directed therapy. Oncotarget 5(6): 1646–56.

[23]   Marcinkiewicz, C., Gałasiński, W. and Gindzieński, A. (1995). EF-1 alpha is a target site for an inhibitory effect of quercetin in the peptide elongation process. Acta Biochim. Pol. 42(3): 347–50.

[24]   Surget, S., Khoury, M.P. and Bourdon, J.-C. (2013, Dec). Uncovering the role of p53 splice variants in human malignancy: A clinical perspective. OncoTargets Ther. 7: 57–68.

[25]   Song, W., Huo, S., Lü, J., Liu, Z., Fang, X., Jin, X. et al. (2009, Apr). Expression of p53 isoforms in renal cell carcinoma. Chin. Med. J. (Engl). 122(8): 921–6.

[26]   Bourdon, J.-C., Khoury, M.P., Diot, A., Baker, L., Fernandes, K., Aoubala, M. et al. (2011). p53 mutant breast cancer patients expressing p53γ have as good a prognosis as wild-type p53 breast cancer patients. Breast Cancer Res. BCR 13(1): R7.

[27]   Khoury, M.P. and Bourdon, J.-C. (2010, Mar). The isoforms of the p53 protein. Cold Spring Harb. Perspect. Biol. 2(3): a000927.

[28]   Avery-Kiejda, K.A., Zhang, X.D., Adams, L.J., Scott, R.J., Vojtesek, B., Lane, D.P. et al. (2008, Mar). Small molecular weight variants of p53 are expressed in human melanoma cells and are induced by the DNA-damaging agent cisplatin. Clin. Cancer Res. Off. J. Am. Assoc. Cancer Res. 14(6): 1659–68.

[29]   Barone, I., Brusco, L. and Fuqua, S.A.W. (2010, May). Estrogen receptor mutations and changes in downstream gene expression and signaling. Clin. Cancer Res. Off. J. Am. Assoc. Cancer Res. 16(10): 2702–8.

[30] Shi, L., Dong, B., Li, Z., Lu, Y., Ouyang, T., Li, J. et al. (2009, Jul). Expression of ER-α36, a novel variant of estrogen receptor α, and resistance to tamoxifen treatment in breast cancer. J. Clin. Oncol. 27(21): 3423–9.

[31] Wang, Z., Wu, Z., Liu, Y. and Han, W. (2017, Feb). New development in CAR-T cell therapy. J. Hematol. Oncol. J. Hematol. Oncol. 10(1): 53.

[32] Sotillo, E., Barrett, D.M., Black, K.L., Bagashev, A., Oldridge, D., Wu, G. et al. (2015, Dec). Convergence of acquired mutations and alternative splicing of CD19 enables resistance to CART-19 immunotherapy. Cancer Discov. 5(12): 1282–95.

[33] Schwerk, C. and Schulze-Osthoff, K. (2005, Jul). Regulation of apoptosis by alternative pre-mRNA splicing. Mol. Cell 19(1): 1–13.

[34] Cascino, I., Papoff, G., De Maria, R., Testi, R. and Ruberti, G. (1996, Jan). Fas/Apo-1 (CD95) receptor lacking the intracytoplasmic signaling domain protects tumor cells from Fas-mediated apoptosis. J. Immunol. Baltim. Md. 1950 156(1): 13–7.

[35] Cheng, J., Zhou, T., Liu, C., Shapiro, J.P., Brauer, M.J., Kiefer, M.C. et al. (1994, Mar). Protection from Fas-mediated apoptosis by a soluble form of the Fas molecule. Science 263(5154): 1759–62.

[36] Ng, K.P., Hillmer, A.M., Chuah, C.T.H., Juan, W.C., Ko, T.K., Teo, A.S.M. et al. (2012, Mar). A common BIM deletion polymorphism mediates intrinsic resistance and inferior responses to tyrosine kinase inhibitors in cancer. Nat. Med. 18(4): 521–8.

[37] Liu, J., Bhadra, M., Sinnakannu, J.R., Yue, W.L., Tan, C.W., Rigo, F. et al. (2017, Sep). Overcoming imatinib resistance conferred by the BIM deletion polymorphism in chronic myeloid leukemia with splice-switching antisense oligonucleotides. Oncotarget. 8(44): 77567–85.

[38] Shkreta, L. and Chabot, B. (2015, Oct). The RNA splicing response to DNA damage. Biomolecules 5(4): 2935–77.

[39] Heath, C.G., Viphakone, N. and Wilson, S.A. (2016, Oct). The role of TREX in gene expression and disease. Biochem. J. 473(19): 2911–35.

[40] Vohhodina, J., Barros, E.M., Savage, A.L., Liberante, F.G., Manti, L., Bankhead, P. et al. (2017, Dec). The RNA processing factors THRAP3 and BCLAF1 promote the DNA damage response through selective mRNA splicing and nuclear export. Nucleic Acids Res. 45(22): 12816–33.

[41] Bauman, J.A., Li, S.-D., Yang, A., Huang, L. and Kole, R. (2010, Dec). Anti-tumor activity of splice-switching oligonucleotides. Nucleic Acids Res. 38(22): 8348–56.

[42] Hong, D., Kurzrock, R., Kim, Y., Woessner, R., Younes, A., Nemunaitis, J. et al. (2015, Nov). AZD9150, a next-generation antisense oligonucleotide inhibitor of STAT3 with early evidence of clinical activity in lymphoma and lung cancer. Sci. Transl. Med. 7(314): 314ra185.

[43] Uehara, H., Cho, Y., Simonis, J., Cahoon, J., Archer, B., Luo, L. et al. (2013, Jan). Dual suppression of hemangiogenesis and lymphangiogenesis by splice-shifting morpholinos targeting vascular endothelial growth factor receptor 2 (KDR). FASEB J. Off. Publ. Fed. Am. Soc. Exp. Biol. 27(1): 76–85.

[44] Dewaele, M., Tabaglio, T., Willekens, K., Bezzi, M., Teo, S.X., Low, D.H.P. et al. (2016, Jan). Antisense oligonucleotide-mediated MDM4 exon 6 skipping impairs tumor growth. J Clin. Invest. 126(1): 68–84.

[45] Ross, S.J., Revenko, A.S., Hanson, L.L., Ellston, R., Staniszewska, A., Whalley, N. et al. (2017, Jun). Targeting KRAS-dependent tumors with AZD4785, a high-affinity therapeutic antisense oligonucleotide inhibitor of KRAS. Sci. Transl. Med. 9(394): eaal5253.

[46] Sangodkar, J., DiFeo, A., Feld, L., Bromberg, R., Schwartz, R., Huang, F. et al. (2009, Dec). Targeted reduction of KLF6-SV1 restores chemotherapy sensitivity in resistant lung adenocarcinoma. Lung Cancer Amst. Neth. 66(3): 292–7.

[47] Goehe, R.W., Shultz, J.C., Murudkar, C., Usanovic, S., Lamour, N.F., Massey, D.H. et al. (2010, Nov). hnRNP L regulates the tumorigenic capacity of lung cancer xenografts in mice via caspase-9 pre-mRNA processing. J. Clin. Invest. 120(11): 3923–39.

[48] Wang, B.-D., Ceniccola, K., Hwang, S., Andrawis, R., Horvath, A., Freedman, J.A. et al. (2017, Jun). Alternative splicing promotes tumour aggressiveness and drug resistance in African American prostate cancer. Nat. Commun. 8: 15921.

[49] Lee, S.C.-W. and Abdel-Wahab, O. (2016, Sep). Therapeutic targeting of splicing in cancer. Nat. Med. 22(9): 976–86.

[50] Sakai, Y., Yoshida, T., Ochiai, K., Uosaki, Y., Saitoh, Y., Tanaka, F. et al. (2002, Oct). GEX1 compounds, novel antitumor antibiotics related to herboxidiene, produced by Streptomyces sp. I. Taxonomy, production, isolation, physicochemical properties and biological activities. J. Antibiot. (Tokyo) 55(10): 855–62.

[51] Mizui, Y., Sakai, T., Iwata, M., Uenaka, T., Okamoto, K., Shimizu, H. et al. (2004, Mar). Pladienolides, new substances from culture of Streptomyces platensis Mer-11107. III. *In vitro* and *in vivo* antitumor activities. J. Antibiot. (Tokyo) 57(3): 188–96.

[52] Salton, M. and Misteli, T. (2016, Jan). Small molecule modulators of Pre-mRNA splicing in cancer therapy. Trends Mol. Med. 22(1): 28–37.

[53] Gao, Y., Trivedi, S., Ferris, R. and Koide, K. (2014, Aug). Regulation of HPV16 E6 and MCL1 by SF3B1 inhibitor in head and neck cancer cells. Sci. Rep. 4: 6098.

[54] Eskens, F.A.L.M., Ramos, F.J., Burger, H., O'Brien, J.P., Piera, A., de Jonge, M.J.A. et al. (2013, Nov). Phase I pharmacokinetic and pharmacodynamic study of the first-in-class spliceosome inhibitor E7107 in patients with advanced solid tumors. Clin. Cancer Res. Off. J. Am. Assoc. Cancer Res. 19(22): 6296–304.

[55] Hong, D.S., Kurzrock, R., Naing, A., Wheler, J.J., Falchook, G.S., Schiffman, J.S. et al. (2014, Jun). A phase I, open-label, single-arm, dose-escalation study of E7107, a precursor messenger ribonucleic acid (pre-mRNA) splicesome inhibitor administered intravenously on days 1 and 8 every 21 days to patients with solid tumors. Invest. New Drugs 32(3): 436–44.

[56] Fukuhara, T., Hosoya, T., Shimizu, S., Sumi, K., Oshiro, T., Yoshinaka, Y. et al. (2006, Jul). Utilization of host SR protein kinases and RNA-splicing machinery during viral replication. Proc. Natl. Acad. Sci. USA 103(30): 11329–33.

[57] Mavrou, A., Brakspear, K., Hamdollah-Zadeh, M., Damodaran, G., Babaei-Jadidi, R., Oxley, J. et al. (2015, Aug). Serine-arginine protein kinase 1 (SRPK1) inhibition as a potential novel targeted therapeutic strategy in prostate cancer. Oncogene 34(33): 4311–9.

[58] Muraki, M., Ohkawara, B., Hosoya, T., Onogi, H., Koizumi, J., Koizumi, T. et al. (2004, Jun). Manipulation of alternative splicing by a newly developed inhibitor of Clks. J. Biol. Chem. 279(23): 24246–54.

[59] Iwai, K., Yaguchi, M., Nishimura, K., Yamamoto, Y., Tamura, T., Nakata, D. et al. (2018, Jun). Anti-tumor efficacy of a novel CLK inhibitor via targeting RNA splicing and MYC-dependent vulnerability. EMBO Mol. Med. 10(6): e8289.

[60] Araki, S., Dairiki, R., Nakayama, Y., Murai, A., Miyashita, R., Iwatani, M. et al. (2015, Jan). Inhibitors of CLK protein kinases suppress cell growth and induce apoptosis by modulating pre-mRNA splicing. PLoS ONE 10(1): e0116929.

[61] Galletti, G., Matov, A., Beltran, H., Fontugne, J., Miguel Mosquera, J., Cheung, C. et al. (2014, Nov). ERG induces taxane resistance in castration-resistant prostate cancer. Nat. Commun. 5: 5548.

[62] Hagen, R.M., Adamo, P., Karamat, S., Oxley, J., Aning, J.J., Gillatt, D. et al. (2014, Oct). Quantitative analysis of ERG expression and its splice isoforms in formalin-fixed, paraffin-embedded prostate cancer samples: Association with seminal vesicle invasion and biochemical recurrence. Am. J. Clin. Pathol. 142(4): 533–40.

11

A Systematic Evaluation of Alternative Splicing and Cancer
A Review

Amrita Bhat,[1] Ruchi Shah[2] and Rakesh Kumar[3],*

1. Introduction

Various studies contain shed light on the intricacy of human transcriptomes in unique depth from last decades [1–3]. This technological advancement has greatly increased our understanding of pre-mRNA splicing patterns, which were found more than four decades ago [4] and were mostly under studied for a extensive period of time. Splicing of Pre-mRNA is a active progression with the intention of involve the deletion of introns and the unification of exons to create mature mRNA, which is afterwards used as a pattern for protein translation.

For decades, scientists have been perplexed as to how a very identical collection of genes which can result in such vastly similar phenotypes. AS, among other things, be revealed to play a significant task in this conclusion [5] and found that the intricacy of alternative splicing differed amongst vertebrates, through primates having the highest level of complexity. Splicing landscapes differ so much between species so as to splicing profiles of the same organs replicate within the species rather than the tissue identity [6]. This is in contrast to tissue-specific gene expression, which is particularly preserved. It's predictable that splicing diversity needs complex regulation.

[1] Institute of Human Genetics, University of Jammu, Jammu and Kashmir, India.
[2] Department of Biotechnology, University of Kashmir, Jammu and Kashmir, India.
[3] School of Biotechnology, SMVDU, Katra, Jammu and Kashmir, India.
* Corresponding author: kumar.rakesh@smvdu.ac.in

1.1 Regulation of Splicing

Pre-mRNAs are not frequently important intended on behalf of protein production in eukaryotic organisms. Awaiting Introns are deleted and the exons are tie mutually to produce mRNAs (Figure 1). More than 95 percent of human genes involve pre-mRNA splicing for expression. The development may also be modified to produce differentially transcribed mRNAs that code for a variety of protein varieties, which is a fundamental process for cellular homeostasis, cell differentiation, and development.

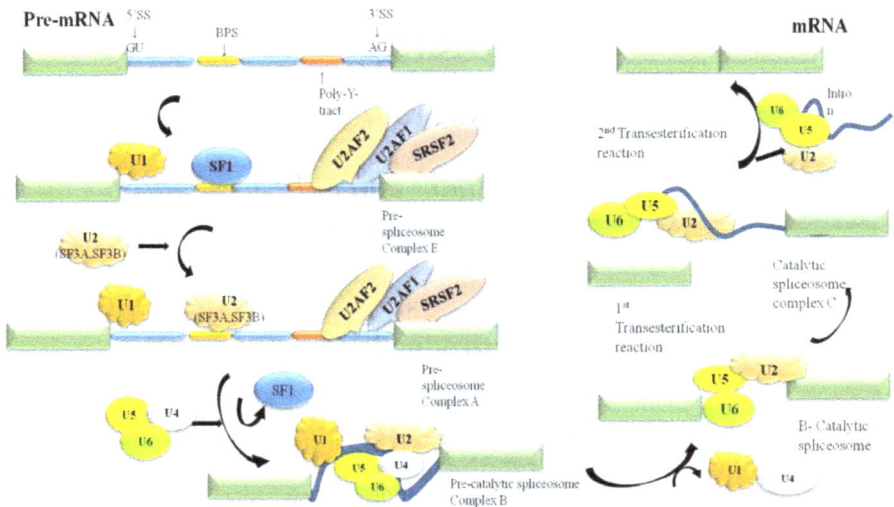

Figure 1: Showing spliceosome assembly.

Splicing changes are now recognized as a cancer hallmark since the splicing pathway and its regulation are so critical to understanding every cancer signature. We go over the splicing process and how it changes in cancer. In this, we will explore the prospects designed meant for the development of original medicinal techniques in experimental oncology and how the alternating splicing play vital role in cancer.

1.2 Main Alternative Splicing Patterns

AS is controlled by the RNA formation [7] and other important rigid processes implicated within gene instruction, such as transcription rate and silencing of gene, in addition to eliminate intronic order [8, 9]. AS events that effect an overall template exon or a fragment of it by activating AS situate in a variation exon, and they can happen in genes with abundant transcription start position or poly adenylation sites. There are five different types of AS patterns like exon skipping, intron preservation, jointly restricted exons, alternative 5' SS, and 3' SS are all examples of exon skipping techniques. Missplicing events have been shown to alter protein utility and cause human diseases, according to growing data as shown in Figure 2.

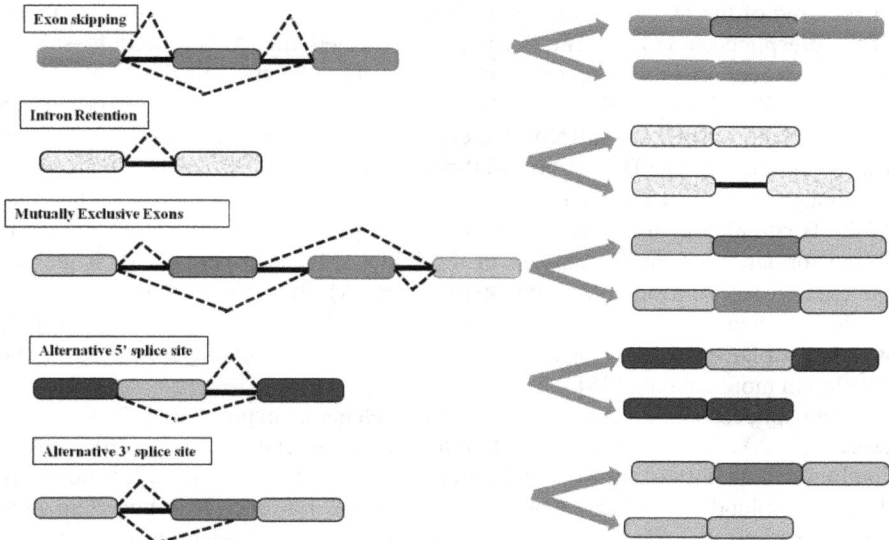

Figure 2: Showing five different types of alternative splicing patterns.

2. Spliceosome

The spliceosome is a massive multi-protein machine that comprises of five little nuclear ribonucleoprotein pieces and also over 150 accessory proteins, that combine pre-mRNA in a stepwise style and appeared rapidly evolving complexes [10]. Exons are joined together and introns are released in the shape of a lariat via two trans esterification processes [11, 12]. AS is linked with a variety of proteins (trans-acting elements) and nucleotide chain to make sure that the appropriate transcript is transcribe and focussed from beginning to end transformation in a tissue-specific fashion. The spliceosome's protein workings also help in the RNA-based active site which is a enzyme multifarious and as well designed for its operation.

2.1 Spliceosome Gathering

Various small nuclear ribonucleoproteins constitute the eukaryote spliceosome (snRNPs). A uridine-rich short nuclear RNA, Sm proteins, and a huge amount of related proteins constitute each snRNP [13, 14]. Around 45 of the molecules are snRNP components, whereas rest be non-snRNP proteins engaged in spliceosome congregation, pre-mRNA unite spot detection, and pre-mRNA binding [15]. They are generally classified into two categories in eukaryotic cells: the major and the minor spliceosome [16, 17]. Apart from those associated with the distinct snRNPs, most spliceosome-related proteins are pooled linked with U2 and U12 spliceosomes [10]. In the early, compound the U1 snRNP is engaged to the 5′ splice site tolerate for required of the U1snRNA [18, 19]. SF1 and U2 auxiliary factor (U2AF), both are non-snRNP factors, which work together with the division end and the 3′ splice site, respectively. The U2 snRNP then join with the branch tip and transfer SF1 by means of required of the U2 snRNA to the division point, consequential in the formation of the prespliceosome [20]. The B complex is formed after a U4/U6.U5 tri-snRNP complex is recruited to the prespliceosome [21]. The U6 snRNA's 5' end base pairs only with 5' splice spot, the U2snRNA outline a duplex by means of the branch point, and the U1 and U4 snRNAs are evacuated, consequentially in Bact complex was being created. The DEAH-box ATP-dependent RNA helicase DHX16 catalyzes the first trans esterification reaction during splicing by catalyzing the configuration of the B complex as of the Bact complex [22] Spliceosome rearrangement results in the formation of the C complex in the initial splicing process [23]. It include the excised lariat-intron and ligated exons (mRNA) after the reaction. The ILS is produced when the DHX8 discharge the merged mRNA from the P complex [24] with help of DEAH-box ATPase DHX15, the ILS detach, allowing the snRNPs to be develop in more splicing [25].

The spliceosome detects 3 different chain elements expressed by 3 associated proteins, U2AF1, U2AF2, and SF1 that are highly altered in cancer [26]. U2AF1 combined to the intron's preserved AG dinucleotide at the 3' end [27]. It is normally deleted in blood cancer and lung adenocarcinoma, however it is uncertain how transformation of a functionally preserved reason with the intention of preserved progression at the 3' splice location lead to tumor development. In haematopoiesis, it has a dissimilar preference designed designed in favour of the nucleotide prior the 3' splice site AG motif, as well as unite variations of predecessor cells in transcript producing RNA-processing reason, ribosomal proteins, and mRNAs of genes such *BCOR* and *KMT2D* with the purpose of changed in MDS and AML [28, 29]. Altered U2AF1 cause aberrant giving out of autophagy-related factor 7 (Atg7) pre-mRNA and an autophagy shortage favors the transformation of mice pro-B cells, implying that such splicing abnormalities may add to cancer expansion [30].

2.2 Functions of AS in Cancer

AS have be related to various studies like Human solid tumors, such as the bladder [31], brain [32], kidney [33], prostate [34], skin [35], stomach [36] thyroid tumors [37], neurological diseases such as Alzheimer's disease (AD) [38], Parkinson's disease (PD) [39], Huntington's disease (HD) [40, 41], as well as haematological malignancies, such as AML, MDS, CML that has been linked to various cancers as shown in Table 1.

ATP-dependent persistent attachment of the U2 snRNP about the branch site is the next step in the spliceosome primary arrangement. Base combination connections connecting 6 nucleotides of U2 snRNA [10] and nucleotides nearby the branch-site adenosine in the pre-mRNA intron are necessary for detection of the branch site by U2 snRNP, just as they are designed for detection of the 5' splice site by U1snRNA. SF3B1 is a input protein component of the U2snRNP that distinguish the spot and the U2snRNA–pre- mRNA helix and assist the rough calculation of the branch-site adenosine to the 5' splice site and the first segment of the splicing reaction by reason a structural alteration persuade by necessary of the pre-mRNA [73, 74].

Table 1: Recurrent splicing-factor mutations in cancer and associated prognosis.

Splicing Factor	Cancer Type	Reference
SF3B1	CLL	[42–44]
	MDS	[45, 46]
	AML	[47–49]
	Breast Cancer	[50–52]
SRSF2	MDS	[53–55]
	CMML	[56]
	AML	[57, 58]
	Primary and secondary myelofibrosis	[59]
U2AF1	MDS	[60–62]
	AML	[63, 64]
	Primary myelofibrosis	[65, 66]
	Lung Carcinoma	[67, 68]
U2AF2	Lung Carcinoma	[69, 70]
	CML	[71]
	Breast cancer	[72]

SF3B1 mutations break up the protein's interface with the splicing factor SUGP1, which consequence in practice of cryptic 3′ splice sites so as near are usually nucleotides ahead of the 3′ splice sites use in wild-type cells. By changing the form or purpose of these specific genes, these modify are likely to give to cancer development.

Members of a variety of families of RNA-binding proteins, such as arginine-serine-rich SR, hnRNP, and RBM protein, support and regulate U1 and U2 snRNPs' detection of splice sites [75]. These factors detect certain regulatory segments in introns or exons and assist or slow up the detection of adjacent splice sites by the core splicing machinery in a position-dependent manner [76]. Modification in AS of pre-mRNAs determined by proto-oncogenes or tumour suppressor genes frequently operate as driver alteration in cancer, resulting in functionally different protein forms or frame shifts [77]. SRSF2, an SR protein so as to attach to assured exonic splicing mediator and induces identification of the beside splice site by U1 or U2 snRNPs, is a classic example with such control [78–80].

SRSF2 is quite often transmogrified in a wide range of myeloid neoplasms which include CMML; at smallest amount one of these transformations alters the comparative similarity for unlike requisite sites in pre-mRNAs, variation in the agreement of exonic splicing improved that are activated, which leads to exon inclusion and change in either SF3B1 or SRSF2 have also been pragmatic to produce MAPKKK7, which stimulate nuclear factor-B signalling. RBM10, for example, increases exon 9 skipping in pre-mRNA programming the Notch regulator NUMB, resulting in higher production of the anti-proliferative isoform. By regulating the alternating look of several oncogenic genes, and splicing factors, AS the stage a

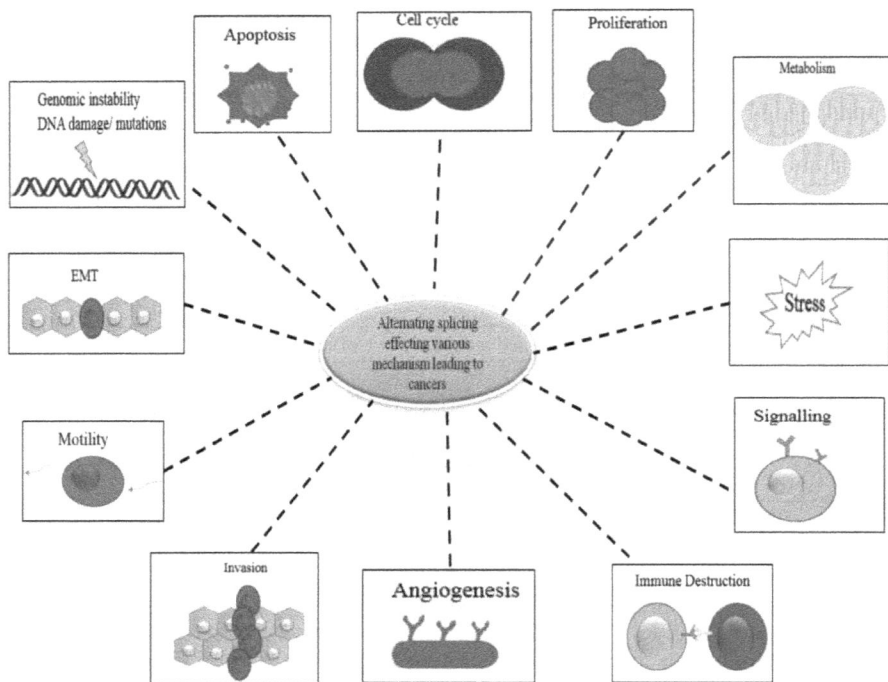

Figure 3: Consequence of AS unregulated on tumor development.

function in the development of production, differentiation, and apoptosis as shown in Figure 3.

2.3 Invasion and Metastasis

These are main important challenge in tumour treatment. Throughout cancer association and repetition, which are dependent comparative on connections linked among tumor cells as well as the microenvironment, the epithelial–mesenchymal transition (EMT) occurs improperly [81]. ESRP1 and hnRNPM may fight for GU-rich binding sites in pre-mRNA and manipulate exon enclosure, influential whether a cell is epithelial or mesenchymal [82]. hnRNPM and ESRP1 have newly been revealed to be crucial managed within the EMT splicing program along with related to breast cancer [83].

2.4 Angiogenesis

It is the development of new blood vessels, is one of the most important aspects of tumor progression. A lot of proteins, such as essential fibroblast growth factor, tumor necrosis factor, and vascular endothelial growth factor (VEGF) [84]. It has both angiogenic and antiangiogenic properties, suggesting that it could be used in antiangiogenic treatments [85]. AS of one-time components in angiogenesis wishes to be extra verified within order to suggest all of the vital nutrients for tumor progression.

2.5 Alternative Splicing in Tumor Metabolism

Due to expansion, tumour cells may face nutrition restriction and hypoxia as the tumor progresses. The conversion of phosphoenol pyruvate to pyruvate, the final stage in glycolysis, to increased by PKM, that is determined by two genes: PKLR and PKM, each of which produces two variations [86]. Although PKLR is mostly uttered in liver and hematopoietic cells, PKM, which codes for PKM1 and PKM2 variations, is found in the majority of organs. PKM2 upregulation is general in a number of cancer and contribute towards a control glycolytic process, shimmering the intonation of the PKM1 and PKM2 ratio by AS in response to metabolic anxiety to choose the choice flanked by aerobic glycolysis and mitochondrial oxidative phosphorylation [87].

The minor spliceosome excises a compartment of introns with a exclusive configuration of splice-site progression in addition to the main spliceosome pathway. U11 and U12 snRNPs, which take part in similar roles to U1 and U2 snRNPs in distinguish into 5' and 3' splice sites, respectively, are part of this machinery.

3. Signal Transduction Driving AS During Tumorigenesis

Though changes in several splicing factors (snRNPs, SR proteins, and hnRNAPs) include, oncogenic signal transduction pathways are a primary driver of dysregulated splicing. Signalling may influence splicing factor production at the transcriptional alteration intensity, together of which alter sub cellular localization to the nucleus or cytoplasm of a cell. Phosphorylation, for example, standardize the movement of SR proteins throughout spliceosome assembly and catalysis [75].

Two groups of kinases disturbed in the parameter of SR-protein phosphorylation are SRPKs in the cytoplasm and CDC-like kinases in the nucleus. Signals from receptors with upstream ligands triggering posterior effectors within the cell are normally sent by a chain of biochemical reactions [88]. During carcinogenesis, abnormal signalling pathways are frequent, and they offer fascinating objective for therapeutic improvement to normalize the growth and survival of malignant cells. Alternative splicing was discovered to be dysregulated as a result of these cancer cell essential changes.

The RAS/RAF/ERK pathway is a characteristic incident throughout lots of epithelia cell-derived malignancies, described with the establishment of the small GTPase RAS and a cascade motivation in various protein kinases. The oncogenic KRAS was linked to an ETS transcription factor-mediated aberrant-splicing regulatory network, according to transcriptomic RNA seq data from colon adenocarcinoma or lung squamous cell carcinoma. The network involved the activation of PTBP1 expression, which was linked in AS of transcripts encoding the small GTPase RAC1, adaptor protein NUMB, and PKM [89] as shown in Figure 4.

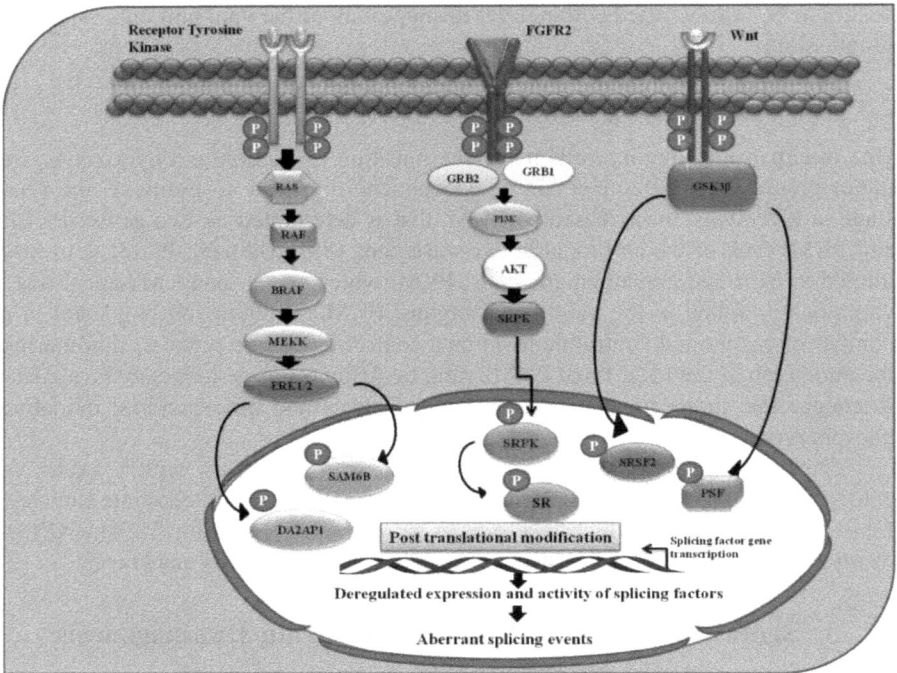

Figure 4: In cancer, oncogenic signalling drives alternative splicing.

This signalling pathway have as well been linked to the phosphorylation of the splicing activator DAZAP1, that is essential designed meant for its translocation from the cytoplasm to the nucleus, as fit as splicing dictatorial action [89, 90]. Furthermore, phosphorylation of splicing factors, such as hint transmission and RNA metabolism, could be caused by MAPK/ERK activation downstream of RAS (SAM68) [91]. Phospho-SAM68 might then combine to and induce an intron in the SRSF1 3'UTR transcript, resulting in a splicing profile switch, as shown in the *RON* transcripts. It's worth noting that enhanced hnRNPA2 expression in hepatocellular carcinoma has been linked to an AS switch that reduces the isoform of A-RAF, resulting in RAF/MEK/ERK pathway activation and cellular transformation [92]. Other essential factor mediating cell endurance and apoptosis in a variety of cancers is the (PI3K)/AKT pathway [93]. The phosphorylation of AKT/SRPK/SR protein by EGF signalling has been shown to affect splicing. The (mTORC1), a crucial control device of cell metabolism and development, is also known to be activated via the PI3K/AKT pathway [94, 95]. The activation and translocation of kinase SRPK2, which further activates SR proteins to engage in the splicing of lipogenesis-related transcripts and promoted by the mTORC1-S6K1 axis signalling [96, 97].

Wnt signalling is renowned for modifiable expansion, as well as its close relationship with a variety of cancers, including (CRC) [98]. Glycogen synthase 3 is a component of the canonical wnt pathway that has be related to the phosphorylation of splicing factors including SRSF2 and PTB related splicing factor (PSF) [99].

Activated signalling has been shown to increase the level of SRSF3 transcripts. There are a lot of more tumor microenvironment derived soluble material, metabolic anxiety circumstances [100], and extracellular matrix-related signals that interact with AS and are not mentioned here [101, 102]. As a result, better investigation among cancer signalling way and splicing abnormalities in cancer are required.

3.1 Diagnosis in Cancer

The process of determining the presence, location, and extent of cancer in a patient's body is known as a diagnosis of cancer. Typically, a combination of clinical examination, imaging, and laboratory tests is used to make the diagnosis. When diagnosing cancer, some common diagnostic procedures include: Tests for images: A variety of imaging procedures, including X-rays, CT scans, MRI scans, and PET scans, can assist in determining the extent and location of cancerous tumors [103] Biopsy: The removal of a small amount of tissue for examination under a microscope is known as a biopsy. A needle biopsy or a surgical biopsy can accomplish this [104]. Blood tests: The levels of certain cancer-associated proteins or markers can be measured with blood tests [105]. Endoscopy: An endoscopy is a procedure in which a camera and light are inserted into the body through a thin, flexible tube to examine the inside of organs or tissues [106]. Further testing may be required to determine the stage and severity of the cancer following a diagnosis. Using this information, a treatment plan that is specific to the patient's needs is created. Early location of disease through customary screenings can build the possibilities of effective treatment and further develop results for patients.

3.2 Prognosis in Cancer

In the context of cancer, the term "prognosis" refers to the anticipated course and outcome of the disease based on a variety of factors, including the patient's overall health, the type and stage of the cancer, and their response to treatment. Patients and their loved ones can learn a lot from a prognosis, which can help them make better decisions about their care and prepare for the future .In cancer, prognosis can be affected by a number of factors, including: Cancer type & stage: Different kinds of cancer respond to treatment in different ways and have different survival rates. The size of the tumor and whether it has spread to nearby lymph nodes or other parts of the body determine the stage of cancer. In most cases, the better the prognosis, the earlier the stage of cancer [107]. Grade: The degree to which cancer cells appear abnormal under a microscope determines the grade of the disease. Cancers with a higher grade typically have a worse prognosis and are more aggressive [108].

3.3 Treatment Response

The prognosis may also be influenced by treatment success. The prognosis of those who respond well to treatment may be better than that of those who do not. A prognosis is an estimate based on the information that is available rather than a precise prediction of what will occur in the future. With their healthcare team, patients should talk about their prognosis, treatment options, and care goals. It is essential to keep in mind that every cancer patient's journey is unique, and many cancer patients are able to lead fulfilling lives for many years after being diagnosed [109].

3.4 Therapeutic Approaches in Cancer

There is a wide variety of treatment options for cancer, including surgery, chemo, radiation, targeted therapy, immunotherapy, and hormone therapy. The type and stage of cancer, as well as the patient's overall health and medical history, influence the choice of treatment. Surgery: Surgery is often used to remove solid tumors or stop cancer from spreading to other parts of the body. It involves removing cancerous tissue [110]. Chemotherapy: The treatment of cancer with drugs is known as chemotherapy. These medications can be given orally or intravenously, and they can be given in cycles to give the body time to heal between treatments [111]. Radiation: Cancer cells are destroyed by high-energy radiation during radiation therapy. This can be given internally by placing radioactive material close to the cancer cells or externally by using a machine that targets the affected area with radiation [112]. Targeted treatment: A type of cancer treatment called targeted therapy makes use of drugs or other substances to target specific proteins or molecules that are involved in the growth and spread of cancer cells. Compared to chemotherapy, this type of treatment may be more precise and have fewer side effects [113].

3.5 Immunotherapy

A type of cancer treatment called immunotherapy uses the body's immune system to fight cancer. This may involve treatments that aid the immune system in recognizing and attacking cancer cells, genetically engineered cells, or drugs that stimulate the immune system [114].

Hormone treatment: Breast and prostate cancer, two types of cancer that are sensitive to hormones, are treated with hormone therapy. This may necessitate the removal of hormone-producing organs or the administration of hormone-blocking medications [115]. In order to achieve the best possible outcome, cancer treatment is frequently individualized and may involve a variety of different therapies in combination. The type and stage of the cancer, the patient's overall health, and their specific requirements and preferences all play a role in the creation of treatment plans. To ensure the best possible outcome, regular monitoring and follow-up care are also essential components of cancer treatment. The treatment process is influenced and informed by all three approaches, which are intimately linked. Prognostic information serves as a guide for treatment decisions, while diagnostic tests provide the information required for both prognostic and therapeutic decisions. Diagnostic tools are then used to monitor treatment response, which can assist in determining the therapeutic approach's efficacy and adjusting the treatment plan if necessary. In general, diagnostic, prognostic, and therapeutic approaches collaborate to assist clinicians in providing cancer patients with the best possible care.

Conclusion and Future Perspectives

In conclusion, AS unfolds a complex network of potential genetic variations arising from a single gene locus, each with distinct roles. The repertoire of such alternatives greatly enriches the diversity of the proteome and forms the genetic underpinnings for cellular functions, disease mechanisms, and the uniqueness of individual organisms. The study of AS enables us to gain insights into the multitudinous cellular phenomena and has thus evolved as a powerful tool to deepen our understanding of the genomic blueprint of life.

Indeed, the dysregulation of AS has been identified as a crucial driver of tumor progression. With an ever-increasing body of evidence linking aberrant splicing patterns to oncogenesis, this field has fast become a vibrant frontier in cancer research. It has cast light upon the complex genetic and epigenetic events shaping the cellular transformation process, offering fresh perspectives on the mechanisms of tumorigenesis and metastasis.

Looking forward, the realm of AS offers an exciting avenue for future research, with the potential to revolutionize cancer management strategies. As we continue to decipher the AS code, novel prognostic markers, and targeted therapeutic strategies are likely to emerge. The challenge ahead lies in expanding our understanding of the splicing machinery, exploring the dynamic interplay between splicing and other gene regulation processes, and translating these findings into viable clinical applications.

Given the vast inter-individual variation in AS patterns, personalized approaches to cancer treatment may well be within our grasp. Furthermore, the prospect of exploiting splicing machinery for therapeutic purposes offers immense potential in the fight against cancer.

In summary, AS stands at the forefront of a new era in cancer research, paving the way for innovative diagnostic and therapeutic approaches. With continued research, the intricate tapestry of alternative splicing may soon be unraveled, ushering in a new age of precision medicine in oncology.

Conflict of Interest

None.

References

[1] Baralle, F.E. and Giudice, J. (2017). Alternative splicing as a regulator of development and tissue identity. Nature Reviews Molecular Cell Biology 18(7): 437–451.

[2] Kahles, A., Lehmann, K.-V., Toussaint, N.C., Hüser, M., Stark, S.G., Sachsenberg, T., Stegle, O., Kohlbacher, O., Sander, C. and Caesar-Johnson, S.J. (2018). Comprehensive analysis of alternative splicing across tumors from 8,705 patients. Cancer Cell 34(2): 211–224. e6.

[3] Cherry, S. and Lynch, K.W. (2020). Alternative splicing and cancer: Insights, opportunities, and challenges from an expanding view of the transcriptome. Genes & Development 34(15-16): 1005–1016.

[4] Chow, L.T., Gelinas, R.E., Broker, T.R. and Roberts, R.J. (1977). An amazing sequence arrangement at the 5′ ends of adenovirus 2 messenger RNA. Cell 12(1): 1–8.

[5]   Gallego-Paez, L.M., Bordone, M.C., Leote, A.C., Saraiva-Agostinho, N., Ascensão-Ferreira, M. and Barbosa-Morais, N. (2017). Alternative splicing: The pledge, the turn, and the prestige. Human Genetics 136(9): 1015–1042.

[6]   Barbosa-Morais, N.L., Irimia, M., Pan, Q., Xiong, H.Y., Gueroussov, S., Lee, L.J., Slobodeniuc, V., Kutter, C., Watt, S. and Colak, R. (2012). The evolutionary landscape of alternative splicing in vertebrate species. Science 338(6114): 1587–1593.

[7]   Jacobs, E., Mills, J.D. and Janitz, M. (2012). The role of RNA structure in posttranscriptional regulation of gene expression. Journal of Genetics and Genomics 39(10): 535–543.

[8]   Luco, R.F., Pan, Q., Tominaga, K., Blencowe, B.J., Pereira-Smith, O.M. and Misteli, T. (2010). Regulation of alternative splicing by histone modifications. Science 327(5968): 996–1000.

[9]   Naftelberg, S., Schor, I.E., Ast, G. and Kornblihtt, A.R. (2015). Regulation of alternative splicing through coupling with transcription and chromatin structure. Annual Review of Biochemistry 84: 165–198.

[10] Wahl, M.C., Will, C.L. and Lührmann, R. (2009). The spliceosome: Design principles of a dynamic RNP machine. Cell 136(4): 701–718.

[11] Bonnal, S.C., López-Oreja, I. and Valcárcel, J. (2020). Roles and mechanisms of alternative splicing in cancer—implications for care. Nature Reviews Clinical Oncology 17(8): 457–474.

[12] Matera, A.G. and Wang, Z. (2014). A day in the life of the spliceosome. Nature Reviews Molecular Cell Biology 15(2): 108–121.

[13] Cvitkovic, I. and Jurica, M.S. (2013). Spliceosome database: A tool for tracking components of the spliceosome. Nucleic Acids Research 41(D1): D132–D141.

[14] Fourmann, J.-B., Schmitzová, J., Christian, H., Urlaub, H., Ficner, R., Boon, K.-L., Fabrizio, P. and Lührmann, R. (2013). Dissection of the factor requirements for spliceosome disassembly and the elucidation of its dissociation products using a purified splicing system. Genes & Development 27(4): 413–428.

[15] Staley, J.P. and Woolford Jr, J.L. (2009). Assembly of ribosomes and spliceosomes: Complex ribonucleoprotein machines. Current Opinion in Cell Biology 21(1): 109–118.

[16] Patel, A.A. and Steitz, J.A. (2003). Splicing double: Insights from the second spliceosome. Nature Reviews Molecular Cell Biology 4(12): 960–970.

[17] Scotti, M.M. and Swanson, M.S. (2016). RNA mis-splicing in disease. Nature Reviews Genetics 17(1): 19–32.

[18] Das, R., Zhou, Z. and Reed, R. (2000). Functional association of U2 snRNP with the ATP-independent spliceosomal complex E. Molecular Cell 5(5): 779–787.

[19] Taylor, J. and Lee, S.C. (2019). Mutations in spliceosome genes and therapeutic opportunities in myeloid malignancies. Genes, Chromosomes and Cancer 58(12): 889–902.

[20] Crisci, A., Raleff, F., Bagdiul, I., Raabe, M., Urlaub, H., Rain, J.-C. and Krämer, A. (2015). Mammalian splicing factor SF1 interacts with SURP domains of U2 snRNP-associated proteins. Nucleic Acids Research 43(21): 10456–10473.

[21] Turunen, J.J., Niemelä, E.H., Verma, B. and Frilander, M.J. (2013). The significant other: Splicing by the minor spliceosome. Wiley Interdisciplinary Reviews: RNA 4(1): 61–76.

[22] De, I., Schmitzová, J. and Pena, V. (2016). The organization and contribution of helicases to RNA splicing. Wiley Interdisciplinary Reviews: RNA 7(2): 259–274.

[23] Krausová, M. and Staněk. D. (2018). snRNP proteins in health and disease. In Seminars in Cell & Developmental Biology. Elsevier.

[24] Yang, H., Beutler, B. and Zhang, D. (2021). Emerging roles of spliceosome in cancer and immunity. Protein & Cell 2021: 1–21.

[25] Will, C.L. and Lührmann, R. (2011). Spliceosome structure and function. Cold Spring Harbor Perspectives in Biology 3(7): a003707.

[26] Yoshida, K., Sanada, M., Shiraishi, Y., Nowak, D., Nagata, Y., Yamamoto, R.,. Sato, Y., Sato-Otsubo, A., Kon, A. and Nagasaki, M. (2011). Frequent pathway mutations of splicing machinery in myelodysplasia. Nature 478(7367): 64–69.

[27] Kandoth, C., McLellan, M.D., Vandin, F., Ye, K., Niu, B., Lu, C., Xie, M., Zhang, Q., McMichael, J.F. and Wyczalkowski, M.A. (2013). Mutational landscape and significance across 12 major cancer types. Nature 502(7471): 333–339.

[28] Seiler, M., Peng, S., Agrawal, A.A., Palacino, J., Teng, T., Zhu, P., Smith, P.G., Caesar-Johnson, S.J., Demchok, J.A. and Felau, I. (2018). Somatic mutational landscape of splicing factor genes and their functional consequences across 33 cancer types. Cell Reports 23(1): 282–296. e4.

[29] Shirai, C.L., Ley, J.N., White, B.S., Kim, S., Tibbitts, J., Shao, J., Ndonwi, M., Wadugu, B., Duncavage, E.J. and Okeyo-Owuor, T. (2015). Mutant U2AF1 expression alters hematopoiesis and pre-mRNA splicing *in vivo*. Cancer Cell 27(5): 631–643.

[30] Park, S.M., Ou, J., Chamberlain, L., Simone, T.M., Yang, H., Virbasius, C.-M., Ali, A.M., Zhu, L.J., Mukherjee, S. and Raza, A. (2016). U2AF35 (S34F) promotes transformation by directing aberrant ATG7 pre-mRNA 3' end formation. Molecular Cell 62(4): 479–490.

[31] Xie, R., Chen, X., Chen, Z., Huang, M., Dong, W., Gu, P., Zhang, J., Zhou, Q., Dong, W. and Han, J. (2019). Polypyrimidine tract binding protein 1 promotes lymphatic metastasis and proliferation of bladder cancer via alternative splicing of MEIS2 and PKM. Cancer Letters 449: 31–44.

[32] Babic, I., Anderson, E.S., Tanaka, K., Guo, D., Masui, K., Li, B., Zhu, S., Gu, Y., Villa, G.R. and Akhavan, D. (2013). EGFR mutation-induced alternative splicing of Max contributes to growth of glycolytic tumors in brain cancer. Cell Metabolism 17(6): 1000–1008.

[33] Liu, F., Dai, M., Xu, Q., Zhu, X., Zhou, Y., Jiang, S., Wang, Y., Ai, Z., Ma, L. and Zhang, Y. (2018). SRSF10-mediated IL1RAP alternative splicing regulates cervical cancer oncogenesis via mIL1RAP-NF-κB-CD47 axis. Oncogene 37(18): 2394–2409.

[34] Fan, L., Zhang, F., Xu, S., Cui, X., Hussain, A., Fazli, L., Gleave, M., Dong, X. and Qi, J. (2018). Histone demethylase JMJD1A promotes alternative splicing of AR variant 7 (AR-V7) in prostate cancer cells. Proceedings of the National Academy of Sciences 115(20): E4584–E4593.

[35] Jensen, M.A., Wilkinson, J.E. and Krainer, A.R. (2014). Splicing factor SRSF6 promotes hyperplasia of sensitized skin. Nature Structural & Molecular Biology 21(2): 189–197.

[36] Ailiken, G., Kitamura, K., Hoshino, T., Satoh, M., Tanaka, N., Minamoto, T., Rahmutulla, B., Kobayashi, S., Kano, M. and Tanaka, T. (2020). Post-transcriptional regulation of BRG1 by FIRΔexon2 in gastric cancer. Oncogenesis 9(2): 1–20.

[37] Lin, P., He, R.-q., Huang, Z.-g., Zhang, R., Wu, H.-y., Shi, L., Li, X.-j., Li, Q., Chen, G. and Yang, H. (2019). Role of global aberrant alternative splicing events in papillary thyroid cancer prognosis. Aging (Albany NY) 11(7): 2082.

[38] Love, J.E., Hayden, E.J. and Rohn, T.T. (2015). Alternative splicing in Alzheimer's disease. Journal of Parkinson's Disease and Alzheimer's Disease 2(2).

[39] Alieva, A.K., Shadrina, M.I., Filatova, E.V., Karabanov, A.V., Illarioshkin, S.N., Limborska, S.A. and Slominsky, P.A. (2014). Involvement of endocytosis and alternative splicing in the formation of the pathological process in the early stages of Parkinson's disease. BioMed Research International 2014.

[40] Fernández-Nogales, M., Santos-Galindo, M., Hernández, I.H., Cabrera, J.R. and Lucas, J.J. (2016). Faulty splicing and cytoskeleton abnormalities in H untington's disease. Brain Pathology 26(6): 772–778.

[41] Hughes, T., Hansson, L., Sønderby, I.E., Athanasiu, L., Zuber, V., Tesli, M., Song, J., Hultman, C.M., Bergen, S.E. and Landén, M. (2016). A loss-of-function variant in a minor isoform of ANK3 protects against bipolar disorder and schizophrenia. Biological Psychiatry 80(4): 323–330.

[42] Wan, Y. and Wu, C.J. (2013). SF3B1 mutations in chronic lymphocytic leukemia. Blood, The Journal of the American Society of Hematology 121(23): 4627–4634.

[43] Te Raa, G., Derks, I., Navrkalová, V., Skowronska, A., Moerland, P., Van Laar, J., Oldreive, C., Monsuur, H., Trbusek, M. and Malcikova, J. (2015). The impact of SF3B1 mutations in CLL on the DNA-damage response. Leukemia 29(5): 1133–1142.

[44] Wang, L., Lawrence, M.S., Wan, Y., Stojanov, P., Sougnez, C., Stevenson, K., Werner, L., Sivachenko, A., DeLuca, D.S. and Zhang, L. (2011). SF3B1 and other novel cancer genes in chronic lymphocytic leukemia. New England Journal of Medicine 365(26): 2497–2506.

[45] Malcovati, L., Stevenson, K., Papaemmanuil, E., Neuberg, D., Bejar, R., Boultwood, J., Bowen, D.T., Campbell, P.J., Ebert, B.L. and Fenaux, P. (2020). SF3B1-mutant MDS as a distinct disease subtype: A proposal from the International Working Group for the Prognosis of MDS. Blood 136(2): 157–170.

[46] Mian, S.A., Rouault-Pierre, K., Smith, A.E., Seidl, T., Pizzitola, I., Kizilors, A., Kulasekararaj, A.G., Bonnet, D. and Mufti, G.J. (2015). SF3B1 mutant MDS-initiating cells may arise from the haematopoietic stem cell compartment. Nature Communications 6(1): 1–14.

[47] Bamopoulos, S.A., Batcha, A., Jurinovic, V., Rothenberg-Thurley, M., Janke, H., Ksienzyk, B., Philippou-Massier, J., Graf, A., Krebs, S. and Blum, H. (2020). Clinical presentation and differential splicing of SRSF2, U2AF1 and SF3B1 mutations in patients with acute myeloid leukemia. Leukemia 34(10): 2621–2634.

[48] van der Werf, I., Wojtuszkiewicz, A., Yao, H., Sciarrillo, R., Meggendorfer, M., Hutter, S., Walter, W., Janssen, J., Kern, W. and Haferlach, C. (2021). SF3B1 as therapeutic target in FLT3/ITD positive acute myeloid leukemia. Leukemia 35(9): 2698–2702.

[49] Dalton, W.B. (2020). The metabolic reprogramming and vulnerability of SF3B1 mutations. Molecular & Cellular Oncology 7(3): 1697619.

[50] Maguire, S.L., Leonidou, A., Wai, P., Marchiò, C., Ng, C.K., Sapino, A., Salomon, A.V., Reis-Filho, J.S., Weigelt, B. and Natrajan, R.C. (2015). SF3B1 mutations constitute a novel therapeutic target in breast cancer. The Journal of Pathology 235(4): 571–580.

[51] Fu, X., Tian, M., Gu, J., Cheng, T., Ma, D., Feng, L. and Xin, X. (2017). SF3B1 mutation is a poor prognostic indicator in luminal B and progesterone receptor-negative breast cancer patients. Oncotarget 8(70): 115018.

[52] Zhou, Z., Gong, Q., Wang, Y., Li, M., Wang, L., Ding, H. and Li, P. (2020). The biological function and clinical significance of SF3B1 mutations in cancer. Biomarker Research 8(1): 1–14.

[53] Rahman, M.A., Lin, K.-T., Bradley, R.K., Abdel-Wahab, O. and Krainer, A.R. (2020). Recurrent SRSF2 mutations in MDS affect both splicing and NMD. Genes & Development 34(5-6): 413–427.

[54] Meggendorfer, M., Roller, A., Haferlach, T., Eder, C., Dicker, F., Grossmann, V., Kohlmann, A., Alpermann, T., Yoshida, K. and Ogawa, S. (2012). SRSF2 mutations in 275 cases with chronic myelomonocytic leukemia (CMML). Blood, The Journal of the American Society of Hematology 120(15): 3080–3088.

[55] Arbab Jafari, P., Ayatollahi, H., Sadeghi, R., Sheikhi, M. and Asghari, A. (2018). Prognostic significance of SRSF2 mutations in myelodysplastic syndromes and chronic myelomonocytic leukemia: A meta-analysis. Hematology 23(10): 778–784.

[56] Patnaik, M.M., Lasho, T.L., Finke, C.M., Hanson, C.A., Hodnefield, J.M., Knudson, R.A., Ketterling, R.P., Pardanani, A. and Tefferi, A. (2013). Spliceosome mutations involving SRSF2, SF3B1, and U2AF35 in chronic myelomonocytic leukemia: Prevalence, clinical correlates, and prognostic relevance. American Journal of Hematology 88(3): 201–206.

[57] Yang, J., Yao, D.-m., Ma, J.-c., Yang, L., Guo, H., Wen, X.-m., Xiao, G.-f., Qian, Z., Lin, L. and Qian, J. (2016). The prognostic implication of SRSF2 mutations in Chinese patients with acute myeloid leukemia. Tumor Biology 37(8): 10107–10114.

[58] Zhang, S.-J., Rampal, R., Manshouri, T., Patel, J., Mensah, N., Kayserian, A., Hricik, T., Heguy, A., Hedvat, C. and Gönen, M. (2012). Genetic analysis of patients with leukemic transformation of myeloproliferative neoplasms shows recurrent SRSF2 mutations that are associated with adverse outcome. Blood, The Journal of the American Society of Hematology 119(19): 4480–4485.

[59] Vannucchi, A.M., Lasho, T., Guglielmelli, P., Biamonte, F., Pardanani, A., Pereira, A., Finke, C., Score, J., Gangat, N. and Mannarelli, C. (2013). Mutations and prognosis in primary myelofibrosis. Leukemia 27(9): 1861–1869.

[60] Shirai, C.L., Tripathi, M., Ley, J.N., Ndonwi, M., White, B.S., Tapia, R., Saez, B., Bertino, A., Shao, J. and Kim, S. (2015). Preclinical activity of splicing modulators in U2AF1 mutant MDS/AML. Blood 126(23): 1653.

[61] Yip, B.H., Steeples, V., Repapi, E., Armstrong, R.N., Llorian, M., Roy, S., Shaw, J., Dolatshad, H., Taylor, S. and Verma, A. (2017). The U2AF1 S34F mutation induces lineage-specific splicing alterations in myelodysplastic syndromes. The Journal of Clinical Investigation 127(6): 2206–2221.

[62] Je, E.M., Yoo, N.J., Kim, Y.J., Kim, M.S. and Lee, S.H. (2013). Mutational analysis of splicing machinery genes SF3B1, U2AF1 and SRSF2 in myelodysplasia and other common tumors. International Journal of Cancer 133(1): 260–265.

[63] Ohgami, R.S., Ma, L., Merker, J.D., Gotlib, J.R., Schrijver, I., Zehnder, J.L. and Arber, D.A. (2015). Next-generation sequencing of acute myeloid leukemia identifies the significance of TP53, U2AF1, ASXL1, and TET2 mutations. Modern Pathology 28(5): 706–714.

[64] Qian, J., Yao, D.-m., Lin, J., Qian, W., Wang, C.-z., Chai, H.-y., Yang, J., Li, Y., Deng, Z.-q. and Ma, J.-c. (2012). U2AF1 mutations in Chinese patients with acute myeloid leukemia and myelodysplastic syndrome.

[65] Tefferi, A., Finke, C.M., Lasho, T.L., Hanson, C.A., Ketterling, R.P., Gangat, N. and Pardanani, A. (2018). U2AF1 mutation types in primary myelofibrosis: Phenotypic and prognostic distinctions. Leukemia 32(10): 2274–2278.

[66] Barraco, D., Elala, Y., Lasho, T., Begna, K., Gangat, N., Finke, C., Hanson, C., Ketterling, R., Pardanani, A. and Tefferi, A. (2016). Molecular correlates of anemia in primary myelofibrosis: A significant and independent association with U2AF1 mutations. Blood Cancer Journal 6(4): e415–e415.

[67] Esfahani, M.S., Lee, L.J., Jeon, Y.-J., Flynn, R.A., Stehr, H., Hui, A.B., Ishisoko, N., Kildebeck, E., Newman, A.M. and Bratman, S.V. (2019). Functional significance of U2AF1 S34F mutations in lung adenocarcinomas. Nature Communications 10(1): 1–13.

[68] Boeckx, B., Shahi, R.B., Smeets, D., De Brakeleer, S., Decoster, L., Van Brussel, T., Galdermans, D., Vercauter, P., Decoster, L. and Alexander, P. (2020). The genomic landscape of nonsmall cell lung carcinoma in never smokers. International Journal of Cancer 146(11): 3207–3218.

[69] Li, J., Cheng, D., Zhu, M., Yu, H., Pan, Z., Liu, L., Geng, Q., Pan, H., Yan, M. and Yao, M. (2019). OTUB2 stabilizes U2AF2 to promote the Warburg effect and tumorigenesis via the AKT/mTOR signaling pathway in non-small cell lung cancer. Theranostics 9(1): 179.

[70] Schnittger, S., Kuznia, S., Meggendorfer, M., Nadarajah, N., Jeromin, S., Alpermann, T., Roller, A., Albuquerque, A., Weissmann, S. and Ernst, T. (2013). Tyrosine kinase inhibitor treated CML patients Harboring Philadelphia-negative cytogenetically aberrant clones show molecular mutations in 31% of cases not present at diagnosis: A high-throughput amplicon sequencing study of 29 genes. Blood 122(21): 611.

[71] Balk, B., Fabarius, A. and Haferlach, C. (2021). Cytogenetics of chronic myeloid leukemia (CML), in Chronic Myeloid Leukemia. Springer, pp. 1–16.

[72] Katayama, H., Boldt, C., Ladd, J.J., Johnson, M.M., Chao, T., Capello, M., Suo, J., Mao, J., Manson, J.E. and Prentice, R. (2015). An autoimmune response signature associated with the development of triple-negative breast cancer reflects disease pathogenesis. Cancer Research 75(16): 3246–3254.

[73] Cretu, C., Schmitzová, J., Ponce-Salvatierra, A., Dybkov, O., De Laurentiis, E.I., Sharma, K., Will, C.L., Urlaub, H., Lührmann, R. and Pena, V. (2016). Molecular architecture of SF3b and structural consequences of its cancer-related mutations. Molecular Cell 64(2): 307–319.

[74] Quesada, V., Conde, L., Villamor, N., Ordóñez, G.R., Jares, P., Bassaganyas, L., Ramsay, A.J., Beà, S., Pinyol, M. and Martínez-Trillos, A. (2012). Exome sequencing identifies recurrent mutations of the splicing factor SF3B1 gene in chronic lymphocytic leukemia. Nature Genetics 44(1): 47–52.

[75] Fu, X.-D. and Ares, M. (2014). Context-dependent control of alternative splicing by RNA-binding proteins. Nature Reviews Genetics 15(10): 689–701.

[76] Wang, E.T., Sandberg, R., Luo, S., Khrebtukova, I., Zhang, L., Mayr, C., Kingsmore, S.F., Schroth, G.P. and Burge, C.B. (2008). Alternative isoform regulation in human tissue transcriptomes. Nature 456(7221): 470–476.

[77] Supek, F., Miñana, B., Valcárcel, J., Gabaldón, T. and Lehner, B. (2014). Synonymous mutations frequently act as driver mutations in human cancers. Cell 156(6): 1324–1335.

[78] Jayasinghe, R.G., Cao, S., Gao, Q., Wendl, M.C., Vo, N.S., Reynolds, S.M., Zhao, Y., Climente-González, H., Chai, S. and Wang, F. (2018). Systematic analysis of splice-site-creating mutations in cancer. Cell Reports 23(1): 270–281. e3.

[79] Singh, B., Trincado, J.L., Tatlow, P., Piccolo, S.R. and Eyras, E. (2018). Genome sequencing and RNA-motif analysis reveal novel damaging noncoding mutations in human tumors. Molecular Cancer Research 16(7): 1112–1124.

[80] Jaganathan, K., Panagiotopoulou, S.K., McRae, J.F., Darbandi, S.F., Knowles, D., Li, Y.I., Kosmicki, J.A., Arbelaez, J., Cui, W. and Schwartz, G.B. (2019). Predicting splicing from primary sequence with deep learning. Cell 176(3): 535–548. e24.

[81] Thiery, J.P., Acloque, H., Huang, R.Y. and Nieto, M.A. (2009). Epithelial-mesenchymal transitions in development and disease. Cell 139(5): 871–890.

[82] Tripathi, V., Shin, J.-H., Stuelten, C. H. and Zhang, Y.E. (2019). TGF-β-induced alternative splicing of TAK1 promotes EMT and drug resistance. Oncogene 38(17): 3185–3200.

[83] Harvey, S.E., Xu, Y., Lin, X., Gao, X.D., Qiu, Y., Ahn, J., Xiao, X. and Cheng, C. (2018). Coregulation of alternative splicing by hnRNPM and ESRP1 during EMT. RNA 24(10): 1326–1338.

[84] Hatcher, J.M., Wu, G., Zeng, C., Zhu, J., Meng, F., Patel, S., Wang, W., Ficarro, S.B., Leggett, A.L. and Powell, C.E (2018). SRPKIN-1: A covalent SRPK1/2 inhibitor that potently converts VEGF from pro-angiogenic to anti-angiogenic isoform. Cell Chemical Biology 25(4): 460–470. e6.

[85] Zhu, H., Carpenter, R.L., Han, W. and Lo, H.-W. (2014). The GLI1 splice variant TGLI1 promotes glioblastoma angiogenesis and growth. Cancer Letters 343(1): 51–61.

[86] Jurica, M.S., Mesecar, A., Heath, P.J., Shi, W., Nowak, T. and Stoddard, B.L. (1998). The allosteric regulation of pyruvate kinase by fructose-1, 6-bisphosphate. Structure 6(2): 195–210.

[87] Takenaka, M., Noguchi, T., Sadahiro, S., Hirai, H., Yamada, K., Matsuda, T., Imai, E. and Tanaka, T. (1991). Isolation and characterization of the human pyruvate kinase M gene. European Journal of Biochemistry 198(1): 101–106.

[88] Zhou, Z. and Fu, X.-D. (2013). Regulation of splicing by SR proteins and SR protein-specific kinases. Chromosoma 122(3): 191–207.

[89] Hollander, D., Donyo, M., Atias, N., Mekahel, K., Melamed, Z., Yannai, S., Lev-Maor, G., Shilo, A., Schwartz, S. and Barshack, I. (2016). A network-based analysis of colon cancer splicing changes reveals a tumorigenesis-favoring regulatory pathway emanating from ELK1. Genome Research 26(4): 541–553.

[90] Choudhury, R., Roy, S.G., Tsai, Y.S., Tripathy, A., Graves, L.M. and Wang, Z. (2014). The splicing activator DAZAP1 integrates splicing control into MEK/Erk-regulated cell proliferation and migration. Nature Communications 5(1): 1–16.

[91] Weg-Remers, S., Ponta, H., Herrlich, P. and König, H. (2001). Regulation of alternative pre-mRNA splicing by the ERK MAP-kinase pathway. The EMBO Journal 20(15): 4194–4203.

[92] Valacca, C., Bonomi, S., Buratti, E., Pedrotti, S., Baralle, F.E. Sette, C., Ghigna, C. and Biamonti, G. (2010). Sam68 regulates EMT through alternative splicing–activated nonsense-mediated mRNA decay of the SF2/ASF proto-oncogene. Journal of Cell Biology 191(1): 87–99.

[93] Shilo, A., Hur, V.B., Denichenko, P., Stein, I., Pikarsky, E., Rauch, J., Kolch, W., Zender, L. and Karni, R. (2014). Splicing factor hnRNPA2 activates the Ras-MAPK-ERK pathway by controlling A-Raf splicing in hepatocellular carcinoma development. RNA 20(4): 505–515.

[94] Shultz, J.C., Goehe, R.W., Wijesinghe, D.S., Murudkar, C., Hawkins, A.J., Shay, J.W., Minna, J.D. and Chalfant, C.E. (2010). Alternative splicing of caspase 9 is modulated by the phosphoinositide 3-kinase/Akt pathway via phosphorylation of SRp30a. Cancer Research 70(22): 9185–9196.

[95] Zhou, Z., Qiu, J., Liu, W., Zhou, Y., Plocinik, R.M, Li, H., Hu, Q., Ghosh, G., Adams, J.A. and Rosenfeld, M.G. (2012). The Akt-SRPK-SR axis constitutes a major pathway in transducing EGF signaling to regulate alternative splicing in the nucleus. Molecular Cell 47(3): 422–433.

[96] Wang, F., Fu, X., Chen, P., Wu, P., Fan, X., Li, N., Zhu, H., Jia, T.-T., Ji, H. and Wang, Z. (2017). SPSB1-mediated HnRNP A1 ubiquitylation regulates alternative splicing and cell migration in EGF signaling. Cell Research 27(4): 540–558.

[97] Lee, G., Zheng, Y., Cho, S., Jang, C., England, C., Dempsey, J.M., Yu, Y., Liu, X., He, L. and Cavaliere, P.M. (2017). Post-transcriptional regulation of *de novo* lipogenesis by mTORC1-S6K1-SRPK2 signaling. Cell 171(7): 1545–1558. e18.

[98] Zhan, T., Rindtorff, N. and Boutros, M. (2017). Wnt signaling in cancer. Oncogene 36(11): 1461–1473.

[99] Hernández, F., Pérez, M., Lucas, J.J., Mata, A.M., Bhat, R. and Avila, J. (2004). Glycogen synthase kinase-3 plays a crucial role in tau exon 10 splicing and intranuclear distribution of SC35: implications for Alzheimer's disease. Journal of Biological Chemistry 279(5): 3801–3806.

[100] Heyd, F. and Lynch, K.W. (2010). Phosphorylation-dependent regulation of PSF by GSK3 controls CD45 alternative splicing. Molecular Cell 40(1): 126–137.

[101] Gonçalves, V., Matos, P. and Jordan, P. (2008). The β-catenin/TCF4 pathway modifies alternative splicing through modulation of SRp20 expression. RNA 14(12): 2538–2549.

[102] Gonçalves, V., Matos, P. and Jordan, P. (2009). Antagonistic SR proteins regulate alternative splicing of tumor-related Rac1b downstream of the PI3-kinase and Wnt pathways. Human Molecular Genetics 18(19): 3696–3707.

[103] Soni, V.D. and Soni, A.N. (2021). Cervical cancer diagnosis using convolution neural network with conditional random field. In 2021 Third International Conference on Inventive Research in Computing Applications (ICIRCA). IEEE.

[104] Vaidyanathan, R., Soon, R.H., Zhang, P., Jiang, K. and Lim, C.T. (2019). Cancer diagnosis: From tumor to liquid biopsy and beyond. Lab on a Chip 19(1): 11–34.

[105] Chen, X., Gole, J., Gore, A., He, Q., Lu, M., Min, J., Yuan, Z., Yang, X., Jiang, Y. and Zhang, T. (2020). Non-invasive early detection of cancer four years before conventional diagnosis using a blood test. Nature Communications 11(1): 3475.

[106] Evans, J.A., Early, D.S., Chandraskhara, V., Chathadi, K.V., Fanelli, R.D., Fisher, D.A.,Foley, K.Q., Hwang, J.H., Jue, T.L. and Pasha, S.F. (2013). The role of endoscopy in the assessment and treatment of esophageal cancer. Gastrointestinal Endoscopy 77(3): 328–334.

[107] Bozzetti, F., Migliavacca, S., Scotti, A., Bonalumi, M., Scarpa, D., Baticci, F., Ammatuna, M., Pupa, A., Terno, G. and Sequeira, C. (1982). Impact of cancer, type, site, stage and treatment on the nutritional status of patients. Annals of Surgery 196(2): 170.

[108] Carriaga, M.T. and Henson, D.E. (1995). The histologic grading of cancer. Cancer 75(S1): 406–421.

[109] Tobore, T.O. (2019). On the need for the development of a cancer early detection, diagnostic, prognosis, and treatment response system. Future Science OA 6(2): FSO439.

[110] Wyld, L., Audisio, R.A. and Poston, G.J. (2015). The evolution of cancer surgery and future perspectives. Nature Reviews Clinical Oncology 12(2): 115–124.

[111] Chu, E. and Sartorelli, A. (2018). Cancer chemotherapy. Lange's Basic and Clinical Pharmacology, 948–976.

[112] Chen, H.H. and Kuo, M.T. (2017). Improving radiotherapy in cancer treatment: Promises and challenges. Oncotarget 8(37): 62742.

[113] Gerber, D.E. (2008). Targeted therapies: A new generation of cancer treatments. American Family Physician 77(3): 311–319.

[114] Zhu, S., Zhang, T., Zheng, L., Liu, H., Song, W., Liu, D., Li, Z. and Pan, C.-x. (2021). Combination strategies to maximize the benefits of cancer immunotherapy. Journal of Hematology & Oncology 14(1): 156.

[115] Mørch, L.S., Løkkegaard, E., Andreasen, A.H., Krüger-Kjær, S. and Lidegaard, Ø. (2009). Hormone therapy and ovarian cancer. Jama 302(3): 298–305.

Index

For Product Safety Concerns and Information please contact our EU
representative GPSR@taylorandfrancis.com
Taylor & Francis Verlag GmbH, Kaufingerstraße 24, 80331 München, Germany